Obesity and Cachexia: Physiological Mechanisms and New Approaches to Pharmacological Control

Edited by

Nancy J. Rothwell
Department of Physiological Sciences
University of Manchester, UK

and

Michael J. Stock
Department of Physiology
St George's Hospital Medical School
London, UK

JOHN WILEY & SONS
Chichester · New York · Brisbane · Toronto · Singapore

Other Wiley Editorial Offices

John Wiley & Sons, Inc., 605 Third Avenue,
New York, NY 10158-0012, USA

Jacaranda Wiley Ltd, G.P.O. Box 859, Brisbane,
Queensland 4001, Australia

John Wiley & Sons (Canada) Ltd, Fourth Floor, 5353 Dundas Road West,
Etobicoke, Ontario M9B 6H8, Canada

John Wiley & Sons (SEA) Pte Ltd, 37 Jalan Pemimpin 05-04,
Block B, Union Industrial Building, Singapore 2057

Library of Congress Cataloging-in-Publication Data:

Obesity and cachexia : physiological mechanisms and new approaches to pharmacological control / edited by N.J. Rothwell and M.J. Stock.
p. cm.
Contains proceedings of the spring symposium of the Biological Council, held in London in 1990.
Includes bibliographical references and index.
ISBN 0 471 93068 7
1. Obesity—Chemotherapy—Congresses. 2. Cachexia—Chemotherapy—Congresses. 3. Body weight—Regulation—Congresses. 4. Appetite—Effect of drugs on—Congresses. I. Rothwell, Nancy. II: Stock, Michael. III. Biological Council.
[DNLM: 1. Cachexia—drug therapy—congresses. 2. Cachexia—physiology—congresses. 3. Obesity—drug therapy—congresses. 4. Obesity—physiology—congresses. WD 210 01144 1990]
RC628.0226 1991
616.3′98061—dc20
DNLM/DLC
for Library of Congress 91-21626
CIP

British Library Catologuing in Publication Data:

A catalogue record for this book is
available from the British Library

ISBN 0 471 93068 7

Typeset by Photo·graphics, Honiton, Devon
Printed in Great Britain by
Courier International Ltd, East Kilbride

The Biological Council and its Coordinating Committee for Symposia on Drug Action are grateful for the financial contribution from the companies listed below. Without their help it would be impossible to hold these symposia.

Beecham Pharmaceuticals
The Boots Company
Glaxo Inc.
ICI Pharmaceuticals
Kabivitrum
Kotobuki Seiyaku
Lilly Industries
MSD Neuroscience Research Centre
Parke-Davis Research Unit
Servier Laboratories
Wyeth Research UK

Contents

Contributors

J.R.S. Arch *SmithKline Beecham Pharmaceuticals, Great Burgh, Yew Tree Bottom Road, Epsom, Surrey KT18 5XQ, UK*

A. Astrup *Research Department of Human Nutrition, Royal Veterinary and Agricultural University, 25 Rolighedsvej, DK-1958 Frederiksberg, Denmark*

W.M. Bennett *Departments of Anatomy and Physiology, University of Dundee, Dundee DD1 4HN, UK*

B. Buemann *Research Department of Human Nutrition, Royal Veterinary and Agricultural University, 25 Rolighedsvej, DK-1958 Frederiksberg, Denmark*

S. Caccia *Istituto di Ricerche Farmacologiche 'Mario Negri', Via Eritrea 62, 20157 Milan, Italy*

M.A. Cawthorne *SmithKline Beecham Pharmaceuticals, Great Burgh, Yew Tree Bottom Road, Epsom, Surrey KT18 5XQ, UK*

N.J. Christensen *Research Department of Human Nutrition, Royal Veterinary and Agricultural University, 25 Rolighedsvej, DK-1958 Frederiksberg, Denmark*

K.A. Coney *SmithKline Beecham Animal Health, Walton Oaks, Dorking Road, Tadworth, Surrey KT20 7NT, UK*

A.A. Connacher *Department of Medicine, University of Dundee, Dundee DD1 4HN, UK*

K.N. Frayn *Sheikh Rashid Diabetes Unit, Radcliffe Infirmary, Oxford OX2 6HE, UK*

S. Garattini *Istituto di Ricerche Farmacologiche 'Mario Negri', Via Eritrea 62, 20157 Milan, Italy*

J. Gelin *Department of Surgery, Sahlgrenska Hospital, S-413 45 Göteborg, Sweden*

B.A. Gusterson *The Institute of Cancer Research, Haddow Laboratories, 15 Cotswold Road, Belmont, Sutton, Surrey SM2 5NG, UK*

A. Hyltander *Department of Surgery, Sahlgrenska Hospital, S-413 45 Göteborg, Sweden*

R.T. Jung *Department of Medicine, University of Dundee, Dundee DD1 4HN, UK*

M.J. Kluger *Department of Physiology, University of Michigan Medical School, Ann Arbor, MI 48109, USA*

S.F. Leibowitz *The Rockefeller University, 1230 York Avenue, New York, NY 10021, USA*

R.A. Little *North Western Injury Research Centre, University of Manchester, Stopford Building, Manchester M13 9PT, UK*

K. Lundholm *Department of Surgery, Sahlgrenska Hospital, S-413 45 Göteborg, Sweden*

T. Mennini *Istituto di Ricerche Farmacologiche 'Mario Negri', Via Eritrea 62, 20157 Milan, Italy*

V. Piercy *SmithKline Beecham Pharmaceuticals, Great Burgh, Yew Tree Bottom Road, Epsom, Surrey KT18 5XQ, UK*

M. Rebuffé-Scrive *Department of Psychology, Yale University, New Haven, CT 06520-7447, USA*

M.J. Rennie *Departments of Anatomy and Physiology, University of Dundee, Dundee DD1 4HN, UK*

N.J. Rothwell *Department of Physiological Sciences, University of Manchester, Stopford Building, Manchester M13 9PT, UK*

R. Samanin *Istituto di Ricerche Farmacologiche 'Mario Negri', Via Eritrea 62, 20157 Milan, Italy*

M.V. Sennitt *SmithKline Beecham Pharmaceuticals, Great Burgh, Yew Tree Bottom Road, Epsom, Surrey KT18 5XQ, UK*

A. Shenkin *University Department of Chemical Pathology, Royal Liverpool Hospital, Prescot Street, PO Box 147, Liverpool L69 3BX, UK*

K. Smith *Department of Medicine, University of Dundee, Dundee DD1 4HN, UK*

S.A. Smith *SmithKline Beecham Pharmaceuticals, Great Burgh, Yew Tree Bottom Road, Epsom, Surrey KT18 5XQ, UK*

M.J. Stock *Department of Physiology, St George's Hospital Medical School, Cranmer Terrace, London SW17 0RE, UK*

D. Stribling *ICI Pharmaceuticals, Alderley Park, Macclesfield, Cheshire SK10 4TG, UK*

M.J. Tisdale *Pharmaceutical Sciences Institute, Aston University, Birmingham B4 7ET, UK*

J. Wallace *The Institute of Cancer Research, Haddow Laboratories, 15 Cotswold Road, Belmont, Sutton, Surrey SM2 5NG, UK*

P.W. Watt *Departments of Anatomy and Physiology, University of Dundee, Dundee DD1 4HN, UK*

D.H. Williamson *Metabolic Research Laboratory, Nuffield Department of Clinical Medicine, Radcliffe Infirmary, Woodstock Road, Oxford OX2 6HE, UK*

S. Wilson *SmithKline Beecham Pharmaceuticals, Great Burgh, Yew Tree Bottom Road, Epsom, Surrey KT18 5XQ, UK*

D.A. York *Pennington Biomedical Research Center, 6400 Perkins Road, Baton Rouge, LA 70808, USA*

Series Preface

The Biological Council, through its Coordinating Committee for Symposia on Drug Action, organizes an annual Symposium on some aspect of drug action in London in the Spring. A number of factors impinge on the choice of topic, which must be one where substantial progress has been made leading to improved therapy and which is of interest to basic and clinical scientists alike. The Symposium serves to provide relevant background material and up-to-date information in a fashion of value to non-specialist and specialist alike. The attendance of post-graduate research students is particularly encouraged.

The present volume contains the proceedings of the 1990 Symposium, which was the 40th in the series. The pharmacological control of appetite and body mass is a topic of abiding concern and currently available drugs have substantial shortcomings. There has been considerable recent progress in our understanding of basic mechanisms which has the promise of opening new avenues for drug treatment and the contents of this volume provide insights into these opportunities.

The Biological Council is greatly indebted to the numerous pharmaceutical companies who contributed towards the substantial costs of organizing the Symposium. Our policy of keeping registration fees low and allowing post-graduate students to attend at no charge requires major financial input and the sustained support of the drug industry is greatly appreciated.

When choosing the topic for the Symposium, the Coordinating Committee also select the main organizers. This year, Nancy Rothwell and Michael Stock shared these responsibilities. They are both well known for their work in this area and their expertise underlies the success of the Symposium and this volume. The Biological Council is grateful to them for their work in organizing the Symposium and editing these proceedings.

John Caldwell

Chapter One

Obesity and Cachexia: Possibilities for an Integrated Approach

Michael J. Stock
Department of Physiology, St George's Hospital Medical School, London SW17 ORE, UK

Obesity is an excessive accumulation of body fat, and in its gross manifestations poses a real threat to health, affecting both morbidity and mortality. Cachexia (from the Greek 'kakos', meaning bad and 'hexis', meaning condition) describes the severe weight loss and emaciation that accompanies many serious and often fatal diseases, and may itself precipitate an early death. Thus, one could not find a greater contrast than that between the physical appearance and pathophysiology of obese and cachectic patients, and yet a symposium has been held to discuss a common approach to both conditions. Many might regard this as a far-fetched and naive attempt to bring the subjects together, but this opening chapter attempts to demonstrate the potential value of considering the two conditions together. Whether that potential is realized will depend very much on the reader's interpretation of the contributions that appear in subsequent chapters.

There are several different ways of defining and describing obesity. One of the most common is to express body weight as a percentage of the 'ideal' weight based on actuarial data. In the clinical context, however, this definition has now been almost completely overtaken by the use of body mass index (BMI = weight/square of height; kg/m^2), where the normal range is 18–25, overweight is 25–30, moderate obesity 30–40 and gross obesity is a BMI in excess of 40. Many of the risks associated with obesity do not become serious until BMI exceeds 30 (although there are exceptions), but given that just over 30% of the adult population in the UK are

Obesity and Cachexia. Edited by N.J. Rothwell and M.J. Stock

overweight, and approximately 12% are obese, the number of people putting their health at risk through this condition could range from 6–19 million, depending on what risks are considered. The risks to health include cardiovascular disease (particularly hypertension), diabetes, hyperlipidaemia, certain cancers, respiratory problems, and other disorders (e.g. gallstones, osteoarthritis, surgical risks and even traffic accidents). Recently, the problem of risk assessment has been complicated by data that indicate that the distribution of body fat has an important influence; upper body, or intra-abdominal fat deposition has a far greater impact on risk than lower body, or subcutaneous fat deposition. Obesity was often thought to be due to nothing more than simple overeating, but more recently there has been a strong revival of the idea that there has to be an additional metabolic component that predisposes to obesity; this is discussed later on.

The weight loss and muscle wasting seen in cachexia is a common feature of many diseases. It is a serious, widespread and often fatal condition, and even relatively modest weight loss can impair immune responses and retard recovery. The most common causes of cachexia are cancer, sepsis, burn, surgery and injury, but there are many other conditions (such as myocardial infarct, cystic fibrosis, Crohn's disease, AIDS and even some mental disorders) where cachexia can cause serious complications. The incidence of cachexia varies tremendously between and within these various diseases, and for example, in cancer cachexia it can range from 8–70%, depending on the tumour type. Given the incidence of many of the diseases associated with cachexia, and the fact that many hospital patients will exhibit weight loss, one can arrive at a very approximate 'guestimate' of 1–2 million cases of cachexia per annum in the UK. A major cause of the weight loss in cachexia is loss of appetite and anorexia, but reduced food intake by itself is often insufficient to account for the loss of body weight. In these cases, and in patients who do not exhibit anorexia, an inappropriate or frankly elevated metabolic rate, together with an increased protein catabolism is another major cause of cachexia.

It should be apparent from these two brief descriptions of obesity and cachexia that the feature common to both is a disruption in the regulation of energy balance resulting from a combination of hyperphagia and hypometabolism in obesity, and a combination of hypophagia and hypermetabolism in cachexia. This is a gross over-simplification, but nevertheless provides a basis for formulating an integrated approach, at least in energetic terms, to the two conditions. However, to take the argument further a more detailed consideration is required, and to do this one has to turn to animal models.

INTAKE EXPENDITURE AND EFFICIENCY

Before considering the animal models, it is perhaps worthwhile to review the relationship between energy intake and expenditure. Normally, a decrease in energy intake results in a compensatory decrease in energy expenditure. This decrease may be brought about by a reduction in work, both external work (i.e. activity) and internal work (i.e. reduced synthesis). Often, however, there is also an increase in the efficiency of energy metabolism such that normal functions are maintained, but at a lower energy cost. This has the effect of attenuating the negative energy balance caused by the reduction of food intake. When energy intake increases above requirements and expenditure does not change, a positive energy balance results and if sustained will obviously lead to obesity. There are exceptions to this where hyperphagia leads to compensatory increases in energy expenditure that inhibit or retard the development of obesity. This adaptive increase in heat production is termed diet-induced thermogenesis, and in animals has been shown to involve sympathetic activation of brown adipose tissue (BAT) thermogenesis, in much the same way as in the non-shivering thermogenesis exhibited by neonates, hibernators and cold-adapted animals. An increase in diet-induced thermogenesis obviously represents a decrease in efficiency of energy utilization—i.e. a greater proportion of the food energy is dissipated as heat, and less is retained in the body.

In pathological situations, such as obesity and cachexia, it is often the efficiency of energy metabolism that is disturbed, and this may not always be apparent from observations of the absolute level of energy intake or energy expenditure. There are several different ways of defining and assessing efficiency, but these are usually inappropriate for clinical assessments, and become somewhat meaningless when there is a negative energy balance. One way out of this problem is to assess not whether metabolic rate is high or low in absolute terms, but to see instead whether it is appropriate for the level of energy intake. For example, if energy intake decreases but expenditure stays the same, this would mean that expenditure was now inappropriately high and there had been a decrease in energetic efficiency. Thus, when assessing the degree of hypo- or hypermetabolism in obesity or cachexia, one approach is to use the ratio of expenditure to intake—i.e. a low ratio representing increased efficiency and hypometabolism, and a high ratio indicating decreased efficiency and hypermetabolism.

ANIMAL MODELS OF OBESITY

There are numerous animal models of obesity, but the most common are shown in Table 1.1 where there are examples of genetic, hypothalamic and dietary forms. It should be noted that, with the exception of the obesity induced by monosodium glutamate, all these experimental models show hyperphagia. However, hyperphagia is not a precondition for obesity, neither in the monosodium glutamate model nor in the other examples of genetic and hypothalamic obesity, because if food intake is held at normal values these animals will still become obese. Thus, the critical factor leading to obesity is the high metabolic efficiency, and all these models show low rates of energy expenditure relative to the level of intake. The only exception to this is the response of young rats and mice to the cafeteria diet. Typically, this diet (which comprises a wide variety of highly palatable food items) induces a profound increase in voluntary food intake, and in young animals this causes a marked stimulation of diet-induced thermogenesis. This diet-induced thermogenesis can compensate entirely for the excess food consumed, resulting in a large increase in heat production (i.e. decreased efficiency) and little or no change in fat deposition. In older animals, however, the capacity for diet-induced thermogenesis is reduced, and hyperphagia leads inevitably to obesity.

The high levels of diet-induced thermogenesis seen in the young cafeteria-fed rat or mouse result mainly from increased sympathetic activation of brown adipose tissue thermogenesis. BAT thermogenesis depends on the presence of a unique uncoupling (proton conductance) pathway in its mitochondria, with fatty acids providing the principle fuel for this uncoupled respiration. Fatty acids are liberated by β-adrenergic activation of lipolysis, and it now appears that the adrenergic receptor involved is an 'atypical' or β_3-adrenoreceptor (see Chapter 15). The mechanisms involved in diet-

Table 1.1 Animal models of obesity

	Energy intake	Energetic efficiency
Genetic		
fa/fa rat	Increased	Decreased
ob/ob mouse	Increased	Decreased
Hypothalamic		
VMH	Increased	Decreased
GTG	Increased	Decreased
MSG	Unchanged	Decreased
Dietary		
Cafeteria—old	Increased	Decreased
Cafeteria—young	Increased	Increased

Energetic efficiency taken as ratio expenditure:intake.
VMH, ventromedial hypothalamic lesions; GTG, gold thioglucose lesions; MSG, monosodium glutamate lesions.

induced thermogenesis are virtually identical to those involved in non-shivering thermogenesis, and in some respects these two forms of thermogenesis resemble the hypermetabolic state seen in cachexia. Another, more obvious, link between obesity and cachexia comes from studies on the genetically obese (fa/fa) Zucker rat, since not only does this animal show a reduced capacity for diet-induced thermogenesis, but also fails to produce a fever in response to endotoxin (see Chapter 2). This suggests that whatever mechanisms are responsible for defective diet-induced thermogenesis in genetic obesity, these same mechanisms may also be important in the hypermetabolic response to infection.

ANIMAL MODELS OF CACHEXIA

Table 1.2 lists examples of cachexia induced in experimental animals, either by cancer, injury or infection. There are a limited number of energy balance studies in these models, but usually there is evidence of hypophagia, and often there is an increase in metabolic rate, although this may be small or transient. However, given that under normal circumstances energy expenditure would fall when intake declines, even these small increases in metabolic rate indicate that energetic efficiency must have decreased, and these animals were hypermetabolic relative to the level of energy intake. The human equivalent of this would be the cachectic patient who shows no decrease in metabolic rate whilst in bed, whereas most hospital patients show a decline in metabolic rate during bedrest. The most obvious examples of hypermetabolism in cachexia are in mice bearing hypernephromas or adenocarcinomas, where intake remains unaltered for much of the time and

Table 1.2 Animal models of cachexia

	Energy intake	Energetic efficiency
Cancer		
Mouse hypernephroma	No change	Increase
Mouse adenocarcinoma	No change	Increase
Mouse sarcoma	Decrease	Increase
Rat leukaemia	Decrease	Increase
Injury		
Rat scald	Decrease	Decrease/increase
Infection		
Rat endotoxin	Decrease	Increase
Mouse malaria	Decrease	Small increase
Guinea-pig pneumonia	Decrease	Increase

the rapid weight loss is due almost entirely to increased sympathetic activation of thermogenesis.

SURVIVAL VALUE OF OBESITY AND CACHEXIA

If one had to give an instant judgement of the survival value, or evolutionary significance of obesity and cachexia, it is likely that most people would argue that a predisposition to obesity was an advantage (i.e. conserving energy) whereas cachexia was a distinct disadvantage (i.e. wasting energy). Surprisingly, arguments have been put forward to suggest quite the opposite. For example, it has been argued that obesity usually develops in later life, well past the age of active reproduction, and cannot therefore influence selection. However, in those who were obese in adolescence or early adult life, the adverse effects of obesity on fertility and the successful outcome of pregnancy would place them at a disadvantage compared with the lean. The only exception to this might be the primitive groups who have survived and evolved in particularly arid regions of the world (e.g. Pima Indians, Aborigines), and who could not have survived, even to the age of maturity unless they had a particularly thrifty and efficient genotype. Unfortunately, when these primitive groups are given *ad libitum* access to food, their energetic thriftiness is rewarded by large numbers of them becoming obese and/or diabetic. Thus, it could be argued that when times are bad, a thrifty genotype and a predisposition to obesity were an advantage, but as food supplies become more readily available the advantage switched to those with less efficient metabolism, or at least with the capacity to vary efficiency flexibly, according to the energy available.

High rates of catabolism, loss of body protein and fever would not seem to most patients or clinicians the most appropriate response to infection, trauma or cancer, but once again it has been argued that cachexia could represent a useful, primitive reflex to mobilize endogenous fuel and inhibit the spread of infection by raising body temperature. Unfortunately, this reflex, like the Pima Indians' thrifty metabolism no longer seems to be an advantage, and we may have reached a point where the primary injury or disease is controllable and no longer fatal, but the cachectic response to that injury or disease is.

FACTORS INFLUENCING INTAKE AND EXPENDITURE

Many of the contributions that follow will be reviewing how far we have got in understanding the basic mechanisms involved in the control of metabolism, and the potential for pharmacological intervention. However, it is perhaps worthwhile to consider some of the possible factors that might influence energy intake and expenditure in obesity and cachexia. Dealing with intake first, the thermostatic theory of food intake control would suggest that an elevated body temperature (i.e. fever) should result in a diminished appetite, but it is more likely that the cytokines involved in the development of fever play a more important role through their central effects on feeding. Moreover, the cytokine influence would not necessarily be confined to forms of cachexia where fever is manifest. Disturbances in electrolyte balance, pain, gastrointestinal discomfort and nausea are all factors which could contribute to anorexia. Oropharyngeal and gastrointestinal tumours obviously make the physical act of eating difficult and painful and may contribute to nausea, but usually it is the radiotherapy and/or chemotherapy that is the major cause of nausea; in these cases the anorexia could be considered to be iatrogenic. Psychological factors cannot be discounted, and not surprisingly many seriously ill patients are depressed and no longer feel like eating, and to this one has to add the impairment in taste experienced by many patients.

When it comes to explanations of overeating in obesity, it is almost impossible to divide serious, scientific suggestions from the apocryphal, and often moralistic assertions that abound in the minds of clinicians and the lay public alike. Furthermore, it has to be remembered that although obesity has to result from overeating, there is no conclusive evidence to show that the obese eat more than the lean. This suggests that their overeating represents a failure to reduce their intake to take account of a lower energy requirement, rather than overt gluttony. Even in the genetic or dietary animal models of obesity, there has been no specific lesion or derangement described that satisfactorily explains the hyperphagia, even though a considerable amount is known about the neuroanatomy and neurochemistry of feeding behaviour in these animals (see Chapter 3). Nevertheless, based on this knowledge one could offer a variety of suggestions as to likely causes. For example, a reduced availability of, or sensitivity to nutrients in the brain areas controlling feeding would result in the CNS being 'nutrient blind', and food intake would increase in an attempt to overcome this. Other possibilities are changes in the synthesis, secretion or turnover of neurotransmitters involved in the control of food intake, or changes in the neuronal population, or the receptor profile of those neurones responding to these neurotransmitters. Many of these hypotheses rely on experiments involving pharmacological manipulations, but it is perhaps worth remembering that just because one can modify feeding behaviour with agonists or antagonists of these neurochemicals, this does not necessarily mean that this is evidence of a specific neurochemical lesion in obesity; no more so than concluding that headaches are due specifically to a deficiency of aspirin.

Given the inability to define any specific physiological defect in appetite control in obesity, psychosocial explanations of overeating abound. While depression may be a cause of anorexia in cachectic patients, it could equally well be responsible for overeating in obese patients. An exaggerated example of this is seen in the seasonal affective disorder syndrome (SADS) where the autumnal and winter depression experienced by these patients is often accompanied by carbohydrate craving, increased energy intake and weight gain, all of which reverse in the spring or early summer. The psychological and social pressures that have been invoked to explain overeating are far too numerous to list here, but in essence they all imply that the obese, for whatever reason, lack, or ignore the physiological signals that control food intake.

If, as in all the animal models, a high energetic efficiency (i.e. hypometabolism) is a major factor in the development of human obesity, there are several components of energy expenditure to consider. First, reduced levels of physical activity would facilitate a positive energy balance, and there is evidence to suggest that, at least in children, there is a strong association between sedentary behaviour (e.g. watching television) and obesity. However, given the relatively low levels of activity in most people in Western countries, it is difficult to believe, or to prove, that this explains the ease with which some people become obese, while others do not.

A considerable amount of research effort has gone into the question of whether the obese have a reduced basal metabolic rate (BMR), but no matter how the data are expressed (per unit body weight, surface area, lean body mass, or simply per individual) the majority of studies suggest that the BMR of the obese is not noticeably different from that of lean subjects; if anything it is higher. However, there have been several studies (see Chapter 10) where a low BMR has been shown to be predictive of a subsequent tendency to increased weight gain. The differences in BMR may be small (approximately 200 kcal per day in adults) but a small daily positive balance of this order, accumulating over two years, could account for a significant amount of fat deposited. The final, and perhaps most controversial suggestion is that the obese have a higher metabolic efficiency due to a reduced level of diet-induced thermogenesis, and that this may be due in part to decreased sympathetic activity, particularly in response to feeding (see Chapter 10). There is a considerable amount of direct and circumstantial evidence consistent with this suggestion, but there are also quite a number of studies that are not.

Increased activity as a contributory factor to cachexia is rarely mentioned, but it might be worth bearing in mind that general restlessness and disturbed sleep in sick patients could add significantly to daily energy requirements. Nevertheless, even in a relaxed, or sedated cachectic patient there is often an increased resting, or basal metabolic rate. If the patient has a fever, a small part of the increased metabolic rate could be ascribed to the Q_{10}

effect; in humans Q_{10} is approximately 2, which means that metabolic rate doubles for every 10°C rise in temperature. In addition, there are the increases in protein catabolism and gluconeogenesis which, although they serve to mobilize endogenous fuels, in themselves involve an increase in heat production. Lastly, there are the elevations in diet-induced and non-shivering thermogenesis, and in some cases, shivering thermogenesis. There is evidence to show that increases in sympathetic activity and thermogenesis occur in many forms of cachexia, and the elevated diet-induced thermogenesis, particularly during parenteral nutrition, can exacerbate the condition by causing further increases in heat production and body temperature.

POSSIBILITIES FOR PHARMACOLOGICAL INTERVENTION

If there is one thing that obesity and cachexia have in common, it is the extraordinary difficulty in trying to treat the two conditions. In both, there would seem to be an inate drive to push energy metabolism into either positive or negative balance, and to prevent it slipping further is not easy, and to reverse it often seems impossible. From simple thermodynamic considerations the treatment would seem obvious—decreasing intake and increasing expenditure in obesity, and vice versa in cachexia. The problem is, however, that our knowledge of the mechanisms controlling intake and expenditure is still far from complete, even under normal conditions, and the techniques available for manipulating energy balance are very largely empirical in origin. One aspect that makes treatment especially difficult is the time course involved, which is quite different in the two conditions. Obesity develops slowly and insidiously, and although there are methods for very rapid weight reduction, most methods rely on producing a relatively slow, but consistent weight loss to bring the subject to target weight. One is therefore talking about a course of treatment which may last for many months or years, and has to be sustained in order to maintain the weight loss. Some cases of cachexia may have a slow development, but most, particularly the serious cases, require urgent treatment. Sometimes the condition develops faster than the clinician can respond, and the choice of treatment has to be made rapidly and often carried out aggressively.

Obesity attracts considerable commercial interest, and it is often said that over a quarter of the UK food industry is devoted to the production of slimming foods. However, in spite of its health implications and the large numbers who are actively trying to lose, or control their weight, the medical profession's interest in obesity is generally lukewarm, and often indifferent. The most charitable explanation for this could be that they are aware that few of the available treatments are particularly successful, even in the short term, and if the long term aim is to maintain the weight loss, the success rate is very low indeed. The methods that have been tried range from behavioural modification to major surgical techniques, together with an entire galaxy of diets and dieting aids.

Unlike obesity, cachexia is something that cannot be easily ignored, although mild cachexia may go unnoticed for some time, thus delaying rehabilitation and increasing the duration and cost of hospitalization. Given that cachexia is always secondary to another serious disease, the priority will be to treat that primary disease, which should in turn reduce the severity of cachexia. However, the rapid deterioration in some patients caused by cachexia can reach a point where it takes precedence. At present, the only methods directed specifically to treating cachexia are either to provide nutritional supplementation (enteral or parenteral nutrition), or to try to reduce metabolic rate by reducing heat loss, such as by maintaining a high ambient temperature and humidity for burns patients. Both approaches are of limited value; parenteral nutrition may raise thermogenesis further, and reducing heat loss, even in burns patients, will not overcome the hypermetabolism. There is clearly scope for other forms of intervention, and pharmacological approaches may prove fruitful. Reverting back to the very simplistic classification of obesity and cachexia given earlier, one needs to look for agents that stimulate intake in cachexia and inhibit it in the obese, whereas one requires thermogenic agonists to treat obesity and thermogenic antagonists for cachexia. In addition, both conditions would benefit from some method for controlling protein catabolism in order to maintain, or increase lean body mass.

It is not the function of this introductory chapter to anticipate in detail the contributions that follow, but it would not be giving too much away to indicate that there have been recent developments in our understanding of the pharmacological control of energy balance which may be of value in treating obesity or cachexia. New serotonergic appetite suppressant drugs are appearing (see Chapter 14), and at least one company is developing orally effective CCK analogues (peptoids) as potential appetite suppressants. Moving beyond ingestion, another company is developing a lipase inhibitor (tetrahydrolipstatin) as a potential method for reducing energy input by inhibiting the digestion and absorption of fat, and increasing faecal excretion. When it comes to possible agents for stimulating appetite in cachectic patients, there do not appear to be any selective appetite stimulants available. However, given the serotonergic influence on satiety, one possible approach would be to look for serotonergic antagonists to stimulate appetite. It is worth noting that the antihistamine cyproheptadine has pronounced serotonergic antagonist properties and has been used clinically to stimulate appetite.

The idea of using thermogenic drugs to treat obesity is not novel. In the past, thyroid hormones, uncoupling agents (e.g. 2,4-dinitrophenol) and ephedrine have been used. Ephedrine, because of its sympathomimetic effects on thermogenesis is still favoured by some, but its side-effects, particularly those on the cardiovascular system, have prevented its widescale use. Recently, there has been considerable interest in developing thermogenic β-agonists, devoid of other sympathomimetic side effects. Several companies have compounds in development, and it appears that they are all selective for the 'atypical', or β_3-adrenoceptor (see Chapter 15). Apart from activating thermogenesis, these compounds also appear to have some other, advantageous effects, such as improvements in insulin sensitivity and retention of nitrogen. It should also be mentioned that some of the serotonergic appetite suppressants (e.g. d-fenfluramine) also appear to have a mild thermogenic activity, which may add to their value in the treatment of obesity. Given the generally high levels of sympathetic activity in cachexia, and the possibility that this activity increases metabolic rate through actions on β_3-adrenoceptors, one might speculate that a selective β_3-antagonist could have a potential role in the treatment of cachexia. There are some recent data showing greater weight gains following chronic treatment after myocardial infarct with the non-selective antagonist propranolol, which suggest that this approach could prove to be an effective way of preventing weight loss in cachexia. Hopefully, alternative approaches will evolve as a result of a better understanding of the thermogenic and pyrogenic actions of the cytokines in cachexia (see Chapters 2 and 13).

So far, it has been argued that an integration of research ideas, at least in energetic terms, could be mutually beneficial to the study and treatment of obesity and cachexia. However, perhaps this argument is no longer tenable when it comes to considering the changes in protein metabolism. Certainly, one cannot find in obesity the opposite of the high rates of protein catabolism and gluconeogenesis seen in many examples of cachexia. Nevertheless, this is probably one area where the two conditions share a common, rather than a conflicting clinical interest, since the preservation of lean body mass is just as important for the obese subject trying to lose weight as it is for the cachectic patient trying to avoid muscle wasting and emaciation. Once again, β-adrenergic agonists may have a potential clinical application since certain β_2-adrenoceptor agonists have quite marked anabolic effects on skeletal muscle protein deposition (see Chapter 15).

Some of the potential developments outlined above are highly speculative, but are intended to show how a better understanding, and improved methods of treatment of obesity, could have spin-offs that could be used to advantage in the treatment of cachexia, and vice versa. This is why obesity and cachexia have been put together in the same symposium, and the reader is now left to see if the contributions that follow add further weight to the idea; if nothing else, the reader gets two symposia for the price of one!

Chapter Two

Autonomic and Central Control of Thermogenesis

Nancy J. Rothwell
Department of Physiological Sciences, University of Manchester, Stopford Building, Manchester M13 9PT, UK

INTRODUCTION

The production of heat is an obligatory result of many biological processes and forms the component of energy expenditure known as basal metabolic rate. However, in addition to this, variable increases in heat production can be induced by both physiological and pathological stimuli. Cold exposure and food ingestion/hyperphagia are both potent activators of thermogenesis eliciting responses known as non-shivering (NST) and diet-induced thermogenesis (DIT) respectively. NST allows maintenance of homeothermy during cold exposure, while DIT is a regulatory component of energy balance which, along with changes in appetite and food intake, allows maintenance or restoration of body weight and composition (see Rothwell and Stock, 1986a,b for reviews). Marked increases in metabolic rate, which cannot be ascribed to muscular activity, also follow stress, infection and injury (see Long, 1977; Little, 1988; Rothwell, 1990a), and if sustained, these hypermetabolic responses will result in cachexia.

EFFECTOR MECHANISMS OF THERMOGENESIS

Studies which have concentrated on the effects of NST and DIT in experimental animals indicate that increases in heat production are mediated

Obesity and Cachexia. Edited by N.J. Rothwell and M.J. Stock

by activation of sympathetic outflow, which results in increased thermogenesis in brown adipose tissue (BAT) (see Rothwell and Stock, 1986a,b; Girardier, 1983; Landsberg and Young, 1983; Himms-Hagan, 1989). Brown adipose tissue is a small and highly specialized tissue which is present in very small amounts (less than 1% of body weight) in most if not all mammals including humans. However, its remarkable capacity for heat production (~500 W/kg) means that BAT can make a dramatic impact on total metabolic rate in small rodents, and even in large mammals can significantly influence heat production (see Girardier, 1983; Rothwell and Stock, 1986). The very high rates of thermogenesis in BAT are induced by interaction of noradrenaline with a β_3-adrenoceptor (see Chapter 15 by Arch), which then stimulates uncoupling of oxidative phosphorylation by a unique mitochondrial conductance pathway (Nicholls and Locke, 1983). Sympathetic activation of BAT has also been implicated in the thermogenic responses to pyrogens (Cooper *et al.*, 1989; Dascombe *et al.*, 1989a; Jepson *et al.*, 1987; Rothwell, 1990a), injury (McCarthy *et al.*, 1990; Rothwell, Little and Rose, unpublished data) and malignant disease (Brooks *et al.*, 1981; Bianchi *et al.*, 1989), and preliminary data indicate that activation of BAT also accompanies these states in humans (Bianchi *et al.*, 1989; Bruce *et al.*, 1990).

The evidence supporting a role for the sympathetic nervous system in non-shivering and diet-induced thermogenesis has been reviewed previously in detail (see Landsberg and Young, 1983; Rothwell and Stock, 1986a,b). Increased metabolic rates of cold adapted, hyperphagic or protein deficient animals can be inhibited or prevented by injection of ganglionic blockers or β-adrenergic antagonists. The β-adrenoceptor antagonist most commonly used is propranolol which has a low affinity for the β_3-adrenoceptor in BAT (Arch *et al.*, 1984) and a short half life in the rat, thus chronic pharmacological inhibition of thermogenesis has not been possible. Surgical denervation of the interscapular BAT depot does inhibit the activity of the tissue in the rat (Rothwell and Stock, 1984a), and in mice, denervation significantly increases body fat content (Dulloo and Miller, 1984). Thermogenic responses to adrenergic agonists such as noradrenaline and isoprenaline are also enhanced in animals exhibiting DIT and NST, due mainly to hypertrophy of BAT (see Rothwell and Stock, 1986a,b).

Sympathetic activity has been assessed most reliably from measurements of noradrenaline turnover, and these have revealed increases in activity during overfeeding, protein deficiency and cold exposure and reductions in food restricted or genetically obese animals (see Young and Landsberg, 1977; Young *et al.*, 1985, 1982; Sakaguchi *et al.*, 1988). Interestingly, changes in sympathetic activity associated with diet and cold show selectivity for brown fat with only very modest changes occurring in other tissues. Physiological or pharmacological activation of the sympathetic nervous system can only cause acute activation of thermogenesis, but also induces the specific protein responsible for uncoupling in the tissue (Cunningham and Nicholls, 1987) and is capable of producing brown adipocyte proliferation (Geloen *et al.*, 1988). Additional evidence for the sympathetic nervous system as a mediator of DIT and NST includes observations of increased circulating concentrations of catecholamines during thermogenesis, activation of BAT by electrical stimulation of the sympathetic nerves innervating the tissue, and the finding that defective thermogenesis in obese or ageing animals is associated with reduced sympathetic activity (see Rothwell and Stock, 1984b).

Increases in sympathetic nervous system activity have also been well documented following injury and illness (see Little, 1988; Aulick and Wilmore, 1983; Young *et al.*, 1983; Wilmore *et al.*, 1974) and noradrenaline turnover in BAT is increased in endotoxaemic animals (Arnold, Little and Rothwell, unpublished). Administration of β-adrenoceptor antagonists significantly inhibits the thermogenic responses to pyrogens such as endotoxin, interleukin-1 (IL-1) and tumour necrosis factor α (Dascombe *et al.*, 1989; Jepson *et al.*, 1988; Rothwell, 1988), infection (Cooper and Rothwell, unpublished) and injury (McCarthy *et al.*, 1989; Little and Rothwell, unpublished data), and BAT activity in these situations is markedly suppressed by surgical denervation in experimental animals.

Studies on diet-induced thermogenesis have indicated an inhibitory effect of the parasympathetic nervous system which was inhibited by atropine and was greater in obese animals with impaired thermogenesis (Rothwell *et al.*, 1981). However, no comparable studies have been performed relating the parasympathetic nervous system to the hypermetabolic responses to illness.

Obviously, the understanding of the importance of the sympathetic nervous system in thermogenesis will be enhanced by the availability of a selective β_3-adrenoceptor antagonist which might also prove clinically advantageous.

AFFERENT SIGNALS

Activation of NST occurs in response to cold exposure which is detected by both peripheral thermoreceptors (mainly on the skin), relaying neural signals to the hypothalamus and by thermosensitive neurones within the brain. Diet-induced thermogenesis, however, is more complex since it can be stimulated by acute ingestion of food, chronic hyperphagia, changes in dietary composition (e.g. low protein diets) and even by the sight, smell and expectation of food (see Rothwell and Stock, 1987a,b). In addition DIT may be modified by the attendant state of energy balance—i.e. the level of body energy stores. In the face of such complexity and variety of stimuli it seems unlikely that a single afferent signal is responsible for the activation of DIT.

Nutrient availability and rate of utilization are obvious candidates for the control of DIT (see Steffens, 1980; Williamson, 1987), but as yet have received little attention. Central injections of glucose stimulate DIT and BAT activity, and the effects are enhanced by insulin (Rothwell and Stock, 1988a). The site of these effects is unknown, but Oomura (1976) has identified glucose sensitive neurones in several hypothalamic regions such as the ventromedial (VMH) and lateral (LH) hypothalamic areas. The potent effects of protein deficiency on energy balance and thermogenesis cannot be ascribed simply to substitution of protein with carbohydrate, since high fat/low protein diets also stimulate BAT and reduce energetic efficiency (Rothwell and Stock, 1987a). An attractive, but unproven hypothesis relates to the effects of protein intake on circulatory concentrations of amino acids, notably tryptophan. Tryptophan is the precursor amino acid for synthesis of the neurotransmitter 5-hydroxytryptamine (5HT) which has been strongly implicated in the central control of appetite (Blundell, 1989) and thermogenesis (Rothwell, 1989b; Rothwell and Stock, 1986b, 1987b, and see later). Uptake of tryptophan into the brain is a rate limiting step for 5HT synthesis and is determined largely by its concentration in plasma relative to that of other competitive large neutral amino acids (Fernstrom, 1986). These processes have been implicated in the control of appetite, and particularly in nutrient selection (Anderson, 1979; Blundell, 1989; Breisch *et al.*, 1976; Leathwood, 1987; Leibowitz, 1986), but might equally relate protein intake to activation of thermogenesis (Fuller *et al.*, 1987; Rothwell and Stock, 1987b).

Some data exist to suggest that psychological factors such as the sight and smell of food and mechanical effects of distension of the stomach may act as signals for DIT (Rothwell *et al.*, 1984). Neural afferents from the stomach and liver appear to modify feeding behaviour and by similar actions could also influence sympathetic outflow and hence thermogenesis. If DIT is related to long-term energy balance as well as to acute fluctuations in energy intake, it is necessary to postulate some signal relating body energy stores (i.e. fat and to a lesser extent protein content). A similar problem has arisen from attempts to explain long-term regulation of energy intake. However, although various hypotheses have been proposed to relate body energy content to the central control of energy intake and expenditure (e.g. Kennedy's lipostatic hypothesis, 1953), they all lack experimental support.

Afferent signals relating stress, injury, infection or illness to the central control of energy balance involve both neural and humoral factors. For example in the thermogenic response to peripheral injury two separate components can be identified. Intramuscular injection of turpentine in rodents induces a sterile abscess and activation of the acute phase response, thus providing a valuable and widely used model to investigate mechanisms of fever and thermogenesis. The early phase of hypermetabolism is attenuated in animals pretreated with capsaicin soon after birth to destroy C-fibre afferent nerves, but the response to IL-1 (endogenous pyrogen) is normal in these animals (Busbridge, Cooper, Kirkman and Rothwell, unpublished data).

Bacterial or viral infections are potent activators of thermogenesis usually resulting in fever (see Cooper, 1987; Dinarello *et al.*, 1988; Kluger, 1989). Peripheral release of a substance known originally as 'endogenous pyrogen' from macrophages was shown to induce fever and this was subsequently identified as the cytokine interleukin-1 (IL-1, see Blatteis, 1988; Dinarello *et al.*, 1988; Kluger, 1989; Sternberg, 1989). Recombinant preparations of the two forms of IL-1 (IL-1α and β) have allowed detailed studies in experimental animals. IL-1 induces fever and activation of thermogenesis when injected peripherally or directly into the cerebroventricular space of rats, mice and rabbits (Dascombe *et al.*, 1989a; Kluger, 1989) and these effects are dependent on the same peripheral effectors as DIT and NST. Interestingly, genetically obese rodents (fatty Zucker rat and obese (ob/ob) mice) which show defective dietary activation of thermogenesis, also exhibit impaired responses to IL-1 (Dascombe *et al.*, 1989b). The main site of action for IL-1 appears to be within the brain, since very low doses are required to induce fever compared with peripheral injection (ng range, Dascombe *et al.*, 1989). However, the means by which IL-1 gains access to hypothalamic sites of action is unknown. Several studies have reported that IL-1 is unable to cross the blood–brain barrier (see Dinarello *et al.*, 1988), while a recent report suggests that there is quite rapid transport of IL-1α into the brain (Banks *et al.*, 1990). Another problem encountered in studies on IL-1 and fever has been the difficulty in detecting significant increases in circulating IL-1 in animals (Busbridge *et al.*, 1989) or humans (Horan *et al.*, 1989) during the development of fever in response to exogenous pyrogen or burn injury (Childs *et al.*, 1989). However, dramatic increases in circulating concentrations of another pyrogenic cytokine IL-6 (Busbridge *et al.*, 1989) have been observed in response to infection, pyrogens, injury and even psychological stress (Childs *et al.*, 1989; Horan *et al.*, 1989; Kluger, 1989; LeMay *et al.*, 1989; Rothwell *et al.*, 1989a). Since IL-1 is a potent releaser of IL-6, the latter may be a more important circulating pyrogen. Nevertheless, the involvement of IL-1 in fever and hypermetabolism is supported by the finding that administration of antibodies to IL-1β or to the IL-1 receptor attenuate the acute and chronic metabolic responses to endotoxin (Rothwell *et al.*, 1989a; Kluger, 1989; Wong *et al.*, 1989) or turpentine abscess (Gershenwald *et al.*, 1989).

There exist a number of cytokines with pyrogenic and thermogenic effects in experimental animals. However, although administration of recombinant forms of these proteins induces hypermetabolism their physiological importance remains uncertain.

Tumour necrosis factor (TNF)α has been proposed as a mediator of cancer cachexia and was originally termed 'cachectin' because it induces wasting (Beutler and Cerami, 1986). TNF also causes activation of thermogenesis when injected centrally or peripherally into rodents (Rothwell, 1988) but the doses required are rather higher than for other cytokines, questioning its physiological importance. Furthermore, there has been a general failure to observe any significant quantities of TNFα in the plasma of cachectic patients (see Chapters 8 and 13).

BRAIN REGIONS INFLUENCING THERMOGENESIS

The hypothalamus is the major area concerned with the control of autonomic functions, particularly feeding behaviour and thermoregulation, and has been the primary focus for studies on thermogenesis. A close association between the ventromedial hypothalamus (VMH) and the regulation of body weight or energy balance was first demonstrated almost 40 years ago, when destruction of the ventromedial nucleus was shown to result in hyperphagia and obesity. The VMH was subsequently referred to as a 'satiety' centre, acting reciprocally with the lateral hypothalamus (LH; known as the hunger centre) to control food intake (Stellar, 1954). However, obesity will develop in response to VMH lesions in the absence of hyperphagia, indicating that energy expenditure is simultaneously reduced (Han, 1965). Perkins *et al.* (1981) reported marked increases in thermogenesis (implied from measurements of temperature) of brown fat in response to VMH stimulation, which were dependent on the sympathetic nervous system, and several groups have confirmed the involvement of the VMH in the control of BAT thermogenesis (Holt *et al.*, 1987; Freeman and Wellman, 1987; LeFeuvre, 1990). VMH stimulation raises brown fat temperature in conscious as well as anaesthetized animals, causes marked increases in noradrenaline turnover (Saito *et al.*, 1987), and blood flow (Iwai *et al.*, 1987) to BAT, and 40% increases in resting oxygen consumption (Atrens *et al.*, 1985).

Inhibition of thermogenesis following surgical or chemical destruction of the VMH has been implied from increases in the efficiency of weight gain (Han, 1965), reduced thermogenic responses to food (Carlisle *et al.*, 1988), noradrenaline turnover and thermogenic activity of brown fat (Seydoux *et al.*, 1981; Hogan *et al.*, 1982) all of which may contribute to the development of obesity. It is difficult to distinguish the influence on thermogenesis of the VMH from that of the ventral noradrenergic bundle (VNAB). Sahakian *et al.* (1983) have shown that depletion of noradrenaline in or around the VNAB stimulates food intake and weight gain, but inhibits BAT activity in the rat.

There is some debate over whether the VMH is involved in the control of NST as well as DIT, and this may again relate to differential effects of the nucleus and of noradrenergic fibres. Hogan *et al.* (1982) reported impaired DIT, but no deficiency of NST in VMH-lesioned rats, whereas Imai-Matsumura *et al.* (1984) demonstrated that injection of lidocaine into the VMH prevented the increase in BAT temperature in response to cooling of the preoptic area in conscious rats.

The VNAB runs along the base of the brain and projects to the paraventricular nuclei (PVN), dorsomedial nuclei (DMN) and the preoptic region of the hypothalamus, all of which have been implicated in the control of brown fat thermogenesis. The preoptic and anterior areas of the hypothalamus have been studied in some detail because of their general involvement in temperature regulation; both can apparently activate a variety of peripheral mechanisms of heat production and conservation. The PVN has been more specifically associated with dietary responses and its destruction results in an obesity which shares some common features with the VMH syndrome (Leibowitz *et al.*, 1984; Weingarten *et al.*, 1985).

Investigations into the involvement of extrahypothalamic areas in the control of thermogenesis are somewhat limited. Activation of thermogenesis in response to conditioning (e.g. at the time of darkness) in laboratory rats and the sight and smell of food (see Rothwell and Stock, 1986a,b) suggests that there is some cortical influence on thermogenesis, presumably via the hypothalamus.

Stereotaxic knife cuts or suction decerebration in the cat or the rat have indicated the presence of a tonic inhibitory centre in the upper pontine region. Decerebration in the prepontine region produces dramatic increases in body temperature in several species (Bignall *et al.*, 1975; Rothwell *et al.*, 1983) which are associated with dramatic increases in metabolic rate (Rothwell *et al.*, 1983). In contrast, premammillary transections prevent or inhibit these responses indicating the presence of a tonic inhibitory system located somewhere between the lower mid-brain and upper pons which controls sympathetic outflow (Benzi *et al.*, 1988; Rothwell *et al.*, 1983).

AMINERGIC PATHWAYS

Several pieces of evidence have implicated central noradrenergic pathways in the control of thermogenesis. Sahakian *et al.* (1983) demonstrated that chemical destruction of noradrenergic pathways in the region of the VNAB resulted in inhibition of BAT activity and obesity. Conversely, Shimazu *et al.* (1986) have observed dramatic increases in body weight gain in rats in which noradrenaline was continuously infused into the VMH over several weeks, but found no changes in body weight in response to noradrenaline infusion into the PVN or the lateral hypothalamus, and only modest obesity in response to adrenaline infusion into the VMH. More recently, Siviy *et al.* (1989) found that infusion of noradrenaline into the PVN inhibited energy expenditure in the rat probably via an α_2-adrenoreceptor. These apparently conflicting findings may represent subtle but crucial differences in the site of infusion and the location and type of receptors affected by experimental manipulation. For example, microinjection of noradrenaline into the PVN can elicit feeding via an α-adrenoceptor mechanism (Leibowitz, 1976), whereas activation of β-adrenoreceptors on the lateral hypothalamus may inhibit feeding (Grossman, 1962). It is likely that the control of thermogenesis is also dependent on both the site of action and receptor subtype.

The reported effects of serotonergic pathways on thermogenesis are slightly less controversial. Central injections of 5-hydroxytryptamine, serotonergic agonists or releasing agents such as D-fenfluramine stimulate metabolic rate and BAT thermogenesis by increasing sympathetic outflow (LeFeuvre *et al.*, unpublished results). Chemical depletion of 5HT pathways by injection of parachlorophenylalanine (PCPA) inhibits BAT activity (Fuller *et al.*, 1987) and causes hyperphagia and obesity in the rat (Breisch *et al.*, 1976). However, central injection of the slightly more selective neurotoxic agent 5,7-dihydroxytryptamine (5,7-DHT), in combination with desipramine to protect noradrenergic pathways, fails to induce obesity but greatly enhances the thermogenic effects of D-fenfluramine, presumably because of increased receptor density or enhanced receptor sensitivity (LeFeuvre *et al.*, 1987). The varied effects of PCPA and 5,7-DHT are probably due to selective destruction of 5HT pathways in different brain regions. Some serotonergic pathways within the Raphe nucleus are not destroyed by 5,7-DHT (Fischette *et al.*, 1987) and these are probably involved in the control of feeding and body weight (Coscina and Magder, 1984).

There are relatively few studies on dopaminergic or cholinergic pathways and thermogenesis, although Shimazu *et al.* (1986) observed no changes in body weight after chronic infusions of acetylcholine into the hypothalamus.

Increasing brain γ-amino-*n*-butyric acid (GABA) concentrations by peripheral administration of GABA transaminase inhibitors inhibits the development of genetic obesity (Coscina and Nobrega, 1984) and stimulates thermogenesis (Horton *et al.*, 1988b). The use of selective GABA agonists and antagonists, injected centrally (into the VMH or third ventricle) has shown that activation of $GABA_B$ receptors, for example by Baclofen, causes very large increases in thermogenesis (Addae *et al.*, 1986), whereas activation of the $GABA_A$ receptor inhibits metabolic rate and BAT activity (Horton *et al.*, 1988a).

PEPTIDES AND THE CENTRAL CONTROL OF THERMOGENESIS

We have identified several peptides which cause dose-dependent increases in metabolic rate in conscious animals when injected into the third ventricle of the brain, and in most cases these have been associated with increased BAT activity (LeFeuvre, 1990; Rothwell, 1989b, 1990b). Particularly notable is the fact that most of these peptides have been demonstrated to exert reciprocal effects on appetite and thermogenesis, inhibiting food intake (see Morley, 1987), although the physiological importance of endogenous peptides in appetite control or thermogenesis has not been elucidated.

One of the major peptide hormones implicated in the central control of energy balance is insulin (see Figlewicz *et al.*, 1987; Rothwell and Stock, 1988a for reviews) which, in high doses peripherally may cause obesity, but within physiological concentrations could have CNS actions to inhibit food intake and stimulate thermogenesis. Central infusion of insulin inhibits weight gain in the rat and cerebroventricular injections enhance the thermogenic actions of glucose (see Rothwell and Stock, 1988 for review). One of the most potent thermogenic peptides is corticotrophin releasing factor (CRF) which is now emerging as an important factor in the central control of energy balance and in the metabolic responses to stress, infection and injury (see Rothwell, 1990a for review).

CRF

CRF is synthesized mainly in the paraventricular nucleus of the hypothalamus although many other sites of synthesis within the CNS and the periphery have been identified (see Emeric-Sauval, 1986; Gillies and Grossman, 1985). The most notable function of CRF is in the control of the synthesis and release of ACTH and other proopiomelanocortin (POMC) derived peptides. However, CRF can also exert actions within the CNS which appear to be independent of the pituitary, and thus functions as a neurotransmitter or neuromodulator. The effects of CRF on thermogenesis, food intake and the sympathetic nervous system are included in this category.

The involvement of CRF in the control of energy balance was first implied from studies on the effects of adrenalectomy in genetically obese rodents (see York, Chapter 6, for detailed review). Briefly, adrenalectomy was found to almost completely normalize the defective energy balance of genetically obese and ageing rats and mice by both suppressing food intake and activating thermogenesis and BAT (Marchington *et al.*, 1983, 1986; Rothwell and Stock, 1984, 1988b; Holt and York, 1982; Yukimura *et al.*, 1978). These effects are reversed by replacement of the animals with corticosterone or dexamethasone and, in the case of obese animals, only very low doses of replacement are required (King and Smith, 1985; Holt and York, 1982). The effects of adrenalectomy are mimicked by hypophysectomy (Holt *et al.*, 1988; Rothwell and Stock, 1985), leading to the suggestion that reduced synthesis, release or actions of CRF might be partially responsible for the impaired energy balance regulation and thermogenesis in obese mutants (Rothwell and Stock, 1985; York, 1989).

Central injection of CRF-41 does indeed cause dose-dependent increases in metabolic rate, BAT blood flow and thermogenic activity which are prevented by inhibition of the sympathetic nervous system (LeFeuvre *et al.*, 1987). Some time ago Brown (1982) reported activation of sympathetic outflow by CRF but failed to identify its site of action within the brain. Genetically obese rats and mice can respond normally to injection of CRF (Carnie *et al.*, 1988), although preliminary reports have indicated reduced concentrations of CRF in specific hypothalamic nuclei (Moore and Routh, 1988), and these mutants may be unable to synthesize and/or release CRF in response to other stimuli (see later). Intracerebroventricular infusion of CRF in lean or obese animals inhibits body weight gain and stimulates BAT and like adrenalectomy, these effects appear greater in the obese mutants (Arase *et al.*, 1988; Rohner-Jeanrenaud *et al.*, 1988).

The actions of CRF on pituitary ACTH release are modified by glucocorticoid status, but this is not the case for the thermogenic effects of CRF which are almost identical in adrenalectomized animals or those treated with high doses of dexamethasone (see Rothwell, 1990b for review). However, glucocorticoids probably influence thermogenesis by inhibiting CRF synthesis. Peripheral or central administration of the glucocorticoid antagonist RU-38486 stimulates thermogenesis and this effect is reversed by pretreatment of the animals with a CRF receptor antagonist (Hardwick *et al.*, 1989). The dissociation between the thermogenic and pituitary effects of CRF is supported by discrepancies in the relative potencies of CRF-like peptides from other species. Sauvagine is a very potent thermogenic agent when administered centrally in the rat but is much less effective in eliciting ACTH release (LeFeuvre *et al.*, 1989). Furthermore, the thermogenic effects of CRF are not modified by pretreatment of animals with a monoclonal antibody to ACTH (LeFeuvre, Rothwell and White, unpublished data). Nevertheless CRF may depend for its thermogenic effects on the release and action of other POMC products, probably within the CNS. Our preliminary data indicate that γMSH and β endorphin might both be involved in thermogenesis (LeFeuvre, Rothwell and White, unpublished data).

CRF could also mediate in the effects discussed earlier of serotonergic agonists on thermogenesis. 5HT causes release of CRF both *in vivo* and *in vitro* (Calogero *et al.*, 1989; Gibbs and Vale, 1983) and results of our studies in conscious rats indicate that thermogenic responses to 5HT and its precursors are attenuated by inhibition of CRF release or action (LeFeuvre and Rothwell, unpublished data).

In addition to its effects on food intake, CRF has many other central actions, for example on behaviour, sleep, peripheral glucose metabolism and immune function (see Rothwell, 1990b for review). CRF could thus be of considerable importance in the central control of these processes in response to stress, injury or illness and this possibility has been enhanced by recent reports that CRF also mediates central effects of immune signals, most notably IL-1.

Berkenbosch *et al.* (1987) and Sapolsky *et al.* (1987) both reported that activation of the pituitary adrenal system by IL-1 involves hypothalamic release of CRF and pointed out that this may explain many of the endocrine responses to infection and injury. In particular, increases in glucocorticoid concentrations form an important feedback mechanism by subsequently inhibiting the metabolic and inflammatory effects of infection (Besedovsky *et al.*, 1985, 1986). It has now also been demonstrated that the central effects of IL-1β on fever and thermogenesis are also dependent on CRF, since they are inhibited by treatment of animals with a CRF receptor antagonist (alphahelical CRF 9–41), or monoclonal or polyclonal antibodies to CRF (Rothwell, 1989a; Busbridge *et al.*, 1989). The central effects of IL-6 are also dependent on CRF release (Rothwell, unpublished data), although the actions of IL-1α and TNFα are independent of CRF. Uehara *et al.* (1987) have reported differential effects of IL-1α and IL-1β on ACTH release in the rat and we have observed that, although genetically obese mutants fail to show significant thermogenic responses to IL-1β, they can respond normally to IL-1α (Busbridge, Carnie, Dascombe, Johnston and Rothwell, unpublished data). This observation supports the suggestion that obese mutants cannot synthesize or release CRF in response to thermogenic stimuli such as IL-1β and strongly questions the dogma that IL-1α and IL-1β act on the same receptor (Dower *et al.*, 1986).

Although IL-1 appears unable to cross the blood–brain barrier, it could be synthesized within the central nervous system. IL-1β has been identified by immunocytochemistry in several areas of the brain and a high density was observed in hypothalamic regions associated with CRF (Breder *et al.*, 1988). Messenger RNA for IL-1β, but not IL-1α, has also been reported in the CNS and receptors have been identified in many brain areas (Farrar *et al.*, 1987). IL-1β appears to be of greater importance in the control of fever and thermogenesis than IL-1α since antibody to the β form but not to IL-1α inhibits these responses to endotoxin in the rat (Kluger, 1990; Rothwell *et al.*, 1989; Wong *et al.*, 1990). It thus seems likely that IL-1β synthesized within the brain causes release of CRF which then exerts a variety of metabolic and hormonal effects. A recent report has demonstrated that the inhibitory effects of IL-1 on food intake are also dependent on CRF release

(Uehara *et al.*, 1989). Thus inhibition of CRF release or action could be of benefit in the clinical management of cachexia.

The site of action of cytokines is not known precisely although *in vitro* effects on hypothalamic slice preparations have been reported (Tsagakaris *et al.*, 1989) and receptors for IL-1 are distributed throughout the hypothalamus (Farrar *et al.*, 1987). Eicosanoids, mainly prostaglandins (PG), are involved in the pyrogenic and thermogenic effects of most cytokines and can induce release of CRF *in vitro* (Bernadini *et al*, 1989). However, PGE_2, a most likely candidate for activation of fever, fails to release CRF *in vitro* (Bernadini *et al.*, 1989) and its central actions on thermogenesis are not prevented by a CRF antagonist (Rothwell, 1990c). $PGF_{2\alpha}$, also a potent pyrogen, can cause release of CRF and its effects on thermogenesis, which are attenuated by inhibiting CRF action (Rothwell, 1990c), may be involved in the actions of IL-1β.

Glucocorticoids inhibit thermogenic responses to diet, stress, injury, infection and exogenous cytokines, and increases in the circulating concentrations or sensitivity to glucocorticoids may be at least partially responsible for impairments in thermogenesis associated with genetic obesity and ageing (Rothwell and Stock, 1987a,b; Sapolsky *et al.*, 1986; York, 1989). The reduced responses to IL-1β in genetically obese Zucker rats are completely normalized by surgical adrenalectomy (Busbridge *et al.*, 1990), and in ageing mice diminished effects of endotoxin and IL-1 on fever and thermogenesis are almost completely restored to normal by the glucocorticoid antagonist RU486 (Rothwell and Strijbos, unpublished data).

Glucocorticoids might inhibit thermogenesis either by direct inhibition of CRF synthesis or by suppressing prostaglandin synthesis. This latter effect is probably responsible for anti-inflammatory actions of steroids and has been ascribed to release of a mediator known as lipocortin(s) (see Flower, 1988). Lipocortin 1 has been identified in the CNS (Smillie *et al.*, 1989) and central injection of recombinant lipocortin 1 suppresses the thermogenic effects of cytokines in the rat (Carey *et al.*, 1990). Conversely administration of antibody to lipocortin can enhance such responses particularly in animals with impaired thermogenesis (Carey *et al.*, 1990). Several years ago we proposed that prostaglandins might be involved in DIT (Rothwell and Stock, 1986c) thus lipocortin might modify dietary as well as inflammatory and pyrogenic responses.

SUMMARY AND CONCLUSIONS

From relative ignorance just a few years ago, considerable advances have now been made in our understanding of the central control of thermogenesis. A number of central pathways and neurotransmitters have now been identified, most notably the hypothalamic serotonergic, noradrenergic and GABAergic pathways and the importance of CRF and cytokines, although details of their sites and mechanisms of action remain somewhat vague. However, the most important advances have concerned the integration and interaction between the control of food intake and thermogenesis. In very many cases stimulation of one parameter is associated with inhibition of the other, tending to maximize effects on body energy content and suggesting an integrated regulation of energy balance. A number of common mechanisms are involved in the central activation of thermogenesis in response to diverse stimuli such as ingestion of food, protein deficiency, stress, infection and injury. Thus research into the central mechanisms of thermogenesis might provide pharmacological means of treating obesity and cachexia by influencing appetite and food intake. Potential targets already identified include agents to modify the release or action of 5HT, prostaglandins, glucocorticoids and CRF.

REFERENCES

Addae, J.I., Rothwell, N.J., Stock, M.J. and Stone, T.W. (1986). Activation of brown fat thermogenesis in rats by baclofen. *Neuropharmacology*, **25**, 627–31.

Anderson, G.H. (1979). Control of protein and energy intake: role of plasma amino acids and brain transmitters. *Can. J. Physiol. Pharmacol.*, **57**, 1043–57.

Arase, K., York, D.A., Shimizu, H., Shargill, N. and Bray, G.A. (1988). Effects of corticotrophin releasing factor on food intake and brown adipose tissue thermogenesis in rats. *Am. J. Physiol.*, **255**, E255–9.

Arch, J.R.S., Ainsworth, A.T., Cawthorne, M.A., Piercy, M.V., Sennitt, M.V., Thody, V.E., Wilson, C. and Wilson, S. (1984). Atypical β adrenoreceptor on brown adipocytes: target for antiobesity drugs. *Nature*, **309**, 163–5.

Atrens, D.M., Sinden, J.D., Penicaud, L., Devos, M. and LeMagnen, J. (1985). *Physiol. Behav.*, **35**, 15–20.

Aulick, L.H. and Wilmore, D.W. (1983). Hypermetabolism in trauma. In: *Mammalian Thermogenesis* (eds L. Girardier and M.J. Stock), Chapman & Hall, London, pp. 259–304.

Banks, W.A., Kastin, A.J. and Durham, D.A. (1990). Bidirectional transport of interleukin-1 alpha across the blood brain barrier. *Brain Res. Bull.* **23**, 439–42.

Benzi, R.H., Shibata, M., Seydoux, J. and Girardier, L. (1988). Prepontine knife cut induced hyperthermia in the rat. *Pflügers Archiv.*, **411**, 593–9.

Berkenbosch, F., Van Oers, J., del Rey, A., Tilders, F. and Besedovsky, H. (1987). Corticotrophin releasing factor producing neurones in the rat—activation by interleukin 1β. *Science*, **238**, 524–8.

Bernadini, R., Chiorenza, A., Cologero, A.E., Gold, P.N.N. and Chronsos, G.P. (1989). Arachidonic acid metabolites modulate rat hypothalamic corticotrophin-releasing hormone secretion *in vitro*. *Neuroendocrinology*, **50**, 708–15.

Besedovsky, H., del Rey, A. and Sorkin, E. (1985). Immunological neuroendocrine feedback circuits. In: *Neural Modulation of Immunity* (eds. R. Guillemin *et al.*) Raven Press, New York, pp. 165–77.

Besedovsky, H., del Rey, A., Sorkin, E. and Dinarello, C.A. (1986). Immunoregulatory feedback between interleukin-1 and glucocorticoid hormones. *Science,* **233**, 652–4.

Beutler, B. and Cerami, A. (1986). Cachectin and tumour necrosis factor as two sides of the same biological coin. *Nature,* **320**, 584–8.

Bianchi, A., Bruce, J., Cooper, A.L., Childs, C., Kohli, M., Morris, I.D., Morris-Jones, P. and Rothwell, N.J. (1989). Increased brown adipose tissue activity in children with malignant diseases. *Horm. Metab. Res.,* **21**, 587–642.

Bignall, K.E., Heggenness, F.W. and Palmer, J.E. (1975). Effect of neonatal decerebration on thermogenesis during starvation and cold exposure in the rat. *Exp. Neurol.,***40**, 174–88.

Blatteis, C.M. (1988). Neural mechanisms in the pyrogenic and acute phase responses to interleukin-1. *Int. J. Neurosci.,* **38**, 223–32.

Blundell, J.E. (1989). Serotonin and appetite. *Neuropharmacology,* **23**, 1537–51.

Breder, C.D., Dinarello, C.A. and Saper, C.B. (1988). Interleukin-1, immunoreactive innervation of human hypothalamus. *Science*, **240**, 321–4.

Breisch, S.T., Zemlan, F.P. and Hoebel, B.G. (1976). Hyperphagic obesity following serotonergic depletion by intraventricular *p*-chlorophenylalanine. *Science,* **192**, 382–4.

Brooks, S.L., Neville, A.M., Rothwell, N.J., Stock, M.J. and Wilson, S. (1981). Sympathetic activation of brown adipose tissue thermogenesis on cachexia. *Biosci. Rep.,* **1**, 509–17.

Brown, M. (1982). Corticotropin releasing factor: central nervous system sites of action. *Brain Res.,* **399**, 10–14.

Brown, M., Fisher, L.A., Speiss, J., Rivier, C., Rivier, J. and Vale, W. (1982). CRF actions on the sympathetic nervous system and metabolism. *Endocrinology,* **111**, 928–31.

Bruce, J., Childs, C.C., Cooper, A.L. and Rothwell, N.J. (1990). Brown adipose tissue activity in children in relation to disease status. *Proc. Nutr. Soc.* **49**, 189A.

Bruch, K. and Zeisberger, E. (1987). Adaptive changes in thermoregulation and their neuropharmacological basis. *Pharmacol. Ther.,* **35**, 163–215.

Busbridge, N.J., Dascombe, M.J., Tilders, F.J.A., van Oers, J.W.A.M., Linton, E.A. and Rothwell, N.J. (1989). Central activation of thermogenesis and fever by interleukin-1β and interleukin-1α involves different mechanisms. *Biochem. Biophys. Res. Commun.*, **162**, 591–6.

Busbridge, N.J., Carnie, J.A., Dascombe, M.J., Johnston, J.A. and Rothwell, N.J. (1990). Adrenalectomy reverses the impaired pyrogenic responses to interleukin 1β in obese Zucker rats. *Int. J. Obesity*, **14**, 809–19.

Calogero, A.E., Bernadini, R., Margioris, A.N., Bagdy, G., Gallucci, W.T., Munson, P.J., Tamarkin, C., Tomai, T.P., Brady, L., Gold, P.W. and Chrousos, G.P. (1989). Effects of serotonergic agonists and antagonists on CRH secretion by explanted hypothalami. *Peptides,* **10**, 189–200.

Carey, F., Forder, R., Edge, M.D., Greene, A.R., Horan, M.A., Strijbos, P.J.L.M. and Rothwell, N.J. (1990). Lipocortin 1 fragment modifies the pyrogenic actions of cytokines in the rat. *Am. J. Physiol.*, **259**, R266–R269.

Carlisle, H.J., Rothwell, N.J. and Stock, M.J. (1988). Thermic responses to food in rats with lesions of the ventromedial hypothalamus. *Proc. Nutr. Soc.,* **47**, 24A.

Carnie, J.A., LeFeuvre, R.A., Linton, E.A., McCarthy, H.D. and Rothwell, N.J. (1988). Thermogenic effect of CRF in genetically obese Zucker rats. *Proc. Nutr. Soc.,* **47**, 165A.

Childs, C., Ratcliffe, R.J., Holt, I. and Hopkins, S.J. (1989). Relationships between interleukin 1, interleukin 6 and pyrexia in burned children. *Cytokine,* **1**, 36.

Cooper, A.L., Fitzgeorge, R.B., Baskerville, A., Little, R.A. and Rothwell, N.J. (1989). Bacterial infection (*Legionella pneumophilia*) stimulates fever, metabolic rate and brown adipose tissue activity in the guinea pig. *Life Sci.*, **45**, 843–7.

Cooper, K.E. (1987). The neurobiology of fever: thoughts on recent developments. *Ann. Rev. Neurosci.*, **10**, 297–394.

Coscina, D.V. and Magder, R.J. (1984). Effects of serotonin depleting midbrain lesions on the defense of hypothalamic obesity. *Physiol. Behav.*, **33**, 575–9.

Coscina, D.V. and Nobrega, J.N. (1984). Anorectic potency of inhibiting GABA transaminase in brain: studies of hypothalamic dietary and genetic obesities. *Int. J. Obesity,* **8**, 181–200.

Cunningham, S.A. and Nicholls, D.G. (1987). Induction of functional uncoupling protein on guinea pigs infused with noradrenaline. *Biochem. J.*, **245**, 485–91.

Dascombe, M.J., Rothwell, N.J., Sagay, B.O. and Stock, M.J. (1989a). Pyrogenic and thermogenic effects of interleukin 1β in the rat. *Am. J. Physiol.*, **256**, E7–11.

Dascombe, M.J., Hardwick, A., LeFeuvre, R.A. and Rothwell, N.J. (1989b). Impaired effects of interleukin 1β in fever, thermogenesis and brown fat in genetically obese rats. *Int. J. Obesity,* **13**, 367–74.

Dinarello, C.A., Cannon, J.G. and Wolff, S.M. (1988). New concepts on the pathogenesis of fever. *Rev. Infect. Dis.*, **10**, 168–87.

Dower, S.K., Kronheim, S.R., Hopp, T.P., Cantrell, M., Deeley, M., Gillis, S., Henney, C.S. and Urdal, D.C. (1986). The cell surface receptors for interleukin-1α and interleukin-1β are identical. *Nature,* **324**, 266–8.

Dulloo, A.G. and Miller, D.S. (1984). Energy balance following sympathetic denervation of brown adipose tissue. *Can. J. Physiol. Pharmacol.*, **62**, 235–40.

Emeric-Sauval, E. (1986). CRF A review. *Psychoneuroendocrinology,* **11**, 277–94.

Farrar, W.L., Kilian, P.C., Ruff, M.R., Hill, J.M. and Pert, C.B. (1987). Visualization and characterization of interleukin-1 receptors in brain. *J. Immunol.*, **139**, 459–63.

Fernstrom, J.D. (1986). Acute and chronic effects of protein and carbohydrate ingestion on brain tryptophan levels and serotonin synthesis. *Nutr. Rev. Suppl.*, **5**, 25–36.

Figlewicz, D.P., Lacour, F., Sipols, A., Porte, D. and Woods, D. (1987). Gastroenteropancreatic peptides and the central nervous system. *Ann. Rev. Physiol.*, **49**, 383–95.

Fischette, C.T., Nock, B. and Renner, K. (1987). Effects of 5,7-hydroxytryptamine on serotonin$_1$ and serotonin$_2$ receptors throughout the rat central nervous system using quantitative autoradiography. *Brain Res.*, **421**, 263–79.

Flower, R.J. (1988). Lipocortin and the mechanisms of action of the glucocorticoids. *Br. J. Pharmacol.*, **65**, 987–1015.

Freeman, P.H. and Wellman, P.J. (1987). Brown adipose tissue thermogenesis induced by low level electrical stimulation of the hypothalamus in rats. *Brain Res. Bull.*, **18**, 7–11.

Fuller, M.J., Stirling, D.M., Dunnett, S., Reynolds, G.P. and Ashwell, M. (1987). Decreased brown adipose tissue thermogeneic activity following a reduction in brain serotonin by intraventricular *p*-chlorophenylalanine. *Biosci. Rep.*, **7**, 121–7.

Geloen, A., Collet, A.J., Guay, G. and Bukowiecki, L.J. (1988). β-Adrenergic stimulation of brown adipocyte proliferation. *Am. J. Physiol.*, **254**, C175–82.

Gershenwald, J.E., Fong, Y., Chizzonite, R., Lowry, S.F. and Moldawer, L.L. (1989). Anti IL-1 receptor monoclonal antibodies attenuate metabolic changes in inflammation. *Cytokine,* **1**, 150.

Gibbs, D.M. and Vale, W. (1983). Effect of the serotonin uptake inhibitor fluoxetine on corticotropin releasing factor and vasopressin secretion into hypophysial portal blood. *Brain Res.,* **280**, 176–9.

Gillies, G. and Grossman, A. (1985). The CRFs and their control: Chemistry, physiology and clinical implications. *Clin. Endocrinol. Metab.,* **14**, 821–45.

Grossman, S.P. (1962). Direct adrenergic and cholinergic stimulation of hypothalamic mechanisms. *Am. J. Physiol.,* **303**, 782–882.

Han, P.W. (1965). Hypothalamic obesity in rats without hyperphagia. *Trans. N.Y. Acad. Sci.,* **30**, 229–43.

Hardwick, A.J., Linton, E.A. and Rothwell, N.J. (1989). Thermogenic effects of the antiglucocorticoid RU436 in the rat: involvement of corticotropin releasing factor and sympathetic activation of brown adipose tissue. *Endocrinology,* **124**, 1684–8.

Himms-Hagen, J. (1989). Role of thermogenesis in the regulation of energy balance in relation to obesity. *Can. J. Physiol. Pharmacol.,* **67**, 394–401.

Hogan, S., Coscina, D.V. and Himms-Hagen, J. (1982). Brown adipose tissue of rats with obesity introducing ventromedial hypothalamic lesions. *Am. J. Physiol.,* **243**, E334–8.

Holt, S., Rothwell, N.J., Stock, M.J. and York, D.A. (1988) Effect of hypophysectomy on energy balance and brown fat activity in obese Zucker rats. *Am. J. Physiol.,* **254**, E162–6.

Holt, S., Wheal, H. and York, D.A. (1987). Hypothalamic control of BAT in Zucker lean and obese rats; effects of electrical stimulation of the VMH and other hypothalamic nuclei. *Brain Res.,* **405**, 227–33.

Holt, S. and York, D.A. (1982). The effect of adrenalectomy on GDP binding to brown adipose tissue mitochondria of obese rats. *Biochem. J.,* **208**, 819–22.

Horan, M.A., Gibbons, L., Hopkins, S.J., Cooper, A., Strijbos, P., Rothwell, N.J. and Little, R.A. (1989). Changes in plasma interleukin-6 during experimentally induced fever in normal humans. *Cytokine,* **1**, 138.

Horton, R.W., LeFeuvre, R.A., Rothwell, N.J. and Stock, M.J. (1988a). Opposing effects of activation of central $GABA_A$ and $GABA_B$ receptors on brown fat thermogenesis in the rat. *Neuropharmacology,* **27**, 363–6.

Horton, R.W., Rothwell, N.J. and Stock, M.J. (1988b). Chronic inhibition of GABA transaminase results in activation of thermogenesis and brown fat in the rat. *Gen. Pharmacol.,* **19**, 403–6.

Imai-Matsumura, K., Matsumura, K. and Nakayama, T. (1984). Involvement of the ventromedial hypothalamus in brown adipose tissue thermogenesis induced by preoptic cooling in rats. *Jap. J. Physiol.,* **34**, 939–43.

Iwai, M., Hell, N.S. and Shimazu, T. (1987). Effects of VMH stimulation on blood flow of BAT in rats. *Pflügers Arch.,* **410**, 44–7.

Jepson, M.M., Cox, M., Bates, P.C., Rothwell, N.J., Stock, M.J., Cady, E.B. and Millward, D.J. (1987). Regional blood flow and skeletal muscle energy status in endotoxemic rats. *Am. J. Physiol.,* **252**, E581–7.

Jepson, M.M., Millward, D.J., Rothwell, N.J. and Stock, M.J. (1988). Involvement of the sympathetic nervous system and brown fat in endotoxin induced fever in the rat. *Am. J. Physiol.,* **255**, E617–23.

Kennedy, G.C. (1953). The role of depot fat in the hypothalamic control of food intake in the rat. *Proc. Roy. Soc. London,* **140**, 578–92.

King, B.M. and Smith, R.L. (1985). Hypothalamic obesity after hypophysectomy or adrenalectomy: dependence on corticosterone. *Am. J. Physiol.,* **249**, 522–6.

Kluger, M.J. (1989). Body temperature changes during inflammation: their mediation and nutritional significance. *Proc. Nutr. Soc.,* **48**, 337–45.

Landsberg, L. and Young, J.B. (1983). Autonomic regulation of thermogenesis. In: *Mammalian Thermogenesis* (eds L. Girardier and M.J. Stock), Chapman & Hall, London, pp. 99–140.

Leathwood, P.D. (1987). Tryptophan availability and serotonin synthesis. *Proc. Nutr. Soc.,* **46**, 143–56.

LeFeuvre, R.A. (1990). *The Hypothalamic Control of Thermogenesis*, PhD Thesis, University of London.

LeFeuvre, R.A., Rothwell, N.J. and Stock, M.J. (1987). Activation of brown fat thermogenesis in response to central injection of CRF in the rat. *Neuropharmacology,* **26**, 1217–21.

LeFeuvre, R.A., Rothwell, N.J. and White, A. (1989). Comparison of the thermogenic effects of CRF, sauvagine and urotensin 1 in the rat. *Horm. Metab. Res.,* **21**, 525–6.

Leibowitz, S.F. (1986). Brain monoamines and peptides: role in the control of eating behaviour. *Fed. Proc.,* **45**, 1396–403.

Leibowitz, S.F., Roland, C.R., Hor, L. and Squillari, V. (1984). Noradrenergic feeding via the paraventricular nucleus is dependent upon circulating corticosterone. *Physiol. Behav.,* **32**, 857–64.

LeMay, L.G., Vander, A. and Kluger, M. (1989). The effects of psychological stress and anesthesia on plasma interleukin 6 (IL-6) activity in rat. *Cytokine,* **1**, 123.

Little, R.A. (1988). Metabolic rate and thermoregulation after injury. In: *Recent Advances in Critical Care Medicine*, (ed. I. Ledingham), Vol. 3, London, Churchill, pp. 159–72.

Long, C.L. (1977). Energy balance and carbohydrate metabolism in infection and sepsis. *Am. J. Clin. Nutr.,* **30**, 301–10.

McCarthy, H.D., O'Shaughnessy, C.T. and Rothwell, N.J. (1989). Endogenous mediators of cerebral ischaemia-induced hypermetabolism in the rat. *Br. J. Pharmacol.*, **99**, 838P.

Marchington, D., Rothwell, N.J., Stock, M.J. and York, D.A. (1983). Energy balance diet-induced thermogenesis and brown adipose tissue in lean and obese (fa/fa) Zucker rats after adrenalectomy. *J. Nutr.,* **113**, 1395–402.

Marchington, D., Rothwell, N.J, Stock, M.J. and York, D.A. (1986). Thermogenesis and sympathetic activity in BAT of overfed rats after adrenalectomy. *Am. J. Physiol.,* **250**, E362–6.

Moore, B.J. and Routh, V.H. (1988). Corticotrophin releasing factor (CRF) levels are altered in the brain of the genetically obese fatty Zucker rat. *Fed. Proc.,* **22**, 5386.

Morley, J.E. (1987). Neuropeptide regulation of appetite and weight. *Endocr. Rev.,* **8**, 256–87.

Nicholls, D.G. and Locke, R. (1983). Thermogenic mechanism in brown fat. *Physiol. Rev.,* **64**, 1–64.

Oomura, Y. (1976). Significance of glucose, insulin and free fatty acid in the hypothalamic and satiety neurones. In: *Hunger: Basic Mechanisms and Diurnal Implications* (eds D. Novin, W. Wyrwicka and G. Bray), Raven Press, New York, pp. 145–57.

Perkins, M.N., Rothwell, N.J., Stock, M.J. and Stone, T.W. (1981). Activation of brown adipose tissue thermogenesis by the ventromedial hypothalamus. *Nature,* **89**, 401–2.

Rohner-Jeanrenaud, F., Walker, C., Greco Perotto, R. and Jeanrenaud, B. (1988). Chronic intracerebroventricular administration of corticotropin releasing factor (CRF) to genetically obese fa/fa rats arrests the further increase of their body

weight. In: *Obesity in Europe '88: Proc. European Congress on Obesity* (eds P. Bjorntorp and S. Rossner), John Libbey, London, pp. 253–8.
Rothwell, N.J. (1988). Central effects of TNFα in the rat. *Biosci. Rep.*, **8**, 345–52.
Rothwell, N.J. (1989a). CRF is involved in the pyrogenic and thermogenic effects of interleukin 1β in the rat. *Am. J. Physiol.*, **256**, E111–5.
Rothwell, N.J. (1989b). Central control of thermogenesis. *Proc. Nutr. Soc.*, **48**, 241–50.
Rothwell, N.J. (1990a). Thermogenesis on obesity and cachexia. In: *Endocrinology and Metabolism—Hormones and Nutrition in Obesity and Cachexia* (ed. M. Muller), Springer, Heidelberg, pp. 77–85.
Rothwell, N.J. (1990b). Central effects of CRF on metabolism and energy balance. *Neurosci. Biobehav. Rev.*, **14**, 263–71.
Rothwell, N.J. (1990c). Central activation of thermogenesis by prostaglandins, dependence on CRF. *Horm. Metab. Res.*, **22**, 616–8.
Rothwell, N.J. and Stock, M.J. (1984). Sympathetic and adrenocorticoid influences on diet-induced thermogenesis and brown fat activity in the rat. *Comp. Biochem. Physiol. [A]*, **79**, 575–9.
Rothwell, N.J. and Stock, M.J. (1984a). Effects of denervating brown adipose tissue on the responses to cold hyperphagia and noradrenaline treatment in the rat. *J. Physiol.*, **355**, 457–63.
Rothwell, N.J. and Stock, M.J. (1984b). Brown adipose tissue. *Rec. Adv. Physiol.*, **10**, 349–85.
Rothwell, N.J. and Stock, M.J. (1985). Thermogenesis and BAT activity in hypophysectomized rats with and without corticotrophin replacement. *Am. J. Physiol.*, **249**, E333–330.
Rothwell, N.J. and Stock, M.J. (1986a). Brown adipose tissue and diet-induced thermogenesis. In: *Brown Adipose Tissue* (eds P. Trayhurn and D.G. Nicholls), Arnold, London, pp. 269–98.
Rothwell, N.J. and Stock, M.J. (1986b). Whither brown fat? *Biosci. Rep.*, **6**, 3–18.
Rothwell, N.J. and Stock, M.J. (1986c). Possible involvement of prostaglandins in diet-induced thermogenesis of cafeteria-fed rats. *Proc. Nutr. Soc.*, **45**, 111A.
Rothwell, N.J. and Stock, M.J. (1987a). Influence of carbohydrate and fat intake on diet-induced thermogenesis and brown fat activity in rats fed low protein diets. *J. Nutr.*, **117**, 1721–6.
Rothwell, N.J. and Stock, M.J. (1987b). Effects of diet and fenfluramine on thermogenesis on the rat: possible involvement of serotonergic mechanisms in the rat. *Int. J. Obesity*, **11**, 319–24.
Rothwell, N.J. and Stock, M.J. (1988a). Insulin and thermogenesis. *Int. J. Obesity*, **1**, 93–102.
Rothwell, N.J. and Stock, M.J. (1988b). Influence of adrenalectomy on age-related changes in energy balance thermogenesis and brown fat activity in the rat. *Comp. Biochem. Physiol.*, **89A**, 265–9.
Rothwell, N.J., Saville, M.E. and Stock, M.J. (1981). Acute effects of 2-deoxy-D-glucose and noradrenaline on metabolic rate and brown adipose tissue in normal and atropinized lean and obese (fa/fa) Zucker rats. *Pflügers Arch.*, **392**, 172–7.
Rothwell, N.J., Stock, M.J. and Thexton, A.J. (1983). Decerebration activates thermogenesis in the rat. *J. Physiol.*, **342**, 15–22.
Rothwell, N.J., Stock, M.J. and Warwick, B.P. (1984). Thermic effects of methyl cellulose. *Proc. Nutr. Soc.*, **43**, 70A.
Rothwell, N.J., Busbridge, N.J., Humphray, H. and Hissey, P. (1989). Central actions of interleukin-1β on fever and thermogenesis. *Cytokine*, **1**, 153.

Sahakian, B.J., Trayhurn, P., Wallace, M., Deeley, R., Winn, P., Robbins, T.W. and Everett, B.B. (1983). Increased weight gain and reduced activity in brown adipose tissue produced by depletion of hypothalamic noradrenaline. *Neurosci. Lett.,* **39**, 321–6.

Saito, M., Minokoshi, Y. and Shimanin, R. (1987). Accelerated norepinephrine turnover in peripheral tissues after ventromedial hypothalamic stimulation in rats. *Brain Res.,* **481**, 298–303.

Sakaguchi, T., Arase, K., Fisler, J.S. and Bray, G.A. (1988). Effect of starvation and food intake on sympathetic activity. *Am. J. Physiol.,* **255**, R284–8.

Sapolsky, R.M., Kray, L.C. and McEwen, B.S. (1986). The neuroendocrinology of stress and ageing: the glucocorticoid cascade hypothesis. *Endocrinol. Rev.,* **7**, 284–301.

Sapolsky, R.M., Rivier, C., Yamamoto, G., Plotsky, G. and Vale, W. (1987). Interleukin 1 stimulates the secretion of hypothalamic corticotrophin releasing factor. *Science,* **238**, 522–4.

Seydoux, J., Rohner-Jeanrenaud, F., Assimacopoulos-Jeannet, F., Jeanrenaud, B. and Girardier, L. (1981). Functional disconnection of brown adipose tissue in hypothalamic obesity in rats. *Pflügers Arch.,* **390**, 1–4.

Shimazu, T., Noma, M. and Saito, M. (1986). Chronic infusion of norepinephrine into the VMH induces obesity in rats. *Brain Res.,* **369**, 215–23.

Siviy, S.M., Kritikos, A., Atrens, D.M. and Shepherd, A. (1989). Effects of norepinephrine infusion in the paraventricular hypothalamus on energy expenditure in the rat. *Brain Res.,* **487**, 79–88.

Smillie, F., Boulton, C., Peers, S. and Flower, R.J. (1989). Distribution of lipocortins I, II and V in tissues of the rat, mouse and guinea pig. *Br. J. Pharmacol.,* **97**, P90.

Steffens, A.B. (1980). The evidence that humoral factors are involved in the regulation of food intake and body weight. *Brain Res. Bull.,* **5**, 13–6.

Stellar, E. (1954). The physiology of motivation. *Psychol. Rev.,* **61**, 522.

Sternberg, E.M. (1989). Monokines, lymphokines and the brain. In *The Year in Immunology* (eds J.M. Cruse and R.E. Lewis), Karger, Basel, pp. 205–17.

Tsagakaris, S., Gillies, G., Rees, L.H., Besser, M. and Grossman, A. (1989). Interleukin 1 directly stimulates the release of corticotropin releasing factor from rat hypothalamus. *Neuroendocrinology,* **49**, 98–101.

Uehara, A., Gottschall, P.E., Dahl, R.R. and Arimura, A. (1987). Stimulation of ACTH release by human interleukin-1 beta but not by interleukin-1 alpha in conscious freely moving rats. *Biochem. Biophys. Res. Commun.,* **146**, 1286–90.

Uehara, A., Sekuya, C., Takasagi, Y., Namiki, M. and Armura, A. (1989). Anorexia induced by interleukin 1: involvement of corticotropin releasing factor. *Am. J. Physiol.,* **257**, R613–7.

Weingarten, H.P., Chang, P. and McDonald, T.J. (1985). Comparison of the metabolic and behavioural disturbances following PVN and VMH lesions. *Brain Res. Bull.,* **14**, 551–9.

Williamson, D.H. (1987). Brain substrates and the effect of nutrition. *Proc. Nutr. Soc.,* **46**, 81–7.

Wilmore, D.W., Long, J.M., Mason, A.D., Skreen, R.W. and Pruitt, B.A. (1974). Catecholamines: mediator of the hypermetabolic response to thermal injury. *Ann. Surg.,* **870**, 653–70.

Wong, N., LeMay, L.G., Otterness, I., Kunkel, S.C., Vander, A.J. and Kluger, M.J. (1990). The roles of IL-1α and β TNF and IL-6 in LPS fever in the rat. *Prog. Leukocyte Biol.*, **10B**, 313–8.

York, D.A. (1987). Neural activity in hypothalamic and genetic obesity. *Proc. Nutr. Soc.,* **46**, 105–17.
York, D.A. (1989). Corticosteroid inhibition of thermogenesis in obese animals. *Proc. Nutr. Soc.,* **48**, 231–5.
Young, J.B. and Landsberg, L. (1977). Stimulation of the sympathetic nervous system during sucrose feeding. *Nature,* **269**, 615–7.
Young, J.B., Fish, S. and Landsberg, L. (1983). Sympathetic nervous system and adrenal medullary responses to ischaemic injury in mice. *Am. J. Physiol.,* **245**, E67–73.
Young, J.B., Kaufman, L.N. and Saville, M.E. (1985). Increased sympathetic nervous system activity in rats fed a low protein diet. Evidence against a role for dietary tyrosine. *Am. J. Physiol.,* **248**, R627–37.
Young, J.B., Saville, M.E., Rothwell, N.J. and Stock, M.J. (1982). Effect of diet and cold exposure on norepinephrine turnover in brown adipose tissue in the rat. *J. Clin. Invest.,* **69**, 1061–71.
Yukimura, Y., Bray, G.A. and Wolfsen, A.R. (1978). Some effects of adrenalectomy in the fatty rat. *Endocrinology,* **103**, 1924–8.

Chapter Three

Brain Neurochemical Systems Controlling Appetite and Body Weight Gain

Sarah F. Leibowitz
The Rockefeller University, 1230 York Avenue, New York, NY 10021, USA

Under normal conditions, animals exhibit consistent as well as distinctive patterns of food intake, in terms of both the quality and quantity ingested. These patterns can be characterized in relation to the animal's ingestion of the three macronutrients, protein, carbohydrate and fat, and also temporally in relation to the circadian cycle. Evidence indicates that neurotransmitters, in specific areas of the hypothalamus and possibly other brain areas, are involved in coordinating the ingestion and metabolism of nutrients at the different phases of the light/dark cycle. Individual differences in natural patterns of nutrient ingestion and body weight gain may be determined by differences in the functional activity of these hypothalamic neurochemical systems. Moreover, appetite and body weight disturbances may reflect abnormalities in the synthesis and release of these neurotransmitters and in the activity of their respective receptor sites.

This review will summarize findings in animals that support a physiological function of these neurotransmitter receptor systems in controlling natural patterns of nutrient ingestion and metabolism. The specific neurotransmitters which have received considerable attention and will be the focus of this review are the monoamines, noradrenaline, serotonin and dopamine, and the peptides, neuropeptide Y and galanin. Each of these potential neurotransmitters exists in large quantities within the hypothalamus, together with their respective receptor sites, and each has profound impact on natural patterns of macronutrient intake and metabolism when administered directly

Obesity and Cachexia. Edited by N.J. Rothwell and M.J. Stock

into the hypothalamus. The diurnal rhythms of these neurochemicals, their relation to circulating hormones and nutrients, and their role in determining body weight gain will be examined. The possibility that these neurochemical systems contribute to the development and maintenance of clinical eating and body weight disorders will be evaluated.

NATURAL FEEDING PATTERNS

The natural feeding activity in the rat is characterized by bursts of eating at the beginning and towards the end of the dark cycle, with a relatively stable trough in the middle of the cycle (Shor-Posner *et al.*, 1991b; Tempel *et al.*, 1989). The burst of food intake at the start of the feeding period is characterized by a strong preference for carbohydrate; this contrasts with the feeding in the later hours of the active cycle when carbohydrate intake dramatically declines and protein and fat are preferred.

It is proposed that, in animals as well as in humans (Armstrong, 1980; Ashley *et al.*, 1982, 1985; DeCastro, 1987; LeMagnen, 1981; Strubbe *et al.*, 1986; Shor-Posner *et al.*, 1991b; Tempel *et al.*, 1989), food ingested during the first portion of the active period, when hepatic glycogen stores and blood glucose levels may actually decline, is used to fulfill immediate energy requirements and then to promote lipogenesis. This is in contrast to later in the cycle when nutrient and energy stores are to a large extent replenished, and feeding is anticipatory in nature, geared towards ingestion of nutrients (fat and protein) that can be stored and later utilized during the inactive phase.

In addition to these temporal traits of natural feeding patterns characteristic of the overall rat population, there exist individual patterns of feeding with specific traits that distinguish certain rats and allow them to be categorized into sub-groups (Shor-Posner *et al.*, 1991a,b). In particular, measurements of nutrient intake demonstrate that rats often exhibit an overall preference for one specific diet, which accounts for 40–50% of their total diet. They prefer either carbohydrate (50% of the population), fat (30%) or protein (20%), and these three sub-populations each display specific differences in their natural feeding patterns. That is, the high-carbohydrate animals consume smaller and more frequent meals, and they also weigh significantly less than the high-fat and high-protein rats. Moreover, exclusively in the high-fat animals, a third peak of food intake, particularly of fat, in addition to the peaks at the beginning and end of the active cycle, is found to occur during the middle of the dark period, closely reflecting the high body weight of these animals. Evidence described below provides some indication that these behavioral patterns may be related to differences in the activity of the various neurochemical–neuroendocrine substrates.

BRAIN NEUROCHEMICAL SYSTEMS

Noradrenaline

It has been proposed (see Leibowitz, 1986, 1988) that the catecholamine, noradrenaline (NA), is active in controlling the first meal of the active feeding cycle, which is characteristically rich in carbohydrate, and, in general, it evokes a state of energy conservation at this particular time when energy stores are particularly low. Hypothalamic injection of NA in brain-cannulated animals elicits feeding behavior in satiated animals, and this phenomenon is mediated via α_2-, rather than α_1-, noradrenergic receptors. Moreover, it is automatically localized to the paraventricular nucleus (PVN) of the medial hypothalamus, and damage to this nucleus, via electrolytic or neurotoxin lesions, disturbs the animal's ability to respond to NA, to regulate normal diurnal patterns of carbohydrate and protein intake, and to respond normally to nutritional imbalances (Leibowitz, 1988; Leibowitz *et al.*, 1981; Shor-Posner *et al.*, 1985, 1986).

The primary effect of NA, or the α_2-receptor agonist clonidine, is to cause a preferential increase in the animal's appetite for carbohydrate, generally at the expense of protein (Leibowitz, 1988). In addition, NA injection into the PVN decreases energy expenditure (Siviy *et al.*, 1989) and increases circulating levels of glucose, corticosterone and vasopressin (Leibowitz *et al.*, 1988), hormones which act synergistically with NA in potentiating feeding and carbohydrate intake (see below). The activity of NA and the α_2-noradrenergic receptors is closely linked to the light/dark cycle, exhibiting a sharp, unimodal peak at the start of the active nocturnal period. This is reflected in microinjection studies of the animal's sensitivity to NA; in microdialysis studies of endogenous PVN NA levels; and in radioligand binding studies of PVN α_2-receptor density (Leibowitz, 1988). This peak in α_2-noradrenergic activity at dark onset occurs in association with a *natural* rise in circulating levels of corticosterone, as well as in the animal's preference for carbohydrate.

Adrenal steroids

Additional evidence suggests that the adrenal steroids, corticosterone (CORT) and aldosterone (ALDO), are critical in determining the action of central neurotransmitters, such as NA, and may by themselves have direct actions on neural function that determines patterns of nutrient intake, as well as nutrient metabolism (Leibowitz, 1988; Tempel and Leibowitz, 1989). It is proposed that the glucocorticoid CORT exhibits a peak of activity at the beginning of the normal feeding cycle, when it naturally rises in blood levels and when it acts both alone and in conjunction with endogenous NA to stimulate carbohydrate intake and restore glucose in the body. The activity of NA and the PVN α_2-receptors is dependent upon and potentiated by the high levels of circulating CORT at the onset of the active feeding cycle. Furthermore, PVN implants of CORT in adrenalectomized rats selectively potentiate carbohydrate intake, only at the onset of the dark cycle when carbohydrate is normally preferred. This effect of the glucocorticoid contrasts with that produced by the mineralocorticoid ALDO, which primarily stimulates fat ingestion and which acts throughout the nocturnal feeding period but is particularly potent at the end of the cycle when fat is more commonly preferred (Shor-Posner *et al.*, 1991b; Tempel *et al.*, 1989).

Serotonin

Brain serotonin (5HT) is known to have an inhibitory effect on feeding behavior. Recent pharmacological and biochemical studies have identified and characterized more precisely the nature of this effect, the brain sites and receptors involved, and the possible physiological role of endogenous 5HT in controlling natural patterns of nutrient ingestion and metabolism (see Leibowitz, 1986, 1990). The medial hypothalamus is a critical brain area in the mediation of 5HT's action. This includes the PVN and ventromedial nucleus, which are known to be involved in controlling energy balance, in addition to the suprachiasmatic nucleus which determines circadian patterns of eating.

Serotonergic stimulation of these nuclei, with exogenous 5HT or drugs (e.g. d-norfenfluramine and fluoxetine) that enhance endogenous 5HT, preferentially reduces carbohydrate intake in naturally feeding animals, specifically through satiety mechanisms involved in the termination of feeding (Leibowitz, 1990). This phenomenon is mediated predominantly by $5HT_{1B}$ receptor sites, which are concentrated in the medial hypothalamic nuclei and are responsive to changes in nutritional status (Leibowitz and Jhanwar-Uniyal, 1990). In addition to inhibiting energy intake, serotonergic stimulation of the PVN increases energy expenditure via brown adipose tissue thermogenesis, thereby linking 5HT to metabolic as well as behavioral systems for control of energy balance (Leibowitz, 1990; Rothwell, 1990).

The activity of serotonergic function in the medial hypothalamus exhibits a circadian rhythm, possibly controlled by the suprachiasmatic nucleus (Leibowitz, 1990). This rhythm is characterized by a peak at the beginning of the active cycle, when motivation to eat is strongest and is triggered by deficits in energy stores. As indicated above, carbohydrate is the naturally preferred macronutrient at this time, and it is suggested that 5HT becomes most activated under these conditions to terminate the carbohydrate-rich meal, possibly by activating 'satiety' neurons localized in the medial hypothalamus.

In this process, 5HT may interact antagonistically with NA and its α_2-noradrenergic receptors, that normally function to enhance carbohydrate intake and reduce energy expenditure at the onset of the feeding cycle (see above). Moreover, while inducing satiety for carbohydrate, 5HT may additionally play a role in switching the animal's preference towards protein. This macronutrient is regulated in close relationship to carbohydrate and is normally preferred in the second meal of the natural feeding cycle (Shor-Posner *et al.*, 1991b; Tempel *et al.*, 1989).

Dopamine

It is believed that other monoamines besides NA and 5HT, possibly active outside the medial hypothalamus, control the ingestion and perhaps the metabolism of the nutrients, fat and protein. For example, there is evidence that dopamine (DA) acts like 5HT in exerting an inhibitory effect on total food intake but that this catecholamine functions very differently from 5HT in terms of its impact on nutrient selection and meal patterns (see Leibowitz, 1980; Leibowitz and Shor-Posner, 1986). Whereas 5HT preferentially reduces carbohydrate intake, DA causes an inhibition of fat and protein intake and no change in carbohydrate intake (Leibowitz *et al.*, 1986). This effect of DA is apparently mediated by the perifornical region of the lateral hypothalamus, which lies just lateral to the PVN and ventromedial nucleus where NA and 5HT control carbohydrate consumption.

The well-known anorectic drug amphetamine, which enhances the release of endogenous catecholamines, appears to act in part through lateral hypothalamic DA to cause a suppression of appetite for protein and fat (Leibowitz, 1980; Leibowitz *et al.*, 1986). Moreover, the importance of local DA receptor sites in this area, for normal control of nutrient intake, is reflected in the finding that catecholamine receptor antagonists, e.g., chlorpromazine, clozapine and haloperidol, produce the opposite effect, namely, stimulate feeding, potentiate the ingestion of protein and fat, and, after chronic administration, increase daily food consumption and body weight gain (Leibowitz and Shor-Posner, 1986). Meal pattern analyses, of these opposing effects produced by amphetamine and the DA antagonists, suggest that lateral hypothalamic DA reduces nutrient intake through the inhibition of a hunger-stimulating mechanism, rather than via enhancement of satiety as proposed for 5HT in the medial hypothalamus.

Amino acids

Work of Wurtman and Fernstrom (see Fernstrom, 1983; Wurtman *et al.*, 1980) suggests that ingestion of specific nutrients may have impact on circulating levels of the amino acid precursors, tyrosine and tryptophan, and that these amino acids may, in turn, affect the synthesis of the monoamines within the brain. Evidence indicates that, while carbohydrate intake preferentially increases circulating tryptophan and consequently 5HT synthesis, protein ingestion under certain circumstances may cause a relative increase in tyrosine, the amino acid precursor for DA. Thus, the activity of DA may be enhanced after consumption of protein, permitting appropriate neurotransmitter control of the quantity of protein to be eaten at the end of the active feeding cycle. This is in contrast to 5HT's activity, which may be enhanced after consumption of carbohydrate, particularly at the beginning of the natural feeding cycle when the ratio of tryptophan to other large neutral amino acids is most likely to rise.

Glucose

There is evidence that circulating glucose, in addition to the steroids and amino acids, may interact with the monoaminergic systems of the PVN to control carbohydrate intake (Leibowitz, 1988). In addition to stimulating carbohydrate intake, the infusion of NA into the PVN increases blood levels of glucose (Leibowitz *et al.*, 1988). In a variety of conditions, a strong positive correlation exists between circulating glucose levels and the density of α_2-receptor sites in the PVN or medial hypothalamus (Jhanwar-Uniyal *et al.*, 1988; Leibowitz, 1988). Moreover, a short delay (30 min) in food presentation at the start of the active cycle is sufficient to cause a significant decline in α_2-receptor sites exclusively in the PVN, in close relation to changes in circulating glucose levels (Jhanwar-Uniyal *et al.*, 1988). These effects are readily reversed either by feeding or by administration or ingestion of glucose, confirming a close association between α_2-noradrenergic activity, circulating glucose levels and natural eating of carbohydrate.

Neuropeptide Y

The peptide, neuropeptide Y (NPY), which like NA stimulates specifically carbohydrate intake and is most potent at dark onset, is believed to participate in the induction of carbohydrate preference and carbohydrate metabolism (see Leibowitz, 1991a). This peptide acts in cooperation with, but independently of NA, primarily under conditions when requirements for energy intake and lipogenesis are particularly strong.

Specifically, NPY, acting through various medial hypothalamic nuclei, is found to have potent effects on a variety of endocrine, physiological and behavioral systems that modulate energy balance. This peptide stimulates the release of various hormones, such as CORT, ALDO and vasopressin, which modulate energy metabolism as well as feeding behavior (Leibowitz *et al.*, 1988). It also has direct impact on energy metabolism through an effect on substrate utilization and lipogenesis. Finally, it has a remarkably potent stimulatory effect on feeding behavior, which is characterized by a selective increase in carbohydrate ingestion (Stanley *et al.*, 1985) that is strongest at the beginning of the active feeding cycle and is dependent upon circulating levels of CORT (Stanley *et al.*, 1989b). This evidence has led to the proposal that NPY exerts anabolic effects to restore energy balance at specific times of energy depletion. Increased NPY activity may occur specifically at the beginning of the active cycle or after an environmentally imposed period of food deprivation.

Galanin

The peptide, galanin (GAL), is known to coexist with NA in the PVN. Recent studies suggest that this peptide, acting via both behavioral and physiological systems, may have a dual function in restoring energy balance (see Leibowitz, 1991b). Its first action is a short-term effect requiring the rapid release of endogenous NA in the PVN and the presence of circulating CORT; this effect involves the stimulation of carbohydrate feeding in conjunction with the enhancement of metabolic processes, e.g., gluconeogenesis, that are geared towards maintaining or replenishing circulating glucose levels specifically at the beginning of the active feeding cycle. The second action of GAL, a more potent and long-term effect, is expressed independently of NA and CORT; this effect involves the stimulation of fat intake, possibly in relation to the natural rise in fat preference and lipogenesis that occurs naturally later in the feeding cycle and possibly in association with the mineralocorticoid ALDO which similarly potentiates fat intake.

Specifically, the initial glucoregulatory function of GAL is supported by the evidence that: (1) GAL, which coexists with NA in neurons of the PVN, acts like this amine in producing a short-duration carbohydrate feeding response, a response that occurs primarily at the beginning of the active feeding period when there is a natural rise in endogenous NA release; (2) PVN GAL administration enhances extracellular levels of endogenous NA; (3) GAL's stimulatory action on carbohydrate feeding, like the effect of NA, is antagonized by α_2-receptor antagonists, NA synthesis inhibitors and PVN lesions; (4) both GAL- and NA-induced carbohydrate feeding is abolished by adrenalectomy and restored by CORT replacement; and (5) GAL and NA injections both reduce energy expenditure and stimulate glucagon secretion, which naturally rises to induce gluconeogenesis at the onset of the dark period.

The second action of GAL, a stimulation of fat consumption, is a more potent and long-term effect (lasting up to 24 h) than the increase in carbohydrate intake that occurs only in the early dark period. This GAL-induced fat response can be observed throughout the dark cycle, although it is strongest at the end of this period. The evidence suggests that GAL may be involved in mediating the natural rise in fat ingestion that rats exhibit in the latter half of the active period, and that its general function is to replenish lipid stores in the body at this time. In this process, this peptide acts independently of endogenous NA release and circulating CORT, which are at relatively low levels during the later portion of the active feeding cycle. However, GAL may function in association with the steroid ALDO, which is also found to potentiate specifically fat ingestion, most potently at the end of the feeding cycle (Tempel and Leibowitz, 1989). It may also involve an inhibition in the release of DA, which would also be expected to cause an enhancement in fat preference (Leibowitz *et al.*, 1986).

BRAIN NEUROTRANSMITTERS IN RELATION TO BODY WEIGHT GAIN

Chronic neurochemical stimulation

In addition to modulating eating patterns and appetite for specific macronutrients, brain neurotransmitters, in conjunction with circulating peptide and steroid hormones, also affect energy metabolism and body weight gain, in some cases leading to the development of obesity. For example, central infusion studies have revealed an increase in body weight with chronic administration of NA or clonidine into the PVN (Lichtenstein *et al.*, 1984; Yee *et al.*, 1987). This effect is strongest under conditions when the animals are given a single, mixed diet. With pure macronutrients, a smaller change in body weight occurs, as the subjects are allowed to overeat selectively their preferred diet, namely, carbohydrate, while maintaining a relatively stable total calorie intake by consuming less of the other nutrients, in particular, protein. Brain site may also be an important factor, since NA chronically infused into the ventromedial hypothalamus (Shimazu *et al.*, 1986) produces a stronger increase in body weight than in the PVN, suggesting differential roles of these two nuclei in controlling, respectively, body weight gain versus patterns of nutrient ingestion (Leibowitz, 1988).

Repeated infusions of NPY into the PVN, however, have been shown to have the most potent stimulatory effect on body weight and body lipid (Stanley *et al.*, 1986, 1989a). In fact, the magnitude of this effect exceeded what would be predicted on the basis of the hyperphagia produced by NPY, indicating that chronic NPY stimulation may, in fact, enhance the deposition of fat, in addition to potentiating consumption of energy-rich foods.

Metabolic effects

Studies examining the effect of NPY on energy metabolism have provided data that support this suggestion and further differentiate the actions of this peptide from those of the coexisting NA. When injected into the PVN, NA reduces energy expenditure, independently of its effects on activity level (Siviy *et al.*, 1989). This is in contrast to NPY (Menendez *et al.*, 1990), which has no effect on energy expenditure but, rather, has a potent stimulatory effect on respiratory quotient, suggesting that NPY potentiates fat synthesis preferentially from carbohydrate. This diversion of metabolism toward carbohydrate utilization and lipogenesis may be related to the preferential stimulatory effect of NPY on carbohydrate ingestion (Stanley *et al.*, 1985) and also to the decrease in PVN NPY content detected in rats maintained on a high-carbohydrate diet (Beck *et al.*, 1990c). These findings agree with recent evidence showing ventricular NPY administration to promote white fat lipid storage while decreasing brown fat thermogenesis (Billington *et al.*, 1991) and to reduce energy expenditure by controlling sympathetic activity to interscapular brown adipose tissue (Egawa *et al.*, 1991).

Neurochemical changes in obese animals

In addition to these studies with NPY, recent investigations have linked α_2-noradrenergic receptor sites in discrete hypothalamic nuclei to patterns of daily food intake and body weight gain (Jhanwar-Uniyal *et al.*, 1989). This link has been demonstrated in rats maintained on a high-fat diet, which in individual rats causes variable changes in body weight ranging from low to severe levels of obesity. Analyses of hypothalamic α_2-receptors in the moderate and severely obese subjects, as compared with the dietary-resistant and chow-fed rats, reveal a significant increase in receptor number in the PVN and dorsomedial nucleus. Studies of Levin (1990a,b), which demonstrate a similar enhancement of α_2-receptor binding sites in the PVN and other hypothalamic sites of dietary obese animals, give some indication that this effect may be more a consequence rather than a cause of the body weight gain. This conclusion is based on the evidence that these receptor changes are not detected in younger animals that are believed to be prone to gaining weight but have not yet exhibited such weight gain.

Similar studies have also been performed in genetically obese Zucker rats, which in addition to enhanced lipogenesis and body weight gain, exhibit impaired thermogenesis, altered circadian rhythms of CORT release, and aggravated hyperphagia with a preference for carbohydrate- and fat-rich foods (Bray, 1989). These obese animals, relative to their lean littermates, are found to have a greater number of available α_2-receptor sites specifically in the PVN (Jhanwar-Uniyal *et al.*, 1989), in addition to reduced concentration of endogenous NA (Cruce *et al.*, 1976). They also exhibit higher levels of NPY content, in the PVN as well as the arcuate and suprachiasmatic nuclei (Beck *et al.*, 1990a), in addition to raised levels of hypothalamic NPY mRNA suggesting enhanced peptide synthesis (Sanacora *et al.*, 1990). This evidence, together with additional findings showing a positive relationship between circulating CORT, hyperphagia, and either hypothalamic α_2-receptors (Jhanwar-Uniyal *et al.*, 1989; Leibowitz, 1988) or NPY synthesis (White *et al.*, 1990), suggests that both of these neurochemical systems may contribute significantly to the development of the abnormal metabolic, endocrine and behavioral patterns characteristic of these obese animals.

Developmental studies

To characterize developmental patterns of nutrient intake and to identify variables that may be predisposing factors in the development of eating and body weight disorders, we have maintained male and female Sprague-Dawley rats on pure macronutrient diets and examined their patterns of nutrient intake and body weight gain from birth to maturity (Lucas *et al.*, 1990; Shor-Posner *et al.*, 1989). Two of the main findings of these studies are that body weight as early as 3 days of age is positively related specifically to preference for carbohydrate and sucrose after weaning, which in turn is related to body weight at maturity, and also that nutrient intake patterns at weaning closely reflect intake patterns at maturity (day 77). In addition, at puberty, female rats are found to exhibit a strong preference for carbohydrate, whereas the males prefer protein, and at maturity, body weight is strongly related to fat intake, appetite for sucrose, and blood levels of insulin.

Analyses of the α_2-noradrenergic receptor sites of these rats at maturity demonstrate that, in the PVN of female rats and dorsomedial nucleus of male rats, the density of these receptors is positively related to total caloric intake, blood glucose levels, body weight gain and responsiveness to the feeding-stimulatory effects of the α_2-receptor stimulant, clonidine. Moreover, analyses of the $5HT_{1B}$ receptor binding sites reveal, in female rats relative to male rats, a higher density of these receptors in the PVN, in association with a greater responsiveness to the feeding-suppressive effect of serotonergic stimulants (Leibowitz and Jhanwar-Uniyal, 1990).

These results reveal a close relationship between nutrient intake and appetite, body weight, circulating hormones and neurotransmitter receptors in the hypothalamus. Taken together, the above findings demonstrate that the α_2-noradrenergic and perhaps NPY receptors in the medial hypothalamus, known to have a role in the control of energy intake as well as energy metabolism, exhibit dramatic changes in relation to dietary and genetic factors that are known to contribute to the onset and maintenance of obesity.

POSSIBLE FUNCTIONS OF BRAIN NEUROTRANSMITTERS FOR CONTROL OF APPETITE IN HUMANS

The ultimate goal of these studies in animals is to determine whether disturbances in brain monoamines, peptides and their receptors contribute to the development and maintenance of clinical disorders of eating behavior, energy metabolism and body weight regulation. Pharmacological studies in both animals and humans show similar effects across species of the antidepressants, antipsychotic agents, stimulants and peptides on eating behavior (Carruba and Blundell, 1986; Johnson *et al.*, 1983; Leibowitz, 1980; Wurtman and Wurtman, 1986). These studies support the suggestion that the hypothalamic neurotransmitter systems mediating the actions of these compounds may, to some extent, act similarly in these different species, in conjunction with circulating hormones, nutrients and amino acids, to control normal food intake, appetite for specific nutrients and body weight gain. The above evidence demonstrates, in animals, that administration of amines and peptides has profound and distinct effects on patterns of eating behavior and metabolism of specific nutrients; it also reveals that the endogenous neurotransmitters and their receptors are significantly altered in animals exhibiting abnormal patterns of eating and body weight gain. It is therefore reasonable to consider that the pathogenesis of clinical eating disorders may, in part, involve specific disturbances in the activity of these hypothalamic neurochemical systems.

This suggestion is consistent with studies which have detected abnormal concentrations of these neurotransmitters in the CSF of patients with anorexia nervosa or bulimia. For example, anorexics, who have partially recovered their body weight, exhibit a significantly reduced concentration of NA in CSF, in association with sustained behavioral abnormalities including disturbed appetite for calorically dense foods and abnormal recovery from episodes of food deprivation (Kaye *et al.*, 1985, 1988c). Furthermore, bulimia, a disorder characterized by compulsive eating but not necessarily obesity (due to self-indulged purging), has been associated with a disturbed release of brain 5HT, which would be expected to produce an increase in meal size and rate of eating, particularly of energy-rich nutrients (Kaye *et al.*, 1984, 1988a,b). Further studies of the pathophysiology of eating disorders have also detected, in bulimic patients, disturbances in CSF content of peptide YY, a pancreatic polypeptide closely related to NPY (Kaye *et al.*, 1990). In particular, a dramatic increase in CSF peptide YY was detected in bulimics who had abstained from bingeing, leading to the proposal that bulimics, when they initiate a binge, may be responding to heightened levels of this peptide.

There is further evidence suggesting a relationship between hypothalamic NPY, the onset of puberty, and the transition towards development of adult eating patterns in young females. Whereas hypothalamic NPY content rises steadily from birth to the onset of puberty, hypophysial-portal NPY levels, specifically in female rats, exhibit a sharp rise specifically on the day of puberty onset, just prior to the prepubertal surge in luteinizing hormone secretion (Sutton *et al.*, 1988). Studies of feeding behavior in pubescent rats (Lucas *et al.*, 1990; Shor-Posner *et al.*, 1989) indicate that female rats also exhibit a natural rise in preference for carbohydrate at this time, in contrast to the preference for protein exhibited by male rats. It is possible that these simultaneous behavioral and endocrine events are related, since central injection of NPY induces luteinizing hormone secretion (Kalra and Crowley, 1984) and preferentially stimulates carbohydrate ingestion (Stanley *et al.*, 1985). Moreover, food deprivation simultaneously increases NPY content of the PVN (Beck *et al.*, 1990b; Sahu *et al.*, 1988), while producing a strong increase in preference for carbohydrate as well as fat (Shor-Posner *et al.*, 1985; Tempel *et al.*, 1989).

It is of interest that anorexia nervosa, a disturbance in appetite for energy-rich foods, occurs predominantly in females, most often around the time of puberty and as a consequence of dieting and malnutrition (Silverman, 1983). Moreover, CSF levels of NPY are altered in anorexic patients before and after weight restoration (Kaye *et al.*, 1990). It is possible that abnormal peptide activity in anorexia nervosa, possibly related to changes in gonadal and adrenal steroids as well as circulating nutrients and amino acids, may provide the link between disturbed appetite and menstrual function, as well as contribute to the vicious cycle of starvation and bingeing characteristic of this eating disorder.

ACKNOWLEDGEMENTS

This research was supported by a USPHS grant, MH 43422.

REFERENCES

Armstrong, S. (1980). A chronometric approach to the study of feeding behavior. *Neurosci. Biobehav. Rev.*, **4**, 27–53.

Ashley, D.V., Barclay, D.V., Chauffard, F., Moennoz, D. and Leathwood, P.D. (1982). Plasma amino acid responses in humans to evening meals of differing nutritional composition. *Am. J. Clin. Nutr.*, **36**., 143–53.

Ashley, D.V.M., Liardon, R. and Leathwood, P.D. (1985). Breakfast meal composition influences plasma tryptophan to large neural amino acid ratios of healthy lean young men. *J. Neural Transm.*, **63**, 271–83.

Beck, B., Burlet, A., Nicolas, J.P. and Burlet, C. (1990a). Hypothalamic neuropeptide Y (NPY) in obese Zucker rats: Implications in feeding and sexual behaviors. *Physiol. Behav.*, **47**, 449–53.

Beck, B., Jhanwar-Uniyal, M., Burlet, A., Chapleur-Chateau, M., Leibowitz, S.F. and Burlet, C. (1990b). Rapid and localized alterations of neuropeptide Y (NPY) in discrete hypothalamic nuclei with feeding status. *Brain Res.*, **528**, 245–9.

Beck, B., Stricker-Krongrad, A., Burlet, A., Nicolas, J.-P. and Burlet, C. (1990c). Influence of diet composition on food intake and hypothalamic neuropeptide Y (NPY) in the rat. *Neuropeptides*, **17**, 197–203.

Billington, C.J., Briggs, J.E. and Levine, A.S. (1991). Effects of intracerebroventricular injection of neuropeptide Y on energy metabolism. *Am. J. Physiol.*, **260**, R321–27.

Bray, G.A. (1989). Genetic and hypothalamic mechanisms for obesity—finding the needle in the haystack. *Am. J. Clin. Nutr.*, **50**, 891–902.

Carruba, M.O. and Blundell, J.E., eds. (1986). *Pharmacology of Eating Disorders: Theoretical and Clinical Developments*, Raven Press, New York.

Cruce, J.A.F., Thoa, N.B. and Jacobowitz, D.M. (1976). Catecholamines in the brains of genetically obese rats. *Brain Res.*, **101**, 165–70.

DeCastro, J.M. (1987). Circadian rhythms of the spontaneous meal pattern, macronutrient intake, and mood of humans. *Physiol. Behav.*, **40**, 437–46.

Egawa, M., Yoshimatsu, H. and Bray, G.A. (1991). Neuropeptide Y (NPY) suppresses sympathetic activity to interscapular brown adipose tissue in rats. *Am. J. Physiol.*, **260**, R328–34.

Fernstrom, J.D. (1983). Role of precursor availability in control of monoamine biosynthesis in brain. *Physiol. Rev.*, **63**, 484–546.

Jhanwar-Uniyal, M., Grinker, J.A., Finkelstein, J.A., Block, W. and Leibowitz, S.F. (1989). Hypothalamic α_2-noradrenergic receptor system relation to dietary, genetic, and hormonally induced obesity. *Ann. N.Y. Acad. Sci.*, **575**, 613–5.

Jhanwar-Uniyal, M., Papamichael, M.J. and Leibowitz, S.F. (1988). Glucose-dependent changes in α_2-noradrenergic receptors in hypothalamic nuclei. *Physiol. Behav.*, **44**, 611–7.

Johnson, C., Stuckey, M. and Mitchell, J. (1983). Psychopharmacological treatment of anorexia nervosa and bulimia. *J. Nerv. Ment. Dis.*, **171**, 524–34.

Kalra, S.P. and Crowley, W.R. (1984). Norepinephrine-like effects of neuropeptide Y on LH release in the rat. *Life Sci.*, **35**, 1173–6.

Kaye, W.H., Berrettini, W., Gwirtsman, H. and George, D.T. (1990). Altered cerebrospinal fluid neuropeptide Y and peptide YY immunoreactivity in anorexia and bulimia nervosa. *Arch. Gen. Psychiat.*, **47**, 548–56.

Kaye, W.H., Ebert, M.H., Gwirtsman, H.E. and Weiss, S.R. (1984). Differences in brain serotonergic metabolism between nonbulimic and bulimic patients with anorexia nervosa. *Am. J. Psychiatry*, **141**, 1598–601.

Kaye, W.H., Gwirtsman, H.E., Brewerton, T.D., George, D.T. and Wurtman, R.J. (1988a). Bingeing behavior and plasma amino acids: a possible involvement of brain serotonin in bulimia nervosa. *Psychiatry Res.*, **23**, 31–43.

Kaye, W.H., Gwirtsman, H.E., George, D.T., Obarzanek, E., Brewerton, T.D., Jimerson, D.C. and Ebert, M.H. (1988b). Altered feeding behavior in bulima:

is it related to mood and serotonin? In: *Eating Behavior In Eating Disorders* (ed. B.T. Walsh) American Psychiatric Press, Washington, pp. 201–216.

Kaye, W.H., Gwirtsman, H.E., Jimerson, D.C., George, D.T., Karoum, F., Ebert, M.H. and Lake, C.R. (1988c). Catecholamine function in anorexia nervosa at low weight and after weight restoration. In: *Progress in Catecholamine Research. Part C: Clinical Aspects* (eds M. Sandler, A. Dahlstrom and R.H. Belmaker), Alan R. Liss, New York, pp. 355–61.

Kaye, W.H., Jimerson, D.C., Lake, C.R. and Ebert, M.H. (1985). Altered norepinephrine metabolism following long-term weight recovery in patients with anorexia nervosa. *Psychiatry Res.*, **14**, 333–42.

Leibowitz, S.F. (1980). Neurochemical systems of the hypothalamus. Control of feeding and drinking behavior and water-electrolyte excretion. In: *Handbook of the Hypothalamus*, Vol. 3, Part A (eds P.J. Morgane and J. Panksepp), Marcel Dekker, New York, pp. 299–437.

Leibowitz, S.F. (1986). Brain monoamines and peptides: role in the control of eating behavior. *Fed. Proc.*, **45**, 1396–403.

Leibowitz, S.F. (1988). Hypothalamic paraventricular nucleus: interaction between α_2-noradrenergic system and circulating hormones and nutrients in relation to energy balance. *Neurosci. Biobehav. Rev.*, **12**, 101–9.

Leibowitz, S.F. (1990). Serotonin in medial hypothalamic nuclei controls circadian patterns of macronutrient intake. In: *Serotonin: From Cell Biology to Pharmacology and Therapeutics* (eds R. Paoletti, P.M. Vanhoutte, N. Brunello and F.M. Maggi), Kluwer Academic Publishers, Dordrecht, pp. 203–11.

Leibowitz, S.F. (1991a). Brain neuropeptide Y: An integrator of endocrine, metabolic and behavioral processes. *Brain Res. Bull.*, **27**(3/4), in press.

Leibowitz, S.F. (1991b). Hypothalamic galanin in relation to feeding behavior and endocrine systems. In: *Galanin: A New Multifunctional Peptide in the Neuroendocrine System* (eds T. Hokfelt and T. Bartfai), in press, Macmillan Press, New York.

Leibowitz, S.F., Hammer, N.J. and Chang, K. (1981). Hypothalamic paraventricular nucleus lesions produce overeating and obesity in the rat. *Physiol. Behav.*, **27**, 1031–40.

Leibowitz, S.F. and Jhanwar-Uniyal, M. (1990). 5-HT_{1A} and 5-HT_{1B} receptor binding sites in discrete hypothalamic nuclei: Relation to circadian rhythm and gender. *Soc. Neurosci. Abstr.*, **16**, 294.

Leibowitz, S.F. and Shor-Posner, G. (1986). Hypothalamic monoamine systems for control of food intake: Analysis of meal patterns and macronutrient selection. In: *Pharmacology of Eating Disorders: Theoretical and Clinical Developments* (eds M.O. Carruba and J.E. Blundell), Raven Press, New York, pp. 29–49.

Leibowitz, S.F., Shor-Posner, G., MacClow, C. and Grinker, J.A. (1986). Amphetamine: Effects on meal patterns and macronutrient selection. *Brain Res. Bull.*, **17**, 681–9.

Leibowitz, S.F., Sladek, C., Spencer, L. and Tempel, D. (1988). Neuropeptide Y, epinephrine and norepinephrine in the paraventricular nucleus: Stimulation of feeding and the release of corticosterone, vasopressin and glucose. *Brain Res. Bull.*, **21**, 905–12.

LeMagnen, J. (1981). The metabolic basis of dual periodicity feeding in rats. *Behav. Brain Sci.*, **4**, 561–607.

Levin, B.E. (1990a). Obesity-prone and -resistant rats differ in their brain ^3H paraminoclonidine binding. *Brain Res.*, **512**, 54–9.

Levin, B.E. (1990b). Increased brain ^{3}H-paraminoclonidine (α_2-adrenoceptor) binding associated with perpetuation of diet-induced obesity in rats. *Int. J. Obesity,* **14**, 689–700.

Lichtenstein, S.S., Marinescu, C. and Leibowitz, S.F. (1984). Chronic infusion of norepinephrine and clonidine into the hypothalamic paraventricular nucleus. *Pharmacol. Biochem. Behav.,* **13**, 591–5.

Lucas, D.J., Jhanwar-Uniyal, M., Leibowitz, K., Jhanwar, Y.S. and Leibowitz, S.F. (1990). Developmental patterns of nutrient intake, specific appetites and body weight: relation to hypothalamic α_2-noradrenergic receptor sites and circulating levels of glucose and insulin. *Soc. Neurosci. Abstr.,* **16**, 910.

Menendez, J.A., McGregor, I.S., Healey, P.A., Atrens, D.M. and Leibowitz, S.F. (1990). Metabolic effects of neuropeptide Y injections into the paraventricular nucleus of the hypothalamus. *Brain Res.,* **516**, 8–14.

Rothwell, N.J. (1990). Central effects of CRF on metabolism and energy balance. *Neurosci. Biobehav. Rev.,* **14**, 263–71.

Sahu, A., Kalra, P.S. and Kalra, S.P. (1988). Food deprivation and ingestion induce reciprocal changes in neuropeptide Y concentrations in the paraventricular nucleus. *Peptides,* **9**, 83–6.

Sanacora, G., Kershaw, M., Finkelstein, J.A. and White, J.D. (1990). Increased hypothalamic content of prepronenropeptide Y messenger ribonucleic acid in genetically obese Zucker rats and its regulation by food deprivation. *Endocrinology,* **127**, 730–7.

Shimazu, T., Noma, M. and Saito, M. (1986). Chronic infusion of norepinephrine into the ventromedial hypothalamus induces obesity in rats. *Brain Res.,* **369**, 215–23.

Shor-Posner, G., Azar, A.P., Insinga, S. and Leibowitz, S.F. (1985). Deficits in the control of food intake after hypothalamic paraventricular nucleus lesions. *Physiol. Behav.,* **35**, 883–90.

Shor-Posner, G., Azar, A.P., Jhanwar-Uniyal, M., Filart, R. and Leibowitz, S.F. (1986). Destruction of noradrenergic innervation to the paraventricular nucleus: deficits in food intake, macronutrient selection, and compensatory eating after food deprivation. *Pharmacol. Biochem. Behav.,* **25**, 381–92.

Shor-Posner, G., Ian, C., Brennan, G., Cohn, T., Moy, H., Ning, A. and Leibowitz, S.F. (1991a). Self-selecting albino rats exhibit differential preferences for pure macronutrient diets: Characterization of three sub-populations. *Physiol. Behav.*, in press.

Shor-Posner, G., Brennan, G., Ian, C., Jasaitis, R., Madhu, K. and Leibowitz, S.F. (1991b). Nocturnal patterns of macronutrient intake in freely feeding rats, submitted manuscript.

Shor-Posner, G., Brennan, G., Jasaitis, R., Ian, C. and Leibowitz, S.F. (1989). Developmental patterns of eating behavior. *Ann. NY Acad. Sci.,* **575**, 622–4.

Silverman, J.A. (1983). Anorexia nervosa clinical and metabolic observations. *Int. J. Eating Disorders,* **2**, 159–66.

Siviy, S.M., Kritikos, A., Atrens, D.M. and Shepard, A. (1989). Effects of norepinephrine infused in the paraventricular hypothalamus on energy expenditure in the rat. *Brain Res.,* **487**, 79–88.

Stanley, B.G., Anderson, K.C., Grayson, M.H. and Leibowitz, S.F. (1989a). Repeated hypothalamic stimulation with neuropeptide Y increases daily carbohydrate and fat intake and body weight gain in female rats. *Physiol. Behav.,* **46**, 173–7.

Stanley, B.G., Daniel, D.R., Chin, A.S. and Leibowitz, S.F. (1985). Paraventricular nucleus injections of peptide YY and neuropeptide Y preferentially enhance carbohydrate ingestion. *Peptides,* **6**, 1205–11.

Stanley, B.G., Kyrkouli, S.E., Lampert, S. and Leibowitz, S.F. (1986). Neuropeptide Y chronically injected into the hypothalamus: A powerful neurochemical inducer of hyperphagia and obesity. *Peptides,* **7**, 1189–92.

Stanley, B.G., Lanthier, D., Chin, A.S. and Leibowitz, S.F. (1989b). Suppression of neuropeptide Y-elicited eating by adrenalectomy or hypophysectomy: reversal with corticosterone. *Brain Res.,* **501**, 32–6.

Strubbe, J.H., Keyser, J., Dijkstra, T. and Alingh Prins, A.J. (1986). Interaction between circadian and caloric control of feeding behavior in the rat. *Physiol. Behav.,* **36**, 489–93.

Sutton, S.W., Mitsugi, N., Plotsky, P.M. and Sarkar, D.K. (1988). Neuropeptide Y (NPY): a possible role in the initiation of puberty. *Endocrinology,* **123**, 2152–4.

Tempel, D.L. and Leibowitz, S.F. (1989). PVN steroid implants: Effect on feeding patterns and macronutrient selection. *Brain Res. Bull.,* **23**, 553–60.

Tempel, D.L., Shor-Posner, G., Dwyer, D. and Leibowitz, S.F. (1989). Nocturnal patterns of macronutrient intake in freely feeding and food-deprived rats. *Am. J. Physiol.,* **256**, R541–8.

White, B.D., Dean, R.G. and Martin, R.J. (1990). Adrenalectomy decreases neuropeptide Y mRNA levels in the arcuate nucleus. *Brain Res. Bull.,* **25**, 711–15.

Wurtman, R.J., Hefti, F. and Melamed, E. (1980). Precursor control of neurotransmitter synthesis. *Pharmacol. Rev.,* **32**, 315–35.

Wurtman, R.J. and Wurtman, J.J. (1986). Carbohydrate craving, obesity and brain serotonin. *Appetite,* **7**(Suppl.), 99–103.

Yee, F., MacLow, C., Chan, I.N. and Leibowitz, S.F. (1987). Effects of chronic paraventricular nucleus infusion of clonidine and α-methyl-para-tyrosine on macronutrient intake. *Appetite,* **9**, 127–38.

Chapter Four

Protein Turnover and Regulation of Muscle Mass in Man

M.J. Rennie, W.M. Bennet, P.W. Watt, A.A. Connacher, K. Smith and R.T. Jung
Departments of Anatomy and Physiology and Medicine, University of Dundee, Dundee DD1 4HN, UK

INTRODUCTION

What is the relevance of the muscle mass to a consideration of the mechanisms of obesity and cachexia? It makes sense that the energy cost of protein turnover in the body as a whole will be a significant fraction of total energy expenditure and for man in energy equilibrium protein synthesis has been calculated to account for between 10 and 20% of BMR (see Waterlow and Millward, 1989 for review). These values are likely to be somewhat inexact since they have been calculated using theoretical stoichiometries of ATP utilized per peptide bond formed and furthermore they usually have been derived from studies in which whole body protein turnover was measured over a relatively short part of the day, usually in the postabsorptive state. There is no doubt that the contribution of protein synthesis to the resting metabolic rate will vary under different circumstances and although the energy cost for catch-up growth in previously malnourished children was estimated at no more than 20% (Waterlow and Millward, 1989), calculations done on the basis of rates of protein synthesis for preterm neonates suggest rates very much higher than this (De Benoist *et al.*, 1984; Pencharz, 1988). Our own figures for protein synthesis in the term human fetus suggest that protein synthesis could be a substantial proportion of the resting metabolic rate (Rennie *et al.*, 1990). In starvation all the evidence

Obesity and Cachexia. Edited by N.J. Rothwell and M.J. Stock

suggests that whole body protein turnover is depressed so that in depleted patients one would expect the contribution of protein turnover to resting metabolic rate to be less than normal and certainly not contribute to any energy deficit.

The situation is less clear for obesity and cachexia. We may hypothesize a deficit in the contribution to total energy expenditure of thermogenesis associated with whole body protein turnover in obesity and an elevated contribution in cachexia: there may be some evidence for the latter but not for the former. Obese individuals have higher lean body masses than normal, probably due to the stimulatory effect of extra weight bearing work on the muscle mass and such evidence as does exist suggests that obese individuals have, if anything, higher rates of whole body protein turnover per kilogram lean body mass than normal weight control subjects (Nair and Halliday, 1985; A.A. Connacher, R.T. Jung and M.J. Rennie, unpublished work). In some wasting conditions, e.g. those associated with burns, severe injury and cancer, all conditions in which the basal metabolic rate may be elevated, we have good evidence to suggest that whole body protein turnover may be elevated in many cases and the elevations are also reflected in protein synthesis (Rennie, 1985). However, it is difficult to make a case for the energy cost of protein synthesis contributing more than normal to the resting metabolic rate.

What about muscle? Because muscle is such a substantial part of the lean body mass it will be a sizeable contributor to whole body protein synthesis and to the proportion of the total energy expenditure due to it. Estimates of the contribution of muscle to whole body protein synthesis vary in the range of 20–50% (Rennie *et al.*, 1982; Halliday *et al.*, 1988; Bennet *et al.*, 1989; Garlick *et al.*, 1989).

The possibilities that rates of muscle protein turnover are altered in obesity and cachexia are interesting in the context of control of the muscle mass, particularly in relationship to alterations of the metabolic and hormonal environment but remarkably little work has been done in these areas. The purposes of the present article are to review some of our more recent work concerning regulation of protein turnover in human skeletal muscle, and to examine what evidence does exist concerning possible alterations in obesity and cachexia.

METHODS

We have recently discussed all the methods available, both direct and indirect, for the measurement of muscle protein turnover and the reader who requires more detail is directed to that work (Smith and Rennie, 1990). For our present purposes it suffices to say that the methods are direct, depending upon incorporation of tracer leucine (provided either by continuous infusion (Rennie *et al.*, 1982; Halliday *et al.*, 1988; Carraro *et al.*, 1990) or by large dose injection (Garlick *et al.*, 1989), or indirect measurement of exchange of tracer amino acids (Rennie *et al.*, 1983; Rennie and Halliday, 1984; Cheng *et al.*, 1985; Gelfand and Barrett, 1987; Bennet *et al*, 1990a) or of tyrosine and 3-methylhistidine (Lundholm *et al.*, 1982; Rennie *et al.*, 1984; Bennegård *et al.*, 1985). The work of Stjernström and Sjölin has demonstrated (Sjölin *et al.*, 1989; Sjölin, 1988) that the extremely critical position we took (Rennie and Millward, 1983) of whole body 3-methylhistidine excretion and of the use of the 3-methylhistidine:creatinine urinary ratio was probably too severe and that, although 3-methylhistidine production is unlikely to give good quantitative estimates of muscle protein turnover, it may still be useful as an index of whole body protein breakdown and, occasionally, of skeletal muscle protein breakdown.

The precursor problem

We would like to discuss one methodological problem since this is of great importance for the interpretation of both the data on leucine incorporation and the arteriovenous tracer exchange methods. This concerns the extent of labelling of the precursor pool (from which leucine is taken into protein synthesis) in relation to the labelling of the plasma pool and to the intracellular pool (which is presumably the pool fed by protein breakdown). Measurements in patients infused with leucine before and during a back operation have shown that human skeletal muscle tRNA is labelled to about the same extent as α-ketoisocaproate (α-KIC) and somewhere between the intracellular and the plasma tRNA leucine enrichment values (Figure 4.1). We have traditionally used the α-KIC labelling as an index of the precursor and so this result was convenient (to say the least) in retrospectively validating the absolute values of muscle protein synthesis we previously obtained. It is possible to criticize our acceptance of the happy conjunction of tRNA and α-KIC labelling on the grounds that the delay in freezing muscle sampled by surgical biopsy might have been sufficient to equilibrate the tRNA pool with another of a higher enrichment (e.g. the extracellular), especially given the known very high activity of aminoacyl tRNA transferase. However, work done by Peter Watt using di-labelled [^{13}C,^{15}N]leucine in the rat suggests that delay in freezing even excised muscle samples taken from rat does little to equilibrate the plasma leucine with leucyl tRNA; furthermore the fact that leucyl tRNA labelling by ^{13}C and ^{15}N leucine are different from each other and substantially lower than the plasma values argues against equilibration during homogenization of the sample (Figure 4.2). Incidentally, the fact that the ^{13}C and ^{15}N labelling in leucyl tRNA is discrepant demonstrates the extent of activity of the branched chain

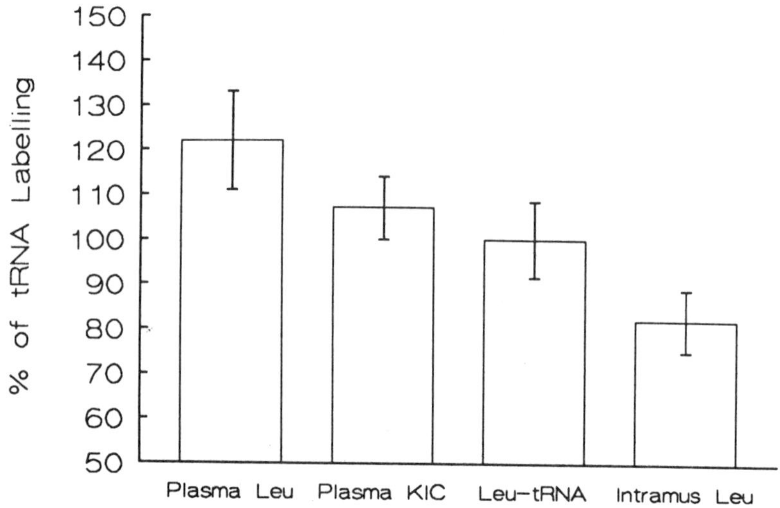

Figure 4.1 Relationship between labelling of leucine metabolites in human skeletal muscle obtained at operation after preoperative infusion of [1-^{13}C]leucine in patients undergoing a back operation

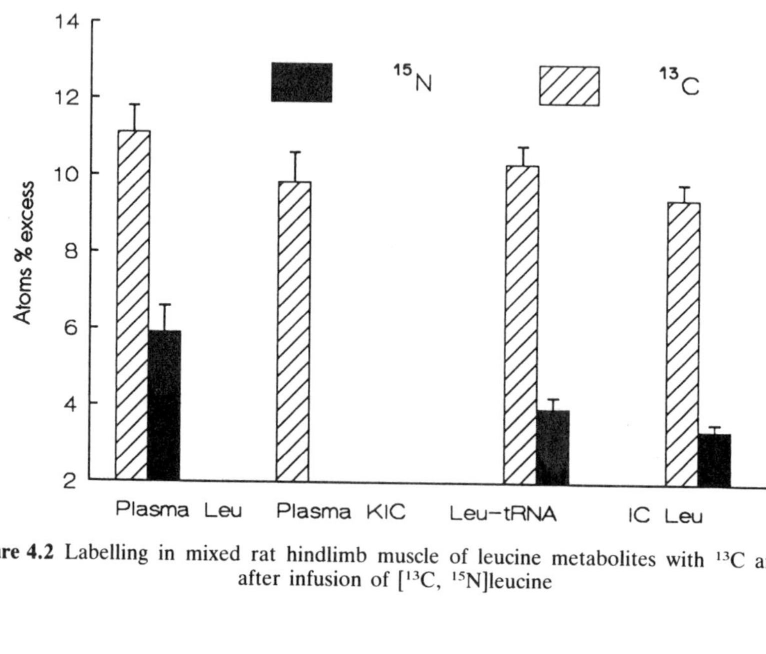

Figure 4.2 Labelling in mixed rat hindlimb muscle of leucine metabolites with ^{13}C and ^{15}N after infusion of [^{13}C, ^{15}N]leucine

amino acid transferase enzyme in skeletal muscle and suggests that the transamination pool which is probably cytoplasmic (Hutson, 1988) and the tRNA charging pool are in close connection.

RESULTS

Age and gender-related influences on muscle protein turnover

The amount of evidence is relatively sparse but for what it is worth (Figure 4.3) rates of muscle protein synthesis determined from incorporation of [^{13}C]leucine into protein sampled from limb, abdominal and back muscle show little change during adulthood. However, the data are rather scattered and the evidence concerning breakdown, from 3-methylhistidine excretion (being mindful of the need to interpret these data carefully), suggests that in man, as in animals, there is the kind of pattern one would expect of high turnover during growth and development with a long drawn out slow fall phase over the period of adult life. It may be that as we collect more data for adults we will discern a gradual fall in muscle protein synthesis but the

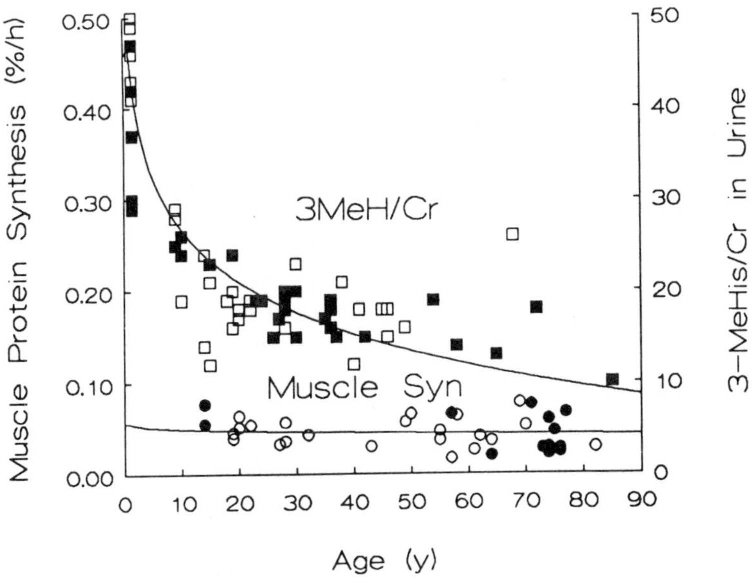

Figure 4.3 Individual values of rates of leucine incorporation into skeletal muscle in persons of different ages and genders (lower data set) and of 3-methylhistidine:creatinine ratios in urine (upper data set) (male ■, ●; female □, ○). Note that the apparently greater negative slope during adulthood of the upper data set is an artefact of the scaling chosen to separate the two data sets (adapted from Smith and Rennie 1990)

extent of this detriment is not likely to be great, given the fact that whole body protein turnover per kilogram lean body mass changes little (Fukagawa *et al.*, 1989) between the second and eighth decades of life. We have been able to determine no gender link differences in muscle protein synthesis, although the 3-methylhistidine:creatinine urinary ratio is slightly less in women (Elia *et al.*, 1981). Estimates derived from 3-methylhistidine production in neonates suggest that muscle protein turnover is very high, possibly approaching values of four times the adult rate (Burgoyne *et al.*, 1982; Pencharz, 1988).

Nutritional modulation

Short-term feeding and fasting appears to modulate muscle protein synthesis (and probably muscle protein breakdown) in ways conforming to common sense, i.e. that synthesis is increased during feeding, with breakdown probably being diminished, and that synthesis seems to be diminished during short-term fasting with breakdown probably being elevated (Rennie *et al.*, 1982; Halliday *et al.*, 1988; Rennie, 1986). Longer term fasting diminishes both (Young *et al.*, 1973; Essén *et al.*, 1989). The results of a number of studies from ourselves and our close collaborators using the leucine incorporation method and the A–V tracer exchange method provide a certain amount of confidence in the consistency of these responses. It may be, however, that as recently discovered by Dave Halliday and Joe Millward (personal communication) for whole body protein turnover, the extent of the nutritional variation very much depends upon the previous protein content of the diet: larger excursions may be expected when the normal dietary intake of protein is in the normal to high range (i.e. above 1.5 g protein/kg body weight per day). Initially when the large dose method was applied to subjects who were refed after an overnight fast no stimulatory effect on muscle protein synthesis could be observed but when food was provided for a longer period (up to 10 h) then a stimulatory effect of about 35% of the already high (0.08%/h) muscle protein synthetic rate could be observed (P.J. Garlick and J. Wernerman, personal communication).

The mechanisms involved in the changes are presently unknown but obvious candidates suggest themselves as mediators. The availability of amino acids, alterations in insulin secretion and possibly alterations in the availability of growth hormone, somatomedin 1, catecholamines, corticosteroids, etc. may be involved. It seems inherently likely that many local paracrine and cytokine effects are involved in addition. So far we have only been able to scratch the surface of the phenomenology by investigating the effects of insulin and amino acids in normal subjects and diabetic patients but the results are encouraging.

Effects of availability of amino acids and insulin on muscle protein turnover

We have been able to show, to our minds convincingly, that incorporation of labelled leucine into human skeletal muscle is enhanced by an increased availability of amino acids (Bennet *et al.*, 1989) since there were no hormonal or metabolic changes other than the increases in amino acids which seemed likely to have affected the response of the muscle. In experiments carried out in pigs (Watt *et al.*, 1991) in which amino acid availability was varied under conditions of fixed insulin and glucagon availability (achieved by use of somatostatin and exogenous infusion of insulin and glucagon), we obtained a similar effect to that seen in man strengthening our feeling of the importance of amino acids in stimulating protein synthesis. More recently, using the A–V tracer exchange method we have confirmed this effect in normal subjects (Bennet *et al.*, 1990a, c) but we do not, as yet, have any idea about the dose-response of the system to amino acids.

None of our data suggest an effect of amino acid on muscle protein breakdown, suggesting that muscle behaves differently from liver in which the availability of alanine, leucine and glutamine appears to be very powerful as a switch in inhibiting proteolysis (Mortimore *et al.*, 1988). The major regulator of protein breakdown appears to be insulin. The dilution across the limb (i.e. A to V) of tracer amino acids (both leucine and phenylalanine) as indices of limb protein breakdown can be shown to be markedly inhibited by insulin provided either systemically or by close arterial injection (Gelfand and Barrett, 1987; Bennet *et al.*, 1990a). These results are in line with those reported for whole body protein turnover (Castellino *et al.*, 1987; Tessari *et al.*, 1987) suggesting that insulin plays an important role in regulating protein breakdown in various tissues of the body. What has been very puzzling has been the difficulty of resolving the effects of insulin, if any, on muscle protein synthesis. Halliday and co-workers (Nair *et al.*, 1984; Pacy *et al.*, 1989) first suggested that in patients with type 1 diabetes (insulin-dependent diabetes) muscle protein synthesis rates were not depressed below normal in the insulin-deprived state and did not respond to replacement of insulin. We found this difficult to fit into our conceptual framework of insulin as an anabolic hormone and hypothesized that the result might be due to a lack of the availability of the full range of amino acids necessary for protein synthesis and that if such availability were to be maintained then an effect of insulin might be demonstrable (Bennet *et al.*, 1990b). In fact we discovered that the amount of amino acid required to maintain the availability was substantially higher than we first thought; provision of exogenous amino acids at infusion rates adequate to maintain blood amino acids did not maintain intramuscular amino acid pools and no stimulatory effect could be observed in the presence of insulin (Bennet *et al.*, 1990b).

However, we have been able to show, using the A–V tracer exchange method both in normal subjects and in diabetic patients that insulin is able to stimulate muscle protein synthesis if amino acid concentrations are elevated to the level seen after a substantial meal (Bennet *et al.*, 1990c; and W.M. Bennet and M.J. Rennie, unpublished results) (Figure 4.4). An intriguing feature of the results of the patients with insulin-dependent diabetes is the suggestion that muscle protein synthesis in them is insulin-resistant, as is known to be the case for glucose uptake.

No dose–response data (for insulin or amino acids and synthesis) are yet available (although we are attempting to remedy this) but our feeling is that the effects of amino acids and insulin probably explain most of the observed changes brought about by feeding, in elevating muscle protein synthesis and decreasing muscle protein breakdown during the switch from the postabsorptive to the fed state.

Mechanisms by which amino acids influence muscle protein synthesis are presently unknown. Since glutamine is an amino acid which is normally present in high concentration in skeletal muscle and shows large pathophysiological fluctuations in concentration we became interested in the idea that glutamine may affect the processes of protein synthesis and breakdown. Work in the isolated perfused rat hindlimb demonstrated that glutamine could stimulate muscle protein synthesis and inhibit muscle protein breakdown

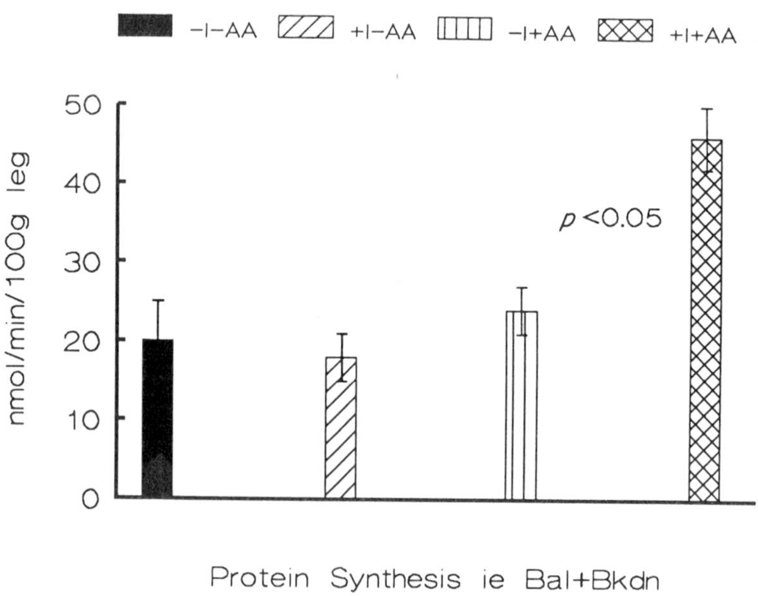

Figure 4.4 Effects of amino acids and insulin on leg phenylalanine uptake (i.e. muscle protein synthesis) in diabetic patients

(MacLennan *et al.*, 1987, 1988) and since then it has been demonstrated that supplementation of parenteral solutions with glutamine peptide diminishes the fall in muscle glutamine and increases whole body nitrogen excretion observed after major surgery (Stehle *et al.*, 1989). Recently we carried out some preliminary experiments in young healthy men infused either with the dipeptide alanyl glutamine or with isonitrogenous amounts of alanine and glutamine and measured incorporation of [1-^{13}C]leucine into muscle protein. Our results suggested a small but significant increase in the rate of incorporation of leucine into muscle protein (J.M. Barua, K. Smith and M.J. Rennie, unpublished work).

Indirect indices of muscle protein synthesis, such as the percentage of total ribosomes present in the polyribosome form, also show some relationship to changes in glutamine concentration in human skeletal muscle (Wernerman *et al.*, 1990). These authors hypothesize that the fall in glutamine in muscle may be the result of a shortage of tricarboxylic acid intermediates in muscle and they have shown that α-ketoglutarate supplementation is as efficient as glutamine dipeptide in preserving muscle glutamine concentration. What is not yet clear is whether or not the α-ketoglutarate supplied is a precursor for glutamine synthesis elsewhere in the body, such as the liver in which it is known that availability of α-ketoglutarate can result in a stoichiometrically equivalent production of glutamine (D. Haussinger, personal communication).

Other hormonal effects

We have not yet been able systematically to investigate the acute effects of any other hormone but there is evidence available in the literature and from our own studies that chronic treatment with hormones causes a considerable modulation of muscle protein turnover in man. High doses of testosterone appear to slightly increase muscle protein synthesis in young men (Griggs *et al.*, 1986, 1989).

In studies of elderly patients with rheumatoid arthritis who normally have slightly elevated rates of muscle protein synthesis (even with higher rates of muscle protein breakdown) we have observed that the therapeutic use of corticosteroids causes a very marked suppression of muscle protein synthesis (Gibson *et al.*, 1990).

Studies by Drs Dennis Bier, Kevin Yarasheski and John Holloszy in St Louis, in which we collaborated, have been unable to demonstrate any additional beneficial effect of growth hormone on muscle protein synthesis in young men undergoing weight-training after 6 weeks of treatment; this finding does not, of course, exclude an early, short-term effect.

We have been able to see acute effects of adrenaline infusion in slightly inhibiting muscle protein breakdown (see below).

Effects of obesity on muscle protein turnover

So far as we are aware there are no published data on the rates of muscle protein synthesis in obese subjects. In an extensive study of whole body and limb protein turnover in lean and obese women studied in the fed and fasted state we found minor increases in the rates of whole body and muscle protein breakdown when expressed per kilogram of lean tissue (Figures 4.5 and 4.6). There was a tendency for rates of protein turnover to be higher in the obese than in the lean, and in obese patients the beneficial effect of adrenaline in decreasing protein breakdown appeared to be smaller. Because the design of the experiments did not allow measurement of protein synthesis it is difficult to interpret the turnover results in terms of energy costs; nevertheless, since protein breakdown consumes a negligible amount of energy, the elevations in the obese are small (and to stay in tissue balance synthesis must have matched breakdown), the results provide no possibility of a deficit in energy consumption in the obese; quite the opposite in fact!

Effects of cachexia on muscle protein turnover

In weight-losing cancer patients we previously observed that rates of protein synthesis (and probably breakdown) were depressed (Emery *et al.*, 1984)

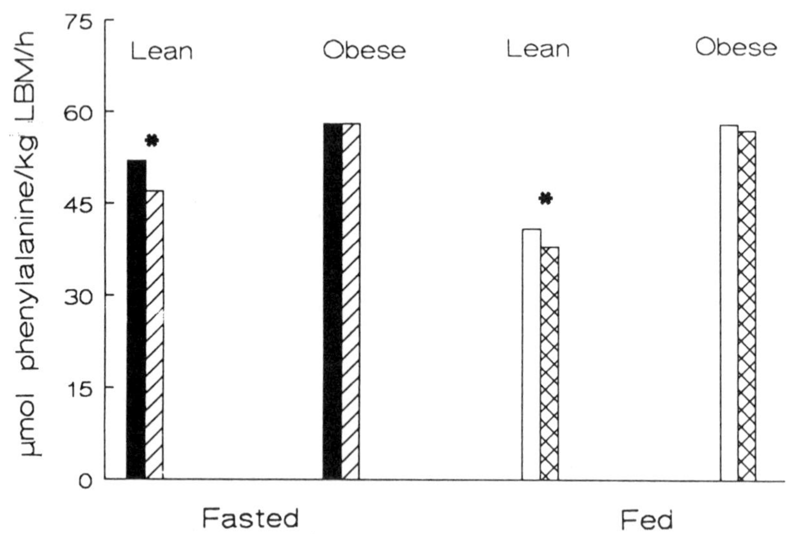

Figure 4.5 Whole body protein breakdown in lean and obese women in the postabsorptive and fed states, in the absence (■, □) and presence (▨, ▩) of infused adrenaline. The values were obtained from plasma flux of [^{15}N]phenylalanine after taking account of amino acid in food. Estimates of protein synthesis were impossible because of our lack of a method for measurement of oxidation

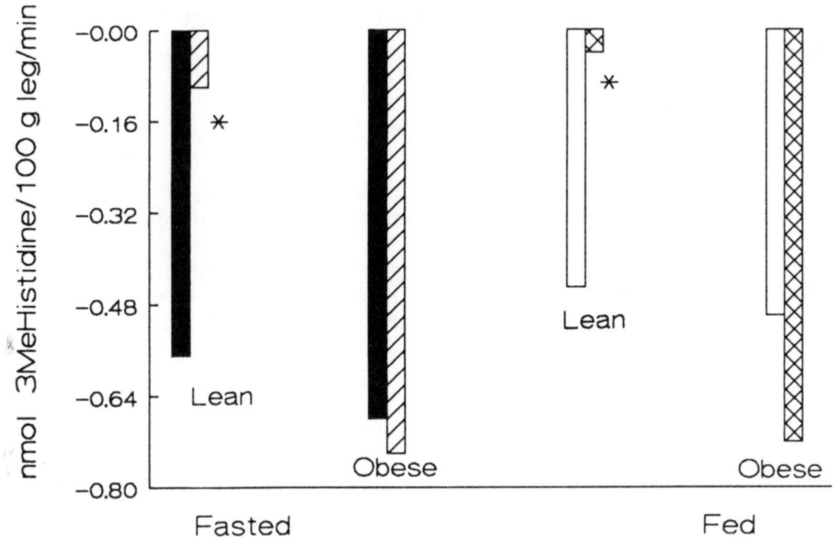

Figure 4.6 Leg efflux of 3-methylhistidine in the patients whose results are presented in Figure 4.5

but more recently Peter Emery and colleagues have discovered that in weight-stable patients with the same carcinoma (presumably pre-cachectic) muscle protein synthesis is not depressed. The general association of a depression of muscle turnover only with muscle wasting (see Rennie, 1985) is thus confirmed in that particular condition.

However, it seems from studies of 3-methylhistidine excretion in trauma, sepsis and burns that muscle protein breakdown is likely to be elevated (Rennie, 1985; Bennegård *et al.*, 1985; Sjölin *et al.*, 1989). Oddly, so far as we know, there are no measurements of muscle protein synthesis postoperatively or in conditions of florid negative nitrogen balance usually associated with early cachexia and we do not know whether synthesis is elevated or depressed compared with normal. It would be our guess that it is actually depressed but we await the appropriate measurements to confirm or refute the position.

ACKNOWLEDGEMENTS

This work was supported by generous grants from Action Research for the Crippled Child, British Diabetic Association, Medical Research Council, Rank Prize Funds, Scottish Home & Health Department, Scottish Hospital Endowment Research Trust, Ajinomoto GmbH, University of Dundee, and

The Wellcome Trust. We are grateful to Mrs D. Watt for typing the manuscript and to our friends and colleagues for constructive criticism.

REFERENCES

Bennegård, C., Edén, E., Emery, P.W., Rennie, M.J. and Lundholm, K. (1985). Efflux of 3-methylhistidine and tyrosine in severe trauma and in patients with septicaemia. *Clin. Nutr.*, **4**, P65.

Bennet, W.M., Connacher, A.A., Scrimgeour, C.M., Smith, K. and Rennie, M.J. (1989). Increase in anterior tibialis muscle protein synthesis in healthy man during mixed amino acid infusion: studies of incorporation of 1-[^{13}C]leucine. *Clin. Sci.*, **76**, 447–54.

Bennet, W.M., Connacher, A.A., Scrimgeour, C.M. and Rennie, M.J. (1990a). The effect of amino acid infusion on leg protein turnover assessed by L-[^{15}N]phenylalanine and L-[^{13}C]leucine exchange. *Eur. J. Clin. Invest.*, **20**, 37–46.

Bennet, W.M., Connacher, A.A., Smith, K., Jung, R.T. and Rennie, M.J. (1990b). Inability to stimulate skeletal muscle or whole body protein synthesis by insulin and glucose during amino acid infusion: studies in type 1 diabetic subjects with tracer L-[^{13}C]-leucine. *Diabetologica,* **33**, 43–51.

Bennet, W.M., Connacher, A.A., Scrimgeour, C.S., Jung, R.T. and Rennie, M.J. (1990c). L-[^{15}N]Phenylalanine and L-[1-^{13}C]leucine leg exchange and plasma kinetics during amino acid infusion and euglycaemic hyperinsulinaemia; evidence for stimulation of muscle and whole body protein synthesis in man by insulin. *Proc. Nutr. Soc.*, **49**, 179A.

Burgoyne, J.L., Ballard, F.J., Tomas, F.M., Dobozy, A., MacLennan, A.H., Fitzgerald, A. and Dahlenburg, G.W. (1982). Measurement of myofibrillar protein breakdown in newborn human infants. *Clin. Science,* **63**, 421–7.

Carraro, F., Stuart, C.A., Hartl, W.H., Rosenblatt, J. and Wolfe, R.R. (1990). Effect of exercise and recovery on muscle protein synthesis in human subjects. *Am. J. Physiol.*, **259**, E470–76.

Castellino, P., Luzi, L., Simonson, D.C., Haymond, M. and DeFronzo, R.A. (1987). Effect of insulin and plasma amino acid concentrations on leucine metabolism in man: Role of substrate availability on estimates of whole body protein synthesis. *J. Clin. Invest.*, **80**, 1784–93.

Cheng, K.N., Dworzak, F., Ford, G.C., Rennie, M.J. and Halliday, D. (1985). Direct determination of leucine metabolism and protein breakdown in humans using L-[1-^{13}C,^{15}N]-leucine and the forearm model. *Eur. J. Clin. Invest.*, **15**, 349–54.

De Benoist, B., Abdulrazzak, Y., Brooke, O.G., Halliday, D. and Millward, D.J. (1984). The measurement of whole body protein turnover in the preterm infant with intragastric infusion of L-[1-^{13}C]leucine and sampling of urinary leucine pool. *Clin. Sci.*, **66**, 155–64.

Elia, M., Carter, A., Bacon, S., Winearls, C.G. and Smith, R. (1981). Clinical usefulness of urinary 3-methylhistidine excretion in indicating muscle protein breakdown. *Br. Med. J.*, **282**, 351–4.

Emery, P.W., Halliday, D., Edwards, R.H.T., Souhami, R.L. and Rennie, M.J. (1984). Protein synthesis in muscle measured *in vivo* in cachectic patients with cancer. *Br. Med. J.*, **289**, 584–6.

Essén, P., McNurlan, M.A., Calder, A.G., Milne, E., Wernerman, J., Garlick, P.J. and Vinnars, E. (1989). The effect of total starvation on protein synthesis rate

in human skeletal muscle determined by the large dose technique. *Clin. Nutr.*, **8**(Suppl.), 90–102.
Fukagawa, N.K., Minaker, K.L., Young, V.R., Matthews, D.E., Bier, D.M. and Rowe, J.W. (1989). Leucine metabolism in aging humans: effect of insulin and substrate availability. *Am. J. Physiol.*, **256**, E288–94.
Garlick, P.J., Wernerman, J., McNurlan, M.A., Essen, P., Lobley, G.E., Calder, G.A. and Vinnars, E. (1989). Measurement of the rate of protein synthesis in muscle of postabsorptive young men by injection of a 'flooding dose' of 1-[^{13}C]leucine. *Clin. Sci.*, **77**, 329–36.
Gelfand, R.A., and Barrett, E.J. (1987). Effect of physiologic hyperinsulinemia on skeletal muscle protein synthesis and breakdown in man. *J. Clin. Invest.*, **80**, 1–6.
Gibson, J.N.A., Poyser, N.L., Morrison, W.L., Scrimgeour, C.M. and Rennie, M.J. (1991). Muscle protein synthesis in patients with rheumatoid arthritis: effect of chronic corticosteroid therapy on prostaglandin $F_{2\alpha}$ availability. *Eur. J. Clin. Invest.* (in press).
Griggs, R.C., Halliday, C., Kingston, W. and Moxley, R.T. (1986). Effect of testosterone on muscle protein synthesis in myotonic dystrophy. *Ann. Neurol.*, **20**, 590–6.
Griggs, R.C., Kingston, W., Jozefowicz, R.F., Herr, B.E., Forbes, G. and Halliday, D. (1989). Effect of testosterone on muscle mass and muscle protein synthesis. *J. Appl. Physiol.*, **66**, E498–503.
Halliday, D., Pacy, P.J., Cheng, K.N., Dworzak, F., Gibson, J.N.A. and Rennie, M.J. (1988). Rate of protein synthesis in skeletal muscle of normal man and patients with muscular dystrophy: a reassessment. *Clin. Sci.*, **74**, 237–40.
Hutson, S.M. (1988). Subcellular distribution of branched-chain aminotransferase activity in rat tissues. *J. Nutr.*, **118**, 1475–81.
Lundholm, K., Bennegård, K., Edén, E., Svaninger, G., Emery, P.W. and Rennie, M.J. (1982). Efflux of 3-methylhistidine from the leg in cancer patients who experience weight loss. *Cancer Res.*, **42**, 4807–11.
MacLennan, P.A., Brown, R.A. and Rennie, M.J. (1987). A positive relationship between intracellular glutamine concentration and protein synthesis in rat skeletal muscle. *FEBS Lett.*, **215**, 187–91.
MacLennan, P.A., Smith, K., Weryk, B., Watt, P.W. and Rennie, M.J. (1988). Inhibition of protein breakdown by glutamine in perfused rat skeletal muscle. *FEBS Lett.*, **237**, 133–6.
Mortimore, G.E., Wert, J.J. Jr. and Adams, C.E. (1988). Modulation of the amino acid control of hepatic protein degradation by caloric deprivation. Two modes of alanine co-regulation. *J. Biol. Chem.*, **263**, 19545–51.
Nair, K.S. and Halliday, D. (1985). Energy and protein metabolism in diabetes and obesity. In: *Substrate and Energy Metabolism* (eds J.S. Garrow and D. Halliday), John Libbey, London p. 195.
Nair, K.S., Read, W.W.C., Ford, G.C. and Halliday, D. (1984). Insulin withdrawal does not reduce skeletal muscle protein synthesis in type 1 diabetic patients. *Diabetes*, **33**, 166A (Abstr.).
Pacy, P.J., Nair, K.S., Ford, C. and Halliday, D. (1989). Failure of insulin infusion to stimulate fractional muscle protein synthesis in type 1 diabetic patients: anabolic effect of insulin and decreased proteolysis. *Diabetes*, **38**, 618–24.
Pencharz, P.B. (1988). Protein metabolism in premature human infants. *Can. J. Physiol. Pharmacol.*, **66**, 1247–52.
Rennie, M.J. (1985). Muscle protein turnover and the muscle wasting due to injury and disease. *Br. Med. Bull.*, **41**, 257–64.

Rennie, M.J. (1986). Metabolic insights from the use of stable isotopes in nutritional studies. *Clin. Nutr.,* **5**, 1–7.

Rennie, M.J. and Halliday, D. (1984). The use of stable isotope tracers as metabolic probes of whole-body and limb metabolism. *Proc. Nutr. Soc.,* **43**, 189–96.

Rennie, M.J. and Millward, D.J. (1983). 3-Methylhistidine excretion and the urinary 3-methylhistidine/creatinine ratio are poor indicators of skeletal muscle protein breakdown. *Clin. Sci.,* **65**, 217–25.

Rennie, M.J., Edwards, R.H.T., Halliday, D., Matthews, D.E., Wolman, S.L. and Millward, D.J. (1982). Muscle protein synthesis measured by stable isotope techniques in man: the effects of feeding and fasting. *Clin. Sci.,* **63**, 519–23.

Rennie, M.J., Babij, P., Sutton, J.R., Tonkins, W.W., Read, W.W., Ford, C. and Halliday, D. (1983). Effects of acute hypoxia on forearm leucine metabolism. *Hypoxia, Exercise, and Altitude: Proc. 3rd Banff Int. Hypoxia Symp.*, Alan R. Liss, New York, pp 317–23.

Rennie, M.J., Bennegård, K., Edén, E., Emery, P.W. and Lundholm, K. (1984). Urinary excretion and efflux from the leg of 3-methylhistidine before and after major surgical operation. *Metabolism,* **33**, 250–6.

Rennie, M.J., Chien, P., Taylor, D.J., Bennet, W.M. and Watt, P.W. (1990). Application of stable isotope tracers in studies of human metabolism. In: *New Techniques in Nutritional Research* (ed. R. Whitehead), Academic Press, San Diego.

Sjölin, J. (1988). Skeletal muscle protein metabolism in human infection: evaluation of 3-methylhistidine as a marker of myofibrillar protein degradation. PhD Thesis. Uppsala: *Acta Universitalis Upsaliensis*, pp. 1–62.

Sjölin, J., Stjernström, H., Henneberg, S., Andersson, E., Martensson, J., Griman, G. and Larsson, J. (1989). Splanchnic and peripheral release of 3-methylhistidine in relation to its urinary excretion in human infection. *Metabolism,* **38**, 23–9.

Smith, K. and Rennie, M.J. (1990). Protein turnover and amino acid metabolism in human skeletal muscle. In: *Baillière's Clinical Endocrinology and Metabolism*, Vol. 4 (eds J.B. Harris and D.M. Turnbull), Baillière Tindall, London, pp. 461–98.

Stehle, P., Zander, J., Mertes, N., Albers, S., Puchstein, C., Lawin, P. and Furst, P. (1989). Effects of parenteral glutamine peptide supplements on muscle glutamine loss and nitrogen balance after major surgery. *Lancet,* **i**, 270–2.

Tessari, P., Inchiostro, S., Biolo, G., Trevisan, R., Fantin, G., Marescotti, M.C., Iori, E., Tiengo, A. and Crepaldi, G., (1987). Differential effects of hyperinsulinemia and hyperaminoacidemia on leucine-carbon metabolism *in vivo*: Evidence for distinct mechanisms in regulation of net amino acid deposition. *J. Clin. Invest.,* **79**, 1062–9.

Waterlow, J.C. and Millward, D.J. (1989). Energy cost of turnover of protein and other cellular constituents. In: *Energy Transformations in Cells and Organisms* (eds W. Wieser and E. Gnaiger), Georg Thieme Verlag, Stuttgart, p. 277.

Watt, P.W., Corbett, M.E. and Rennie, W.J. (1991). Stimulation of protein synthesis in pig skeletal muscle by infusion of amino acids during constant availability of insulin. *Am. J. Physiol.* Submitted.

Wernerman, J., Hammarqvist, F. and Vinnars, E. (1990). Alpha-ketoglutarate and postoperative muscle catabolism. *Lancet,* **i**, 701–3.

Young, V.R., Haverberg, L.N., Bilmazes, C. and Munro, H.N. (1973). Potential use of 3-methylhistidine excretion as an index of progresssive reduction in muscle protein catabolism during starvation. *Metabolism,* **22**, 1429–36.

Chapter Five

Regulation of Adipose Tissue Mass and Metabolism

Dermot H. Williamson
Metabolic Research Laboratory, Nuffield Department of Clinical Medicine, Radcliffe Infirmary, Woodstock Road, Oxford OX2 6HE, UK

INTRODUCTION

The major function of white adipose tissue is to store fat for use in situations associated with decreased energy intake. The body fat content represents a balance between deposition of lipid as triacylglycerols and mobilization in the form of non-esterified fatty acids (NEFA). There are three sources of lipid for deposition in adipose tissue: (a) chylomicrons derived from dietary fat, secreted by the intestinal cells and transported via the lymphatic system; (b) very-low-density lipoproteins (VLDL) secreted by the liver, and to a lesser extent the intestine; (c) fatty acids synthesized *de novo* (lipogenesis) within adipose tissue from endogenous precursors, mainly glucose. The lipid contained in VLDL may originate from *de novo* synthesis or NEFA derived from adipose tissue or from delipidated lipoproteins (remnants) returned to the liver. Uptake and deposition of plasma triacylglycerols (chylomicrons or VLDL) in adipose tissue requires their initial conversion to the constituent NEFA and free glycerol in a reaction catalysed by lipoprotein lipase (EC 3.1.1.34) (Figure 5.1).

The amount of NEFA released to the bloodstream from stored triacylglycerol within adipose tissue is mainly dependent on the rate of lipolysis which in turn is a function of the activity of the hormone-sensitive lipase (EC 3.1.1.3). In addition, the NEFA liberated by the action of this

Obesity and Cachexia. Edited by N.J. Rothwell and M.J. Stock

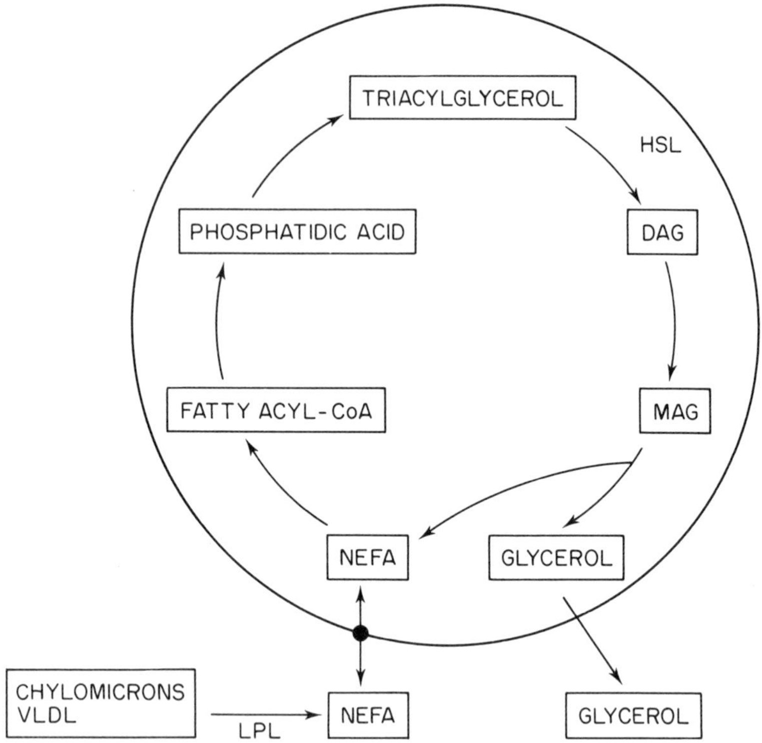

Figure 5.1 Simplified scheme for exogenous lipid deposition in adipose tissue. LPL, lipoprotein lipase; HSL, hormone-sensitive lipase; DAG, diacylglycerol; MAG, monoacylglycerol; NEFA, non-esterified fatty acids

intracellular enzyme can be re-esterified to triacylglycerol and thus escape release to the plasma; this energy-consuming process is referred to as the 'lipolysis–re-esterification' cycle (Figure 5.1).

The aim of this chapter is to discuss the physiological regulation of adipose tissue metabolism, particularly the control by hormones, and to compare this with changes which may occur in the pathological situations of cachexia and obesity. The adipose tissue mass can be altered by changes in cell number in addition to alteration of lipid content (Cryer and Cryer, 1990), but discussion of this is outside the remit of the present contribution. The differences in metabolic response of the various adipose tissue depots and the effects of gender is discussed by M. Rebuffé-Scrive in Chapter 11 of this book. It is not intended to give a detailed bibliography with a historical perspective but rather to cite key recent references and, in particular, reviews of the area. For more detailed and expert accounts of normal adipose tissue metabolism the reader is referred to Saggerson (1985) and

Vernon and Clegg (1985), and for particular information on human adipose tissue, to Galton (1971). The effects of cancer and obesity on adipose tissue lipid metabolism have been reviewed by Kern and Norton (1988) and Bray and York (1979) respectively.

METHODOLOGICAL ASPECTS

Animal studies

There are considerable amounts of data, obtained over several decades, on adipose tissue metabolism and its regulation *in vitro*. This includes measurements of enzyme activities in adipose tissue extracts, substrate fluxes in incubations of adipose tissue pieces or isolated adipocytes and hormonal responses in long-term cultures of adipocytes (see Saggerson, 1985; Vernon and Clegg, 1985). In addition, the use of cell lines capable of differentiation (e.g. 3TL1 pre-adipocytes) has provided insight into maturation and other longer-term processes (Cryer and Cryer, 1990).

Less information is available on adipose tissue metabolism and its regulation *in vivo*, in part because of technical difficulties and lack of methods to obtain quantitative information. However, it is possible to measure rates of fatty acid synthesis *in vivo* by the incorporation of ^3H from ^3H_2O into the fatty acid moiety (Jungas, 1968; Stansbie *et al.*, 1976). Similarly, the rate of acylglycerol glycerol synthesis (fatty acid esterification) can be measured by the rate of incorporation of ^3H from ^3H_2O into the glycerol moiety (Gandemer *et al.*, 1982; Brooks *et al.*, 1983). The rate of net accumulation of ^{14}C-labelled lipid in adipose tissue after oral administration of ^{14}C-labelled fatty acid (as a triacylglycerol) gives an indication of the functional activity of lipoprotein lipase and the ability of the tissue to 'capture' the ^{14}C-labelled NEFA released extracellularly from ^{14}C-labelled chylomicrons (Haggarty *et al.*, 1986; Oller do Nascimento and Williamson, 1986). This approach assumes that little of ^{14}C-labelled NEFA esterified to triacylglycerol within the tissue is released by lipolysis during the experiment. The advantages of these techniques is that they are non-invasive and do not depend on information on blood flow. They are, however, unsuitable for human studies.

Human studies

Recently, renewed interest in the study of human adipose tissue *in vivo* has been generated by the development of two techniques: (a) micro-dialysis (Lönnroth *et al.*, 1988; Hagström *et al.*, 1990; Jansson *et al.*, 1990); (b) measurements of arterial–venous differences across the subcutaneous abdominal fat (Frayn *et al.*, 1989). Both these methods are invasive and for provision of quantitative data require knowledge of the mass of the adipose tissue depot dialysed or perfused and the blood flow to it. Information on the latter can be obtained with ^{133}Xe washout (Rosell and Belfrage, 1979; Coppack *et al.*, 1990a). A disadvantage of the micro-dialysis technique is that it is not suitable for studying lipid (NEFA, triacylglycerol) metabolism of adipose tissue, whereas measurements of triacylglycerol extraction by human adipose tissue by the catheterization technique have already been reported (Coppack *et al.*, 1989, 1990a).

TRIACYLGLYCEROL DEPOSITION

As already indicated in the Introduction the two processes which contribute to deposition of triacylglycerol in white adipose tissue are uptake of NEFA derived from plasma triacylglycerols (chylomicrons or VLDL) and *de novo* synthesis from exogenous precursors. *In vivo* measurements in experimental animals indicate that uptake of pre-formed NEFA is the more important process (Table 5.1) except when the diet is free of fat and this conclusion is likely to be true also for the human (Marin *et al.*, 1987; Coppack *et al.*, 1990a). The key components of the pathway for deposition of exogenous lipid are: (a) availability and species of plasma triacylglycerol; (b) the activity of lipoprotein lipase; (c) the transport of NEFA into the adipocyte; (d) the rate of esterification.

Species of triacylglycerol

It is well established that the rate of lipolysis catalysed by lipoprotein lipase is related to the size of the triacylglycerol substrate (Grundy and Mok, 1976; Nicolls and Lewis, 1980). This is reflected in a very short half-life for chylomicrons as compared to hepatic VLDL. Thus in the post-prandial phase on diets containing long-chain lipid, chylomicrons will be the preferred substrate for lipoprotein lipase leading to the accumulation in the plasma of hepatic VLDL (Robins *et al.*, 1987). The factors regulating the secretion of hepatic VLDL have recently been reviewed (Gibbons, 1990). Diets in which the lipid is as medium-chain triacylglycerols will not result in secretion of chylomicrons and therefore lipid deposition is low (Bach and Babayan, 1982).

Lipoprotein lipase

Lipoprotein lipase, the initiating enzyme for triacylglycerol metabolism is synthesized in adipocytes, but its site of action is the luminal surface of the

Table 5.1 Relative uptake and synthesis of triacylglycerol fatty acid *in vivo* in rat white adipose tissue

Type of rat	Adipose depot	Triacylglycerol fatty acid uptake (nmol fatty acid/min per g)	Triacylglycerol fatty acid synthesis (nmol fatty acid/min per g)	Uptake/synthesis	References
Fed male	Epididymal	268	17–39	6.9–15.8	Hartman (1977); Stansbie *et al.* (1976); Rothwell *et al.* (1983)
Fed female	Parametrial	41	5	8.2	Oller do Nascimento and Williamson (1986)
Fed female + oral glucose	Parametrial	—	24	—	Agius *et al.* (1979)
Fed lactating	Parametrial	19	5.5	3.45	Agius *et al.* (1979); Scow *et al.* (1977)
Fed lactating unsuckled rat	Parametrial	90–188	23.6	3.8–7.9	

vascular endothelium (see Cryer, 1981; Eckel, 1987, 1989; Robinson and Speake, 1989). Consequently, its activity can be modulated at a number of sites: synthesis within the adipocyte; transport to the endothelial cell; degradation (Figure 5.2). The synthesis of lipoprotein lipase in the endoplasmic reticulum initially yields an inactive precursor of the mature active enzyme (Chajek-Shaul *et al.*, 1985a, b; Olivecrona *et al.*, 1985, 1987; Vannier *et al.*, 1985). Maturation is achieved in the Golgi apparatus by progressive *N*-glycosylation of the inactive precursor (Amri *et al.*, 1986). There is evidence in adipocyte cultures that lipoprotein lipase exists in an active dimeric form in the Golgi apparatus (Vannier and Ailhaud, 1989; Vannier *et al.*, 1989). The active enzyme is assumed to be transported in

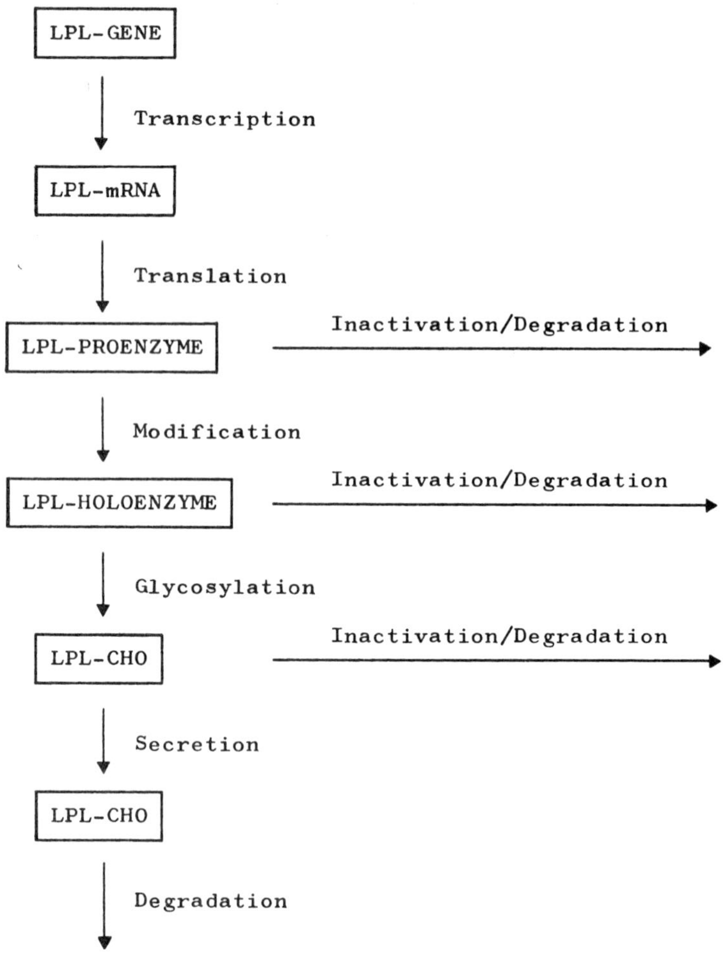

Figure 5.2 Sites for regulation of lipoprotein lipase activity. LPL, lipoprotein lipase

vesicles to the adipocyte plasma membrane where it is bound by heparin sulphate proteoglycans. How lipoprotein lipase is then transferred to the luminal surface of the endothelial cells is by no means certain (Olivecrona and Bengtsson-Olivecrona, 1987; Robinson and Speake, 1989). It has been suggested that after release from the adipocyte membrane the enzyme is internalized in the endothelial cell and then transported in vesicles to the luminal surface (Robinson and Speake, 1989).

The tissue activity of lipoprotein lipase can change rapidly and although this is in part due to changes in synthesis, alterations in degradation play an important role. Like many secretory proteins, lipoprotein lipase can undergo degradation within the adipocyte (Speake *et al.*, 1985a; Vannier *et al.*, 1985; Cupp *et al.*, 1987; Semb and Olivecrona, 1987), indeed this may be a means of regulating the intracellular pool of enzyme (Figure 5.2). Binding of the enzyme to membrane proteoglycans appears to be necessary (Cisar *et al.*, 1989) and the site of degradation is lysosomal (Vannier and Ailhaud, 1989). In addition, the enzyme released from the luminal surface of the endothelial cell is taken up by the liver and degraded. Hepatic degradation ensures a low concentration of lipoprotein lipase in the plasma and decreases the possibility of the enzyme binding to the endothelium of tissues where its synthesis is not expressed (Olivecrona and Bengtsson-Olivecrona, 1987).

The nutritional state of the animal greatly effects the activity of lipoprotein lipase in adipose tissue (Robinson, 1970; Eckel, 1989); the activity decreases on fasting and is rapidly restored on refeeding, high carbohydrate diets being particularly effective (Eckel, 1987). The major changes in activity occur in the endothelial cells rather than the adipocytes. The rapid decrease in lipoprotein lipase activity on fasting illustrates the problem in interpreting measurements in biopsies of human adipose tissue which are usually collected after a fast of 12–15 h. An additional uncertainty is that in the majority of studies the whole tissue activity is measured and this does not necessarily indicate the proportion at the endothelial cell surface. As discussed under Methodological Aspects, one approach to obtaining information on the functionally active enzyme in experimental animals is to measure the accumulation in adipose tissue of radioactivity derived from injected ^{14}C-labelled chylomicrons (Cryer *et al.*, 1976) or produced from an oral load of ^{14}C-labelled lipid (Oller do Nascimento and Williamson, 1986). The present consensus from work on experimental animals is that there is good agreement between measured accumulation of ^{14}C-labelled lipid *in vivo* and the tissue activity *in vitro*.

The close relationship between the nutritional state and the activity of lipoprotein lipase in adipose tissue has led to the view that insulin is the key external signal which regulates activity (Robinson, 1970; Eckel, 1987) (Figure 5.2). A positive correlation between adipose tissue lipoprotein lipase activity and plasma insulin has been reported (Cryer *et al.*, 1976). Administration of insulin *in vivo*, either as a single dose or as a hyperinsulinaemic clamp, results in an increase in adipose tissue lipoprotein lipase activity (Sadur and Eckel, 1982; Eckel, 1987).

There is convincing evidence for a direct effect of insulin *in vitro* to stimulate synthesis of the enzyme (Vydelingum *et al.*, 1983; Speake *et al.*, 1985b). In primary cultures of rat adipocytes insulin rapidly and specifically increased the rate of lipoprotein lipase synthesis and this was accompanied by a two-fold increase in mRNA for the enzyme (Ong *et al.*, 1988); there was no change in the rate of degradation. In contrast, the lipoprotein lipase of human adipocytes (Kern *et al.*, 1985) or adipose tissue (plus endothelium) (Elkeles, 1974) was not increased by physiological concentrations of insulin *in vitro*. Similarly, comparison of lipoprotein lipase activity with the immunoreactive mass of the enzyme has shown that on refeeding a carbohydrate-rich meal (which results in hyperinsulinaemia) the activity increased in human adipose tissue, but the mass did not alter, nor was there any increase in lipoprotein lipase mRNA, suggesting post-translational activation of previously inactive proenzyme (Figure 5.2) (Ong and Kern, 1989). Similar conclusions have been reached in respect of lipoprotein lipase activity of rat adipose tissue (Doolittle *et al.*, 1990). Support for this mechanism of action of insulin comes from studies with differentiated 3T3-L1 adipocytes (Semenkovich *et al.*, 1989), whereas the differentiation-dependent induction of lipoprotein lipase in this cell line is likely to be due to increased accumulation of its mRNA (Semenkovich *et al.*, 1989). Insulin may also act via increased transfer of lipoprotein lipase to its site of action. For example, in fully differentiated 3T3-L1 adipocytes the hormone causes rapid release of the enzyme (Spooner *et al.*, 1979). This may be via the action of a specific phospholipase which hydrolyses a glucosyl phosphatidylinositol molecule responsible for anchoring the lipoprotein lipase to the plasma membrane (Chan *et al.*, 1988). Thus insulin can increase the activity of lipoprotein lipase in adipose tissue by a number of different mechanisms and it is likely that the predominant site of action may vary depending on the experimental situations or the species. Apart from insulin, the only other hormones to have a stimulatory effect on lipoprotein lipase activity are the glucocorticoids which act in the longer term by increasing the rate of transcription (Figure 5.2).

As might be expected, adrenaline and other hormones (glucagon, noradrenaline) which increase cyclic AMP in adipocytes, decrease the activity of lipoprotein lipase (see Eckel, 1987). It is thought that they act via stimulation of degradation because the effects *in vitro* are independent of changes in protein synthesis, insulin or glucose concentrations (Ball *et al.*, 1986).

Transport of fatty acids into adipose tissue

The fatty acids released from plasma triacylglycerols by the action of lipoprotein lipase at the surface of the endothelial cell have two fates: (a) they can enter the plasma NEFA pool and thus be transported from the adipose tissue via the venous blood to other tissues (muscle, liver) or (b) they can enter the adipose tissue and be re-esterified to triacylglycerol for storage. There is considerable variation in the estimates of the proportion of NEFA released from triacylglycerol which enters the adipose tissue. For example, in perfused fat pads only 55% was taken up (Scow, 1977), whereas *in vivo* virtually complete uptake has been reported in the case of VLDL (Wolfe and Durkot, 1985). It would seem likely that the proportion of NEFA released which is 'captured' by adipose tissue will vary in different physiological and pathological situations and will depend on the blood flow through the tissue, the rate of transport of NEFA into the adipocyte and the rate of esterification within the cell (Figure 5.1).

An adipocyte membrane transport system for NEFA has been described (Abumrad *et al.*, 1984) and it appears that this is stimulated five to ten-fold by adrenaline via a mechanism involving cAMP and protein kinase (Abumrad *et al.*, 1986a and b). Insulin antagonizes the activation of NEFA transport by adrenaline (Abumrad *et al.*, 1986b) and this is likely to be due to a lowering of intracellular cAMP (Abumrad *et al.*, 1988). At first sight it might seem surprising that insulin should inhibit adrenaline-stimulated NEFA transport, because a high plasma insulin is associated with increased activity of adipose tissue lipoprotein lipase (Cryer *et al.*, 1976) and lipid deposition. However, the inhibitory effect of insulin on stimulated NEFA transport is presumably related to its ability to inhibit intracellular triacylglycerol hydrolysis (see below) rather than the deposition of exogenous fatty acids.

The presence in human plasma of a protein (acylation stimulating protein), particularly after an oral fat load (Cianflone *et al.*, 1989a), which stimulates triacylglycerol synthesis *in vitro* without any effect on *de novo* fatty acid synthesis or triacylglycerol hydrolysis (Cianflone, 1989b) may provide a means of ensuring more effective capture of dietary fat.

The esterification pathway

Once inside the adipocyte the major fate of NEFA is esterification to triacylglycerol, because the rate of fatty acid oxidation is low in this tissue (Harper and Saggerson, 1976). Knowledge of the regulation of the esterification pathway is still rudimentary, in part because none of the enzymes of the pathway (Figure 5.3) have been fully purified and characterized (Saggerson, 1985). Alterations in the *in vitro* measured activities of some of the enzymes of the pathway have been reported

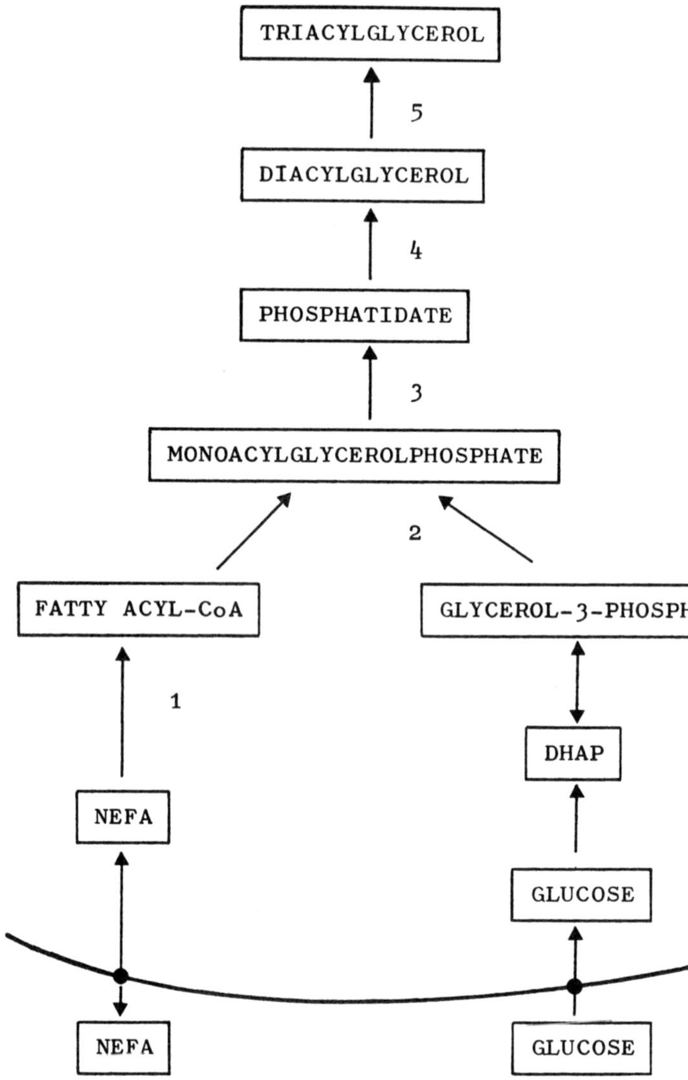

Figure 5.3 The fatty acid esterification pathway. Enzymes: 1, fatty acyl-CoA synthetase; 2, glycerol-phosphate acyltransferase; 3, glycerolphosphate acyltransferase; 4, phosphatidate phosphohydrolase; 5, diacylglycerol acyltransferase

(Lawson *et al.*, 1981) but it must be emphasized that the changes are relatively small and slow compared with those for lipoprotein lipase.

The activity of the initiating enzyme for the pathway, fatty acyl-CoA synthetase (EC 6.2.1.3) decreases slightly (20%) on 24 h starvation (Lawson *et al.*, 1981) and this effect can be reproduced *in vitro* by a 10 min incubation with adrenaline (Soorana and Saggerson, 1978). The decrease is prevented by insulin or a beta-blocker, propranolol. The recent cloning of the long-chain fatty acyl-CoA synthetase from rat liver (Suzuki *et al.*, 1990) is likely to provide information on the possible regulation of this enzyme at the level of the gene.

Glycerol phosphate acyltransferase (EC 2.3.1.15) activity also decreases in starvation and this is reversed by refeeding (Aas and Daae, 1971; Jamdar and Osborne, 1982). It has been reported that the adipocyte enzyme is inactivated by cAMP-dependent protein kinase and re-activated by phosphatase treatment in a cell-free system (Nimmo and Houston, 1978). However, attempts to show that this mechanism is an explanation for the inhibitory effects of adrenaline have not been substantiated in isolated adipocytes (Rider and Saggerson, 1983). The activity of this enzyme may also be regulated by the supply of glycerol 3-phosphate which in turn is dependent on the entry of glucose into the adipocyte (Figure 5.3; see below). Lactate and pyruvate are poor precursors for glyceride-glycerol (Katz and Wals, 1974; Palacin *et al.*, 1988).

Phosphatidate phosphohydrolase (EC 3.1.3.4) activity in adipose tissue is also altered with change in nutritional state (Moller *et al.*, 1977; Jamdar and Osborne, 1982). Its activity is decreased by noradrenaline and this effect can be reversed by insulin (Cheng and Saggerson, 1978). Thus there is a tendency for the esterification pathway to decrease in activity in a concerted way in starvation and the action of insulin is to oppose these changes. However, there is little firm information as to how these changes in activity are brought about.

De novo **synthesis of fatty acids**

The other source of NEFA for net deposition as triacylglycerols is *de novo* synthesis (lipogenesis). The major substrate for lipogenesis in white adipose tissue is glucose, although lactate can make an appreciable contribution, both *in vivo* and *in vitro* (Katz and Wals, 1974; Palacin *et al.*, 1988). The pathway for the conversion of glucose to fatty acid, together with the sites where hormonal regulation may occur is shown in Figure 5.4. However, it must again be emphasized that except on fat-free diets the rate of *de novo* synthesis of NEFA in adipose tissue is low.

It has long been known that glucose entry into adipocytes is stimulated by insulin. More recent work has shown that this increased rate of glucose

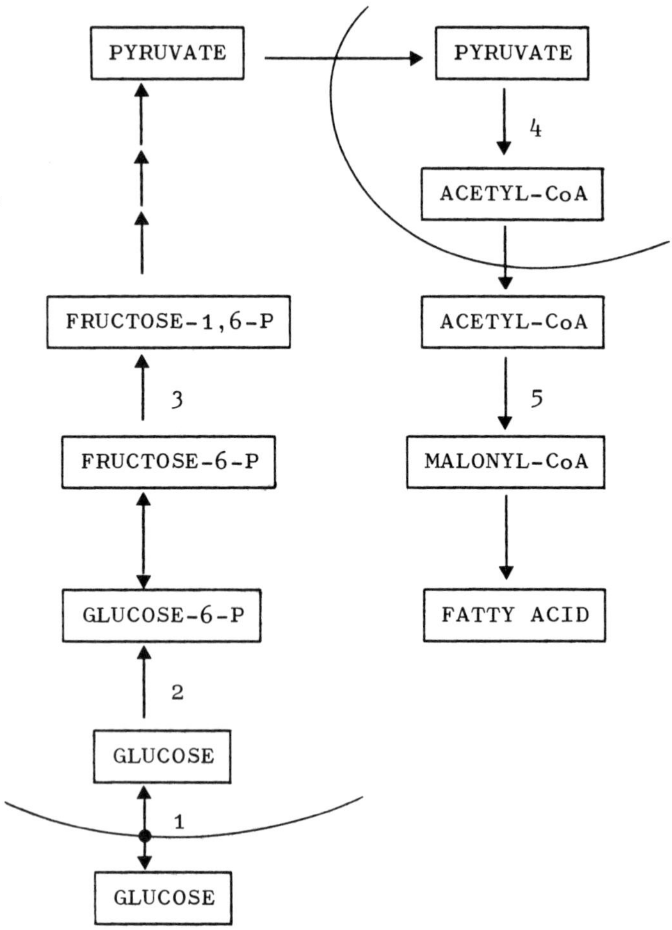

Figure 5.4 The *de novo* fatty acid synthesis pathway. Sites of regulation: 1, glucose transport; 2, hexokinase; 3, phosphofructokinase; 4, pyruvate dehydrogenase; 5, acetyl-CoA carboxylase

transport is due to an increase in the number of glucose transporters in the plasma membrane (see Simpson and Cushman, 1986). The mechanism appears to be as follows: insulin binds to its receptor on the plasma membrane and generates a putative signal. This promotes the translocation of glucose transporters from intracellular vesicles and their insertion in the plasma membrane (Karnieli *et al.*, 1981b). Thus chronic hyperinsulinaemia increased the concentration of transporters, whereas starvation or streptozotocin-diabetes caused a decrease (Karnieli *et al.*, 1981a). The effects of insulin deficiency are more complex because not only is the amount of transporter protein decreased (Berger *et al.*, 1989) but also the steady state concentration of messenger RNA for the transporter (Sivitz *et al.*, 1989). As might be expected these changes can be reversed by refeeding (Kahn *et al.*, 1989) or insulin administration.

The stimulation of glucose transport by insulin can be decreased by lipolytic hormones (catecholamines, glucagon) or agents such as isoproterenol which act via increased adenylate cyclase activity (Smith *et al.*, 1984). However, the overall effect of adrenaline *in vitro* is to increase glucose uptake and glycolytic flux, at least at the level of triosephosphates (Halperin and Denton, 1969) and thus increase glycerol 3-phosphate availability for fatty acyl-CoA formation (Figure 5.4).

The regulatory enzymes of the glycolytic pathway, hexokinase (EC 2.7.1.1), phosphofructokinase (EC 1.7.1.11) and pyruvate kinase (EC 2.7.1.40) exhibit the same control properties characteristic of these enzymes in other tissues. Insulin stimulates their activity, but this effect is likely to be less important than its effects on glucose transport and the succeeding steps after pyruvate kinase.

Pyruvate dehydrogenase (EC 1.2.4.1) is the key enzyme in deciding the conversion of glucose on lactate carbon to lipogenesis (Figure 5.4). It exists in a phosphorylated inactive form which can be activated by dephosphorylation and these interconversions are brought about by an intrinsic kinase and phosphatase respectively (see Randle, 1986). In addition, the active enzyme can be inhibited by the end-products of the reaction (NADH and acetyl-CoA). Incubation of adipose tissue (fat pads or adipocytes) with insulin results in a rapid and stable activation of pyruvate dehydrogenase (see Denton *et al.*, 1989) and *in vivo* alteration of the plasma insulin causes a parallel change in the activation state (Stansbie *et al.*, 1976). The mechanism whereby insulin activates pyruvate dehydrogenase in adipose tissue is not known, although it appears to be due to increased phosphatase activity. The key question is how is the phosphatase activated? Changes in mitochondrial Ca^{2+}, an activator of the phosphatase, have been excluded as the signal (Denton *et al.*, 1989). Another possibility is the release of a putative mediator (e.g. Seals and Jarett, 1980) from the plasma membrane which activates the phosphatase. However, the methodology for assay of the effects of such mediators has recently been criticized (Denton *et al.*, 1989). Denton and colleagues (1989) have suggested insulin may cause a change in the interaction of the pyruvate dehydrogenase system with a large molecular weight component within the mitochondrial inner membrane.

Adrenaline can decrease the proportion of pyruvate dehydrogenase which is in the active form in isolated adipocytes. It is likely that this effect is secondary to the increased lipolysis and raised intracellular concentration of NEFA (Saggerson, 1985). These fatty acids are oxidized and consequently increase the mitochondrial $NADH:NAD^+$ and acetyl-CoA:CoASH ratios leading to inactivation of pyruvate dehydrogenase. Insulin plus glucose can overcome the inhibitory effect of adrenaline and other lipolytic agents, presumably by stimulation of esterification. In contrast, noradrenaline increases pyruvate dehydrogenase activity of brown adipose tissue *in vivo*, but lipogenesis is not stimulated because of inactivation of acetyl-CoA carboxylase (Gibbins *et al.*, 1985).

Acetyl-CoA carboxylase catalyses the committed step in lipogenesis in which acetyl-CoA derived from pyruvate is carboxylated to malonyl-CoA (Figure 5.4). It can be regulated in the short-term by both allosteric regulation and covalent modification (phosphorylation–dephosphorylation; Hardie, 1989). Activation and polymerization is promoted by citrate, whereas long-chain acyl-CoA causes disaggregation and inactivation. These changes in activity make physiological sense, in that citrate can be viewed as a feed-forward activator and long-chain acyl-CoA as a feedback inhibitor to decrease lipogenesis when intracellular NEFA increases (e.g. on stimulation of triacylglycerol hydrolysis). However, the concentrations of citrate occurring in adipose tissue appear to be too low to stimulate enzyme, and much of the fatty acyl-CoA is protein bound, therefore it seems more likely that covalent modification by phosphorylation–dephosphorylation is the predominant form of short-term regulation of acetyl-CoA carboxylase (Hardie, 1989).

Understanding the regulation of acetyl-CoA carboxylase by phosphorylation–dephosphorylation is by no means as simple as one might expect. It was originally thought that physiological inactivation of the enzyme occurred by phosphorylation with cAMP-dependent protein kinase (Lent *et al.*, 1978), but more recent work has indicated that the primary kinase is AMP-dependent (Hardie, 1989; Hardie *et al.*, 1989; Haystead *et al.*, 1990). Apparently, this kinase is activated by AMP and stimulated by nanomolar concentrations of fatty acyl-CoA (Carling *et al.*, 1987). Thus higher fatty acyl-CoA levels in adipose tissue due to increased lipolysis would be expected to potentiate inactivation of acetyl-CoA carboxylase via increased phosphorylation.

Changes in the activation state of acetyl-CoA carboxylase in adipose tissue *in vivo* occur in response to alterations in plasma insulin (Stansbie *et al.*, 1976) and incubation of fat pads with the hormone increases the initial activity of the carboxylase (Halestrap and Denton, 1974). This activation appears to be due to increased polymerization of the enzyme which is accompanied by increased phosphorylation (Borthwick *et al.*, 1987). The site (I) phosphorylated in the presence of insulin is distinct from those sites (A) phosphorylated after addition of adrenaline which leads to inactivation (Brownsey and Denton, 1982). There is considerable speculation as to the nature of the protein kinase which phosphorylates acetyl-CoA carboxylase at the I-site, and some degree of controversy as to whether phosphorylation at this site is essential for activation of the enzyme or whether it is a so-called 'silent phosphorylation'. Indeed, evidence is accumulating that the major action of insulin, at least in liver, is to bring about dephosphorylation of acetyl-CoA carboxylase (Witters *et al.*, 1988; Mabrouk *et al.*, 1990) which suggests that insulin action must either inhibit AMP-dependent protein kinase or stimulate a protein phosphatase.

Adrenaline (Halestrap and Denton, 1974) and glucagon (Zammit and Corstophine, 1982) decrease the activity of acetyl-CoA carboxylase in rat fat pads or adipocytes. Whether this inactivation involves a primary action of cAMP-dependent protein kinase, as was originally thought, is now an open question (Hardie, 1989; Hardie *et al.*, 1989).

LIPID MOBILIZATION

Lipolysis

A specific function of adipose tissue is to release NEFA to the bloodstream to maintain caloric homeostasis and their mobilization from stored triacylglycerol is dependent on the activity of hormone-sensitive lipase or triacylglycerol lipase (EC 3.1.1.3) (Hales *et al.*, 1978; Belfrage, 1985) (Figure 5.1). This enzyme is distinct from lipoprotein lipase in its site of action (within the adipocyte) and its mode of regulation (see below). It catalyses the hydrolysis of both long-chain triacylglycerols and diacylglycerols (the product of the first step), but the rate of reaction with the latter is considerably faster. This lipase, like pyruvate dehydrogenase and acetyl-CoA carboxylase, exists in interconvertible forms; the phosphorylated form is active, whereas removal of covalently bound phosphate leads to inactivation. Catecholamines and glucagon, which stimulate the activity of cyclic AMP-dependent protein kinase, activate the lipase via phosphorylation and increased lipolysis (Strälfors and Belfrage, 1982, 1983). Recent experiments have shown that the purified enzyme in human adipose tissue is also inactivated by cAMP-dependent protein kinase (Holm *et al.*, 1989). Conversely, exposure to insulin results in dephosphorylation of the lipase and inhibition of lipolysis (Strälfors *et al.*, 1984) and decreased flux of NEFA from adipose tissue. The action of insulin to decrease the latter has recently been elegantly demonstrated by measurement of arteriovenous differences across human adipose tissue (Coppack *et al.*, 1989b). The identity of the phosphatases which lead to loss of phosphate from the basal and regulatory sites is known (Strälfors *et al.*, 1984; Olsson and Belfrage, 1987), but how insulin increases the activity of these phosphatases is not known. The pronounced anti-lipolytic effects of insulin may be mediated by other mechanisms, e.g. a decrease in cAMP-dependent protein kinase secondary to a decrease in cAMP (Londos *et al.*, 1983). The latter may be brought about by the activation of a low K_m phosphodiesterase (see Denton *et al.*, 1981). It has recently been shown that the AMP-dependent kinase phosphorylates purified bovine hormone-sensitive lipase without activating the enzyme; the site phosphorylated is adjacent to that phosphorylated by cAMP-dependent protein kinase (Garton *et al.*, 1989). The regulatory significance of this phosphorylation remains to be identified.

LIPOLYSIS–RE-ESTERIFICATION CYCLES

Intracellular cycle

The presence of the lipolysis and esterification pathways in adipose tissue means that a portion of the NEFA released from stored triacylglycerol by the action of hormone-sensitive lipase can be re-esterified rather than released into the plasma (Steinberg, 1963) (Figures 5.1 and 5.5). Thus the activity of this intracellular triacylglycerol–fatty acid cycle is important in the regulation of the net amount of NEFA released in response to a hormonal signal. As already discussed, hormones can effect the rates of lipolysis and re-esterification and therefore they can modulate the rate of the cycle. Indeed, Newsholme (1980) has suggested that substrate cycling allows the amplification of the response of substrate flux to a change in concentration of a regulator (e.g. hormone). Re-esterification of fatty acids requires ATP (8 high-energy phosphate bonds per mole of triacylglycerol recycled; Elia *et al.*, 1987) and therefore the operation of the cycle must be regulated to prevent excessive consumption of energy. Intracellular rates of re-esterification have been measured in adipose tissue *in vivo* by incorporation of ^{3}H into the glycerol moiety of triacylglycerols, assuming a negligible contribution of NEFA from the activity of lipoprotein lipase (see above) (Brooks *et al.*, 1983). Stimulation of lipolysis both *in vivo* (Brooks *et al.*, 1983) and *in vitro* (Brooks *et al.*, 1982) with β-adrenergic agents (noradrenaline, isoprenaline) or glucagon results in an increased rate of cycling.

Extracellular cycle

The NEFA released from adipose tissue has two main fates: (1) it can be oxidized by peripheral tissues (mainly muscle); (2) it can be taken up by the liver and oxidized (completely or partially to ketone bodies) or re-esterified to triacylglycerols. These can be secreted as VLDL and the NEFA component utilized by peripheral tissues as a source of fuel or be taken up by adipose tissue and deposited as triacylglycerol. The latter is an extracellular cycle of lipolysis and re-esterification (Figure 5.5) which represents a larger energy demand than the intracellular cycle because of the involvement of

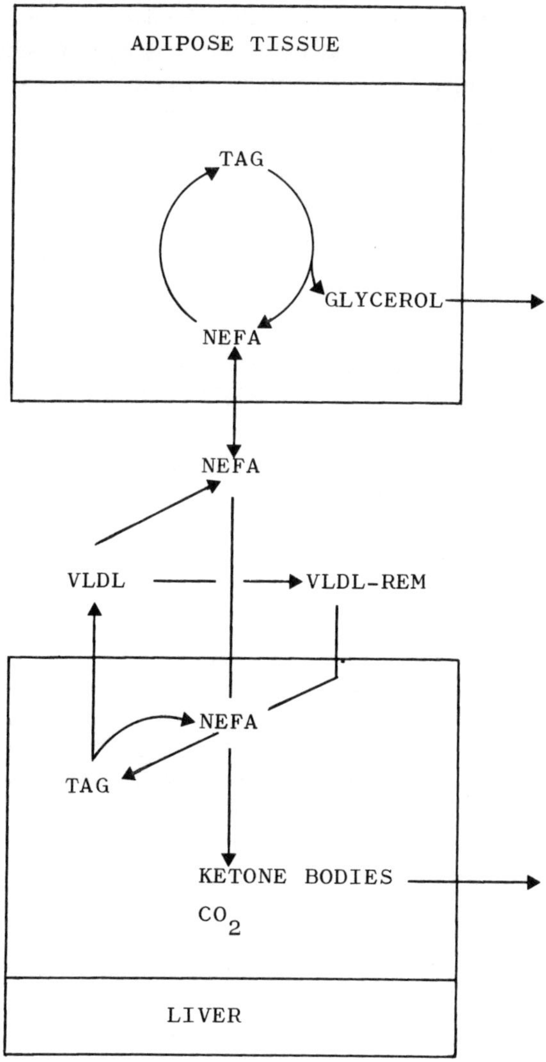

Figure 5.5 Simplified scheme of the lipolysis–re-esterification cycles. TAG, triacylglycerol; NEFA, non-esterified fatty acids; VLDL, very-low-density lipoproteins; VLDL-REM, very-low-density lipoprotein remnants

two esterification steps (liver and adipose tissue). The situation may be more complex because of the possibility that NEFA released from VLDL at the adipose tissue endothelial cells by the action of lipoprotein lipase may return to the liver rather than enter adipose tissue. In addition, the partially delipidated VLDL remnants returned to the liver may undergo a cycle of lipolysis and re-esterification within the liver (Mayes, 1976). By use of stable isotopes of glycerol and palmitate it is now possible to estimate the rates of intracellular and extracellular triacylglycerol–fatty acid cycling in the human (Wolfe and Peters, 1987; Wolfe *et al.*, 1987; 1990). Generally, the measured rate of the intracellular cycle is less than 20% of the whole body cycle (intracellular plus extracellular). Adrenaline stimulates the intracellular cycle and lipolysis (Miyoshi *et al.*, 1988), whereas glucose infusion also increases the cycle but inhibits lipolysis (Wolfe and Peters, 1987).

PHYSIOLOGICAL ALTERATIONS IN LIPID METABOLISM IN ADIPOSE TISSUE

Before discussing the changes in lipid metabolism in adipose tissue in cachexia and obesity it is worthwhile considering two different physiological situations in which there are marked changes, namely lactation and starvation.

Lactation

Lactation in the rat is characterized by marked hyperphagia (two- to three-fold increase), yet the adipose tissue mass remains relatively constant or may even decrease (Williamson, 1980; Vernon and Flint, 1983). The simplistic answer to the question why does the adipose tissue mass not increase in lactation, even on high fat diets (Steingrimsdottir *et al.*, 1980), is that the extra energy consumed during lactation is directed to the gland for the synthesis of macronutrients of milk (lactose, lipid, protein). However, as available substrates are 'seen' by the adipose tissue this implies that lipid deposition (exogenous and endogenous) is depressed in this tissue. The changes in lipid metabolism during lactation are listed in Table 5.2; the net result is increased flux of NEFA from adipose tissue. These alterations in adipose tissue are the mirror image of those occurring in the mammary gland (Table 5.3; Williamson, 1980; Vernon and Flint, 1983) and are rapidly reversed on cessation of lactation, when there is a period of lipid repletion in adipose tissue to replace losses during lactation (Oller do Nascimento and Williamson, 1986).

The key question is the signal(s) which depresses deposition of lipid in adipose tissue during lactation. A putative candidate would be prolactin, which increases in plasma in response to the suckling stimulus, but there appear to be no prolactin receptors on rat adipose tissue plasma membranes (Vernon and Flint, 1983), consequently any effect of this hormone must be indirect. Suppression of prolactin secretion with bromocryptine results in increased lipoprotein lipase activity (Flint *et al.*, 1981) and lipogenesis (Agius

Table 5.2 Changes in lipid metabolism in white adipose tissue of the rat during lactation

Parameter	Change	Reference
Deposition of exogenous lipid	Decreased	Oller do Nascimento and Williamson (1986)
Lipoprotein lipase	Decreased	Scow and Chernick (1987)
Lipogenesis	Decreased and less sensitive to insulin	Agius *et al.* (1979) Burnol *et al.* (1983)
Glucose utilization	Decreased *in vivo* Unchanged *in vitro*	Burnol *et al.* (1987) Burnol *et al.* (1986)
Pyruvate dehydrogenase	Decreased concentration and sensitivity to insulin	Kilgour and Vernon (1987)
Acetyl-CoA carboxylase	Decreased concentration	Sinnett-Smith *et al.* (1980)
Esterification	Decreased	Cubero *et al.* (1983)
Lipolysis–re-esterification cycle	Decreased	Hansson *et al.* (1987)
Lipolysis	Increased	Smith and Walsh (1976)

Table 5.3 Changes in lipid metabolism in mammary gland of the rat during lactation

Parameter	Change	Reference
Uptake of exogenous lipid	Increased	Oller do Nascimento and Williamson (1986)
Lipoprotein lipase	Increased	Scow and Chernick (1987)
Lipogenesis	Increased and more sensitive to insulin	Robinson *et al.* (1978) Burnol *et al.* (1983)
Glucose utilization	Increased	Williamson and Robinson (1977)
Pyruvate dehydrogenase	Increased	Coore and Field (1974)
Acetyl-CoA carboxylase	Increased	Howanitz and Levy (1965)

et al., 1979) in adipose tissue of suckled lactating rats and this effect could be reversed by administration of ovine prolactin. Maintenance of endogenous prolactin secretion by teat-sealing, which decreases the metabolic activity of mammary gland, resulted in a switch-on of uptake of exogenous lipid by adipose tissue (Oller do Nascimento and Williamson, 1989). One interpretation of these findings is that prolactin is essential for an active mammary gland and that the latter generates a signal which can depress adipose tissue lipid metabolism. Some support for this view is the finding that the mammary gland can act as an endocrine organ, e.g., expression of a calcium-mobilizing parathyroid hormone-like peptide (Thiede and Rodan, 1988). An alternative explanation is that an active mammary gland decreases the concentration of a known signal which has a positive effect on adipose tissue lipid metabolism. The obvious candidate is insulin (Table 5.4) (Williamson, 1989).

It is now well established that, despite the hyperphagia, the plasma insulin concentration in the fed state is low during lactation (about 50%) compared with non-lactating rats (Williamson, 1980; Vernon and Flint, 1983). There are a number of explanations for this unexpected finding but the most likely scenario is that glucose utilization in mammary gland is highly insulin sensitive (Burnol *et al.*, 1983, 1986, 1987), and therefore removal of glucose by the mammary gland lowers the plasma concentration which in turn decreases insulin secretion; in addition there is increased removal of insulin by the lactating gland via increased receptor binding (Jones *et al.*, 1984a, b). On removal of the suckling stimulus there is an increase (within 24 h) in plasma insulin to values higher than in virgin rats (Agius *et al.*, 1979; Oller do Nascimento and Williamson, 1988). Thus the transient hyperinsulinaemia (and continuing moderate hyperphagia) on cessation of lactation is likely to be involved, via increased activity of lipoprotein lipase and switch-on of lipogenesis, for the restoration of adipose tissue mass (Oller do Nascimento and Williamson, 1986; Williamson, 1989). However, the decrease in plasma insulin during lactation is not as great as that on starvation (see below) and therefore it is difficult to accept that it is the sole factor, particularly as there is no decrease in insulin receptor numbers

Table 5.4 Summary of effects of hormones on lipid metabolism in white adipose tissue

Parameter	Insulin	Catecholamines	Glucagon
Lipid deposition	Increased	Decreased	Decreased
Lipoprotein lipase	Increased	Decreased	Decreased
Lipogenesis	Increased	Decreased	Decreased
Lipolysis	Decreased	Increased	Increased
Lipolysis–re-esterification cycle	Increased (?)	Increased	Increased

in lactation (Flint *et al.*, 1979). A factor which may potentiate the effects of the lowered plasma insulin is the unchanged plasma glucagon and therefore a decrease in the plasma insulin/glucagon ratio (Robinson *et al.*, 1978). Glucagon is an inhibitor of lipogenesis in isolated adipocytes and thus the prevailing insulin/glucagon ratio in lactation would favour triacylglycerol mobilization rather than fat deposition (Table 5.4) (Robinson *et al.*, 1978; Zammit, 1985). In this context it is appropriate that lactating mammary gland has no glucagon receptors (Robson *et al.*, 1984) and therefore glucagon cannot antagonize the action of insulin in this tissue, which may contribute to the increased sensitivity to the latter.

Starvation

In many respects the alterations in adipose tissue lipid metabolism on starvation resemble those seen in lactation but are more marked (compare Tables 5.2 and 5.5). The decrease in plasma insulin and in the insulin/glucagon ratio are likely to be the key factors which bring about the changes (Table 5.5) and which are rapidly reversed on refeeding (Oller do Nascimento and Williamson, 1988). An important point (see later) is that despite the depression of adipose tissue lipid deposition neither lactation nor starvation is associated with increased plasma triacylglycerol (hypertriglyceridaemia). In the case of lactation this is because of their removal by the mammary gland (Scow *et al.*, 1977; Oller do Nascimento and Williamson, 1986; Scow and Chernick, 1987), whereas in starvation it is presumably the muscle mass which is responsible for the continued utilization of triacylglycerols in the form of VLDL (Williamson, 1989).

LIPID METABOLISM IN ADIPOSE TISSUE IN CANCER

Malignancy in experimental animals and man is associated with progressive involuntary weight loss (cachexia), part of which is due to the accompanying

Table 5.5 Changes in lipid metabolism in white adipose tissue of the rat on starvation

Parameter	Change
Deposition of exogenous lipid	Decreased
Lipoprotein lipase	Decreased
Lipogenesis	Decreased
Lipolysis–re-esterification cycle	Unchanged (?)
Lipolysis	Increased

anorexia, but in many instances there is accelerated loss of fat (see Kern and Norton, 1988). Thus although under-nutrition plays a role in the loss of fat, changes in lipid metabolism of adipose tissue of tumour-bearing animals can occur without any alteration of dietary intake (Evans and Williamson, 1988). These changes are listed in Table 5.6 and the net result is decreased lipid deposition and increased mobilization of stored triacylglycerols. Comparison of Tables 5.2 and 5.6 indicates the similarity of the pattern of change in tumour-bearing and lactating rats; the essential difference is the occurrence of hypertriglyceridaemia in malignancy. As one might predict from this comparison, rapid tumour growth during lactation has little further effect on the already depressed lipid metabolism of adipose tissue, more interestingly it does not decrease exogenous lipid uptake or lipogenesis in mammary gland (Evans and Williamson, 1988). On cessation of lactation, however, the presence of tumour depresses the normal rapid (24 h) physiological switch-on of adipose tissue lipid metabolism (Evans and Williamson, 1988) indicating the tissue-specific nature of the effect. This depression occurs without any apparent change in plasma insulin or food intake (Evans and Williamson, 1988).

How does the presence of tumour bring about these marked changes in lipid metabolism (Table 5.6)? Lanza-Jacoby and colleagues (1984) have shown that the decrease in lipoprotein lipase activity of adipose tissue of tumour-bearing rats is correlated with a decrease in plasma insulin which might explain the change in enzyme activity. On the other hand, the findings with tumour-bearing lactating rats discussed above (Evans and Williamson, 1988) would suggest that at least in this experimental model changes in plasma insulin are not necessarily involved. Similarly, measurements of heparin releasable-lipoprotein lipase in human patients have shown that

Table 5.6 Effects of tumour growth on lipid metabolism in adipose tissue of experimental animals

Parameter	Change	Reference
Deposition of exogenous lipid	Decreased	Evans and Williamson (1988)
Lipoprotein lipase	Decreased	Thompson *et al.* (1981) Lanza-Jacoby *et al.* (1984) Evans and Williamson (1988)
Lipogenesis	Decreased	Thompson *et al.* (1981) Evans and Williamson (1988)
Lipolysis	Increased	Thompson *et al.* (1981)
Lipolysis–re-esterification	Decreased	Ookhtens *et al.* (1986) Evans and Williamson (1988)
Plasma triacylglycerols	Increased	Lanza-Jacoby *et al.* (1984) Evans and Williamson (1988)

cancer-associated decreases in activity of the enzyme are accompanied by either normal insulin levels or increased levels (the latter in patients with marked suppression of activity and profound weight loss; Vlassara *et al.*, 1986). The alternative is that the tumour or its presence produces a factor(s) which can depress adipose tissue lipid metabolism either directly or via inhibition of the anabolic effects of insulin. It is in this area where much of the current interest lies.

There have been reports of the presence of lipid mobilizing factors in serum of tumour-bearing animals (Kitada *et al.*, 1980) or in extracts of tumours which cause cachexia (Beck and Tisdale, 1987; see Tisdale, Chapter 8) but these factors have not been fully characterized. Other potential candidates as mediators of the cachetic response to tumour growth are a family of cytokines, polypeptides produced by lymphocytes, macrophages or endothelial cells as part of the inflammatory response to malignancy or infection. One of these cytokines, tumour-necrosis factor-α (cachectin) is considered to play a key role in cachexia and sepsis (Beutler and Cerami, 1988) and another, interleukin-1 (see Dinarello, 1988), may act synergistically. The overall effect of these cytokines is to depress lipid deposition in adipose tissue, mainly by a decrease in lipoprotein lipase activity, and to promote lipolysis (Table 5.7; see Evans *et al.*, 1989).

The effects of these cytokines on lipoprotein lipase are likely to be direct because they decrease activity in differentiating 3T3-L1 adipocytes (Cornelius *et al.*, 1988; Zechner *et al.*, 1988). Tumour necrosis factor decreases the mRNA for lipoprotein lipase (Cornelius *et al.*, 1988; Zechner *et al.*, 1988), whereas interleukin-1 does not and therefore must act at some other step in the maturation of enzyme activity (Figure 5.2). Surprisingly, tumour necrosis did not suppress lipoprotein lipase activity in primary cultures of human adipocytes, although the cytokine was bound to the plasma membrane (Kern, 1988). More recently, depression of lipoprotein lipase activity by tumour necrosis factor has been observed in adipose tissue pieces from humans (Fried and Zechner, 1989) and rats (Mackay *et al.*, 1990). The differences between the effects of the cytokine on isolated adipocytes (Kern,

Table 5.7 Summary of effects of cytokines on lipid metabolism in adipose tissue

Parameter	Tumour necrosis factor	Interleukin-1
Deposition of exogenous lipid	Decreased	Decreased
Lipoprotein lipase	Decreased (3T3-L1 cells) Decreased *in vivo* in rat	Decreased (3T3-L1 cells) No change *in vivo* in rat
Lipogenesis	No change *in vivo* in rat	No change *in vivo* in rat
Plasma triacylglycerol	Increased	Increased

1988, Mackay *et al.*, 1990) or adipose tissue pieces (Fried and Zechner, 1989; Mackay *et al.*, 1990) may be due to (a) loss of cytokine receptor during isolation of adipocytes; (b) differences in the hormonal milieu on culture; (c) the presence of endothelial cells in the adipose tissue pieces.

The ability of tumour-necrosis factor to induce cachexia in nude mice has been elegantly demonstrated by injection of non-malignant tumour cells which had been genetically engineered to produce a constant low level of circulating cytokine (Oliff *et al.*, 1987).Tumour necrosis factor was found to be present in the serum of cachexic, tumour-bearing rats (Stovroff *et al.*, 1989) and the levels correlated directly with tumour burden and inversely with food intake and body weight change. In this model, the toxic effects of tumour necrosis factor could be reversed by concurrent administration of insulin (Fraker *et al.*, 1989) but no measurements of adipose tissue metabolism were made. Similarly, in sarcoma-bearing mice administration of anti-tumour necrosis factor antibodies attenuated the development of cachexia and decreased the elevated plasma triacylglycerol (Sherry *et al.*, 1989). In summary, there is reasonable evidence that tumour necrosis factor has a role in the development of cachexia, particularly the loss of fat, and that this is reversible with insulin. What is by no means clear is whether tumour necrosis factor or other cytokines can act synergistically with other insulin antagonists (Table 5.4) of lipid metabolism to produce a greater inhibitory effect. It is unlikely that tumour necrosis factor is the sole mediator of cachexia, for example, a protein produced by a melanoma cell line associated with cancer cachexia has been shown to inhibit lipoprotein lipase activity in adipocytes (Mori *et al.*, 1989). Interestingly, this factor may be identical with human leukaemia inhibitory factor (Gough *et al.*, 1988).

LIPID METABOLISM IN ADIPOSE TISSUE IN OBESITY

Obesity in the human may have a genetic, dietary or hypothalamic basis (or a combination of these) and there are experimental animal models for all three types (see Bray and York, 1979). The most studied is the Zucker obese rat (see Argilés, 1989) in which the massive increase in adiposity is now considered to be secondary to central mechanisms (see Martin *et al.*, 1986; Bray *et al.*, 1989) rather than a specific genetic defect in adipose tissue. The Zucker rat, in common with other experimental models of obesity, exhibits hyperphagia and hyperinsulinaemia (York *et al.*, 1972), unlike the increased food intake of lactation where plasma insulin is low. These changes are accompanied by an increase in lipoprotein lipase activity and lipogenesis in adipose tissue (Table 5.8). *In vitro*, basal, but not stimulated lipolysis is increased but as there is also an increase in the intracellular lipolysis–re-esterification cycle net efflux of NEFA may not be

Table 5.8 Effects of obesity on lipid metabolism in adipose tissue of the obese Zucker rat

Parameter	Change	Reference
Deposition of exogenous lipid	Increased	Maggio and Greenwood (1982) Haggarty *et al.* (1986)
Lipoprotein lipase	Increased	De Gasquet *et al.* (1973) Gruen *et al.* (1978) Horwitz *et al.* (1984)
Lipogenesis	Increased	Godbole and York (1978)
Lipolysis	Increased	York and Bray (1973)
Lipolysis–re-esterification	Increased	York and Bray (1973)
Plasma triacylglycerols	Increased	Zucker (1965)

higher on a unit weight basis (York and Bray, 1973) but is likely to be increased on a total mass of tissue basis. Thus the adipose tissue of the Zucker rat has a high capacity for lipid deposition (Table 5.8). Despite this there is hypertriglyceridaemia (Zucker, 1965; Barry and Bray, 1969), which extends to all lipoprotein classes, but there are particularly marked increases in VLDL (Schonfeld *et al.*, 1974). The latter may be due to a combination of hypersecretion of a VLDL fraction from the liver with a high triacylglycerol and low apoprotein B content and decreased removal of these altered lipoproteins (Witzum and Schonfeld, 1979).

It is reasonable to conclude that the major cause of the increased lipid deposition in adipose tissue in obesity is the accompanying hyperinsulinaemia, particularly as this is one of the earliest abnormalities detected during the development of experimental obesity (Jeanrenaud, 1979) and would explain most of the metabolic changes (Table 5.8). Although hyperphagia often accompanies the raised plasma insulin of animal obesity syndromes, it is clear that the hyperinsulinaemia is not simply the result of overstimulation of the pancreatic β-cells by excessive nutrient intake (Jeanrenaud, 1979). Similarly, it has been demonstrated that experimentally induced hyperinsulinaemia in the rat increases body fat despite strict control of food intake and physical activity (Torbay *et al.*, 1985). If the raised plasma insulin is solely responsible for the obesity of the Zucker rat, it is intriguing that other aspects of insulin action, namely on carbohydrate metabolism in liver, muscle and to a lesser extent adipose tissue, show resistance to the effects of the hormone, particularly in the later phases of obesity (Jeanrenaud, 1979; see Frayn, Chapter 7). This insulin resistance has been attributed to the 'down regulation' of insulin receptors by the hyperinsulinaemia (Jeanrenaud, 1979). However, in normal rats, depending on the period of experimental hyperinsulinaemia, glucose transport and metabolism may be initially increased (Wardsala *et al.*, 1985). Another facet of obesity is increased concentrations of plasma NEFA (Opie and Walfisch, 1963) which is likely to be the consequence of the increased adipose tissue mass (Flatt, 1972). The greater availability of NEFA results in increased oxidation (Golay *et al.*, 1984) and may lead to glucose intolerance and insulin insensitivity (Randle *et al.*, 1963).

A key question is whether lowering plasma insulin in obese states can reverse the metabolic changes in adipose tissue. Food intake restriction or pair-feeding returned plasma insulin to normal levels but did not prevent the relative obesity (Bray *et al.*, 1973; Godbole and York, 1978). However, streptozotocin treatment to lower plasma insulin resulted in decreased adiposity in the obese Zucker rats and a decrease in fatty acid synthesis in parametrial and subcutaneous adipose tissue (Godbole and York, 1978).

Apart from the putative role of hyperinsulinaemia in promoting obesity discussed above, there is another possibility which may play an auxiliary role. For many years, particularly amongst physiologists, there has been the concept of feedback control of the body fat mass (see Hervey and Duggan, 1990). This feedback may be brought about by an effect of body fat on food intake (see Scharrer and Langhans, 1990). Thus restricted feeding or starvation leads to a loss of body fat, but on refeeding there is a transient hyperphagia which contributes to the recovery of the fat mass. Alternatively, the body fat may produce a signal which can either stimulate (decreased fat mass) or inhibit (increased fat mass) lipid deposition in adipose tissue. From studies on parabiotic rats (one of which is made obese) evidence has been obtained for humoral 'lipostatic' factors which act on satiety centres (Hervey and Duggan, 1990) or cause depletion of adipose tissue lipid (Harris and Martin, 1986). More recently a crude preparation from serum of overfed obese rats has been shown to inhibit lipogenesis in epididymal adipocytes from *ad libitum* fed rats (Harris *et al.*, 1989). It is tempting to speculate that there is a decrease in the production of these 'lipostatic' factors in genetic obesity or that they are still produced but are not effective on their target systems (satiety and/or lipid deposition).

THE FUTURE

The main conclusion from the present incomplete survey of the regulation of adipose tissue mass is that a key factor is the activity of lipoprotein lipase, and in particular, that portion of the enzyme which is active at the luminal surface of the adipose tissue endothelial cells. Indeed, lipoprotein lipase has been likened to a metabolic 'gatekeeper' (Greenwood, 1985). Clearly, insulin plays an important role in maintaining high activity of lipoprotein lipase. The recent cloning and characterization of the human lipoprotein lipase gene (Kirchgessner *et al.*, 1989; Deeb and Peng, 1989) will allow insight into the molecular basis of human genetic disorders of lipid metabolism. Similarly, the advances in cell biology at the level of intracellular and intercellular protein movement should help to unravel the complexities of the hormonal regulation of the activity of lipoprotein lipase (Doolittle *et al.*, 1990).

The discovery that cytokines, and particularly tumour necrosis factor, can inhibit lipid deposition in adipose tissue has raised a number of related questions. What is the nature of the transmembrane-signalling system whereby the cytokines exert their effect? Can these effects be opposed by insulin or enhanced by lipolytic hormones? Will intervention with anti-tumour necrosis factor prevent the hypertriglyceridaemia and loss of adipose tissue in cachectic patients? Answers to these questions may assist in the decision as to whether 'derivatives' of the cytokines have potential as agents to deplete lipid in obesity.

The application of modern techniques of protein/peptide separation and of molecular endocrinology is likely to provide information on the so-called lipostatic factors, which when identified may be of value in controlling obesity. However, loss of adipose tissue may not necessarily solve the problem of obesity. A recent study has shown that the high activity of lipoprotein lipase in the adipose tissue of obese humans is not decreased after a programme of weight reduction (Kern *et al.*, 1990). On the contrary, the activity increased further, as did the amount of immunoreactive protein and the level of lipoprotein lipase messenger RNA, thereby potentially enhancing lipid storage rather than depletion. What is not known is whether this is the normal physiological response to food restriction or not.

At the physiological level, the development of metabolic studies on human abdominal adipose tissue *in vivo* (Frayn *et al.*, 1989; Coppack *et al.*, 1990a, b) has been an important advance and is likely to provide considerable information. Work on small animals, including models of obesity or changes in adipose tissue mass, will allow further insight into the physiological integration between adipose tissue and other organs, as well as the effects of nutritional modifications. For example, meal-feeding of rats results in increased lipogenesis (Pallardo and Williamson, 1989) and increased exogenous lipid deposition (Tedstone *et al.*, 1991) in adipose tissue during consumption of a meal. This may be in response to the under-nutrition brought about by meal-feeding rather than the meal pattern itself.

ACKNOWLEDGEMENTS

I am grateful to Mrs M. Barber for patient preparation of the manuscript. The author's work is supported by the Medical Research Council (UK).

REFERENCES

Aas, M. and Daae, L.N.W. (1971). Fatty acid activation and acyl-transfer in organs from rats in different nutritional states. *Biochim. Biophys. Acta*, **239**, 208–16.

Abumrad, N.A., Park, J.H. and Park, C.R. (1984). Permeation of long-chain fatty acid into adipocytes. *J. Biol. Chem.*, **259**, 8945–53.

Abumrad, N.A., Park, C.R. and Whitesell, R.R. (1986a). Catecholamine activation of the membrane transport of long-chain fatty acids in adipocytes is mediated by cyclic AMP and protein kinase. *J. Biol. Chem.*, **261**, 13082–6.

Abumrad, N.A., Perry, P.R. and Whitesell, R.R. (1986b). Insulin antagonizes epinephrine activation of the membrane transport of fatty acids. Potential site for hormonal suppression of lipid mobilization. *J. Biol.Chem.*, **261**, 2999–3001.

Abumrad, N.A., Harman, C.M., Barnela, U.S. and Whitesell, R.R. (1988). Insulin antagonism of catecholamine stimulation of fatty acid transport in the adipocyte. Studies on its mechanism of action. *J. Biol. Chem.*, **263**, 14678–83.

Agius, L., Robinson, A.M., Girard, J.R. and Williamson, D.H. (1979). Alterations in the rate of lipogenesis *in vivo* in maternal liver and adipose tissue on premature weaning of lactating rats. A possible regulatory role for prolactin. *Biochem. J.*, **180**, 689–92.

Amri, E.-Z., Vannier, C., Etienne, J. and Ailhaud, G. (1986). Maturation and secretion of lipoprotein lipase in cultured adipose cells. II. Effects of tunicamycin on activation and secretion of the enzyme. *Biochem. Biophys. Acta*, **875**, 334–43.

Argilés, J.M. (1989). The obese Zucker rat: a choice for fat metabolism 1968–1988: Twenty years of research on the insights of the Zucker mutation. *Prog. Lipid Res.*, **28**, 53–66.

Bach, A. and Babayan, V.K. (1982). Medium chain triglycerides: an update. *Am. J. Clin. Nutr.*, **36**, 950–62.

Ball, K.L., Speake, B.K. and Robinson, D.S. (1986). Effect of adrenaline on the turnover of lipoprotein lipase in rat adipose tissue. *Biochim. Biophys. Acta*, **877**, 399–405.

Barry, W.S. and Bray, G.A. (1969). Plasma triglycerides in genetically obese rats. *Metabolism*, **18**, 833–9.

Beck, S.A. and Tisdale, M.J. (1987). Production of lipolytic and proteolytic factors by a murine tumor-producing cachexia in the host. *Cancer Res.*, **47**, 5919–23.

Belfrage, P. (1985). Hormonal control of lipid degradation. In: *New Perspectives in Adipose Tissue* (ed. A. Cryer and R.L.R. Van) London, Butterworths, pp. 121–44.

Berger, J., Biswas, C., Vicario, P.P., Strout, H.V., Saperstein, R. and Pilch, P.F. (1989). Decreased expression of the insulin-responsive glucose transporter in diabetes and fasting. *Nature*, **340**, 70–2.

Beutler, B. and Cerami, A. (1988). Tumor necrosis, cachexia, shock, and inflammation: a common mediator. *Ann. Rev. Biochem.*, **57**, 505–18.

Borthwick, A.C., Edgell, N.J. and Denton, R.M. (1987). Use of rapid gel-permeation chromatography to explore the interrelationships between polymerization, phosphorylation and activity of acetyl-CoA carboxylase. Effects of insulin and phosphorylation by cyclic AMP-dependent protein kinase. *Biochem. J.*, **241**, 773–82.

Bray, G.A. and York, D.A. (1979). Hypothalamic and genetic obesity in experimental animals: an autonomic and endocrine hypothesis. *Physiol. Rev.*, **59**, 779–809.

Bray, G.A., York, D.A. and Fisher, J.S. (1989). Experimental obesity: A homeostatic failure due to defective nutrient stimulation of the sympathetic nervous system. *Vitam. Horm.*, **45**, 1–125.

Bray, G.A., York, D.A. and Swerloff, R.S. (1973). Genetic obesity in rats. I. The effects of food restriction on body composition and hypothalamic function. *Metabolism*, **22**, 435–43.

Brooks, B.J., Arch, J.R.S. and Newsholme, E.A. (1982). Effects of hormones on the rate of the triacylglycerol/fatty acid substrate cycle in adipocytes and epididymal fat pads. *FEBS Lett.*, **146**, 327–30.

Brooks, B.J., Arch, J.R.S. and Newsholme, E.A. (1983). Effect of some hormones on the rate of the triacylglycerol/fatty acid substrate cycle in adipose tissue of the mouse *in vivo*. *Biosci. Rep.*, **3**, 263–7.

Brownsey, R.W. and Denton, R.M. (1982). Evidence that insulin activates fat cell acetyl-CoA carboxylase by increased phosphorylation at a specific site. *Biochem. J.*, **202**, 77–86.

Burnol, A.-F., Leturque, A., Ferré, P. and Girard, J. (1983). Glucose metabolism during lactation in the rat: quantitative and regulatory aspects. *Am. J. Physiol.*, **245**, E351–8.

Burnol, A.-F., Guerre-Millo, M., Lavau, M. and Girard, J. (1986). Effect of lactation on insulin sensitivity of glucose metabolism in rat adipocytes. *FEBS Lett.*, **194**, 292–6.

Burnol, A.-F., Ferré, P., Leturque, A. and Girard, J. (1987). Effect of insulin on *in vivo* glucose utilization in individual tissues of anaesthetized lactating rats. *Am. J. Physiol.*, **252**, E183–8.

Carling, D., Zammit, V.A. and Hardie, D.G. (1987). A common bicyclic protein kinase cascade inactivates the regulatory enzyme of fatty acid and cholesterol biosynthesis. *FEBS Lett.*, **223**, 217–22.

Chajek-Shaul, T., Friedman, G., Knobler, H., Stein, O., Etienne, J. and Stein, Y. (1985a). Importance of different steps of glycosylation for the activity and secretion of lipoprotein lipase in rat preadipocytes studied with monensin and tunicamycin. *Biochim. Biophys. Acta*, **837**, 123–34.

Chajek-Shaul, T., Friedman, G., Stein, O., Etienne, J. and Stein, Y. (1985b). Endogenous plasma lipoprotein lipase activity in fed and fasting rats may reflect the functional pool of endothelial lipoprotein lipase. *Biochim. Biophys. Acta*, **837**, 271–8.

Chan, B.L., Lisanti, M.P., Boulan-Rodriguez, E. and Saltiel, A.R. (1988). Insulin-stimulated release of lipoprotein lipase by metabolism of its phosphatidylinositol anchor. *Science*, **241**, 1670–2.

Cheng, C.H.K. and Saggerson, E.D. (1978). Rapid antagonistic actions of noradrenaline and insulin on rat adipocyte phosphatidate phosphohydrolase activity. *FEBS Lett.*, **93**, 120–4.

Cianflone, K., Vu, H., Walsh, M., Baldo, A. and Sniderman, A. (1989a). Metabolic response of Acylation Stimulating Protein to an oral fat load. *J. Lipid Res.*, **30**, 1717–33.

Cianflone, K., Walsh, M. J., Vu, H., Gagnon, J., Sniderman, A.D. and Rodrigues, M.A. (1989b). Ramification and characterization of ASP. *J. Biol. Chem.*, **265**, 426–30.

Cisar, L.A., Hoogewerf, A.J., Cupp, M., Rapport, C.A. and Bensadoun, A. (1989). Secretion and degradation of lipoprotein lipase in cultured adipocytes. Binding of lipoprotein lipase to membrane heparin sulfate proteoglycans is necessary for degradation. *J. Biol. Chem.*, **264**, 1767–74.

Coore, H.G. and Field, B. (1974). Properties of pyruvate dehydrogenase of rat mammary tissue and its changes during pregnancy, lactation and weaning. *Biochem. J.*, **142**, 87–95.

Coppack, S.W., Frayn, K.N. and Humphreys, S.M. (1989a). Plasma triacylglycerol extraction in human adipose tissue *in vivo*: effects of glucose ingestion and insulin infusion. *Eur. J. Clin. Nutr.*, **43**, 493–6.

Coppack, S.W., Frayn, K.N., Humphreys, S.M., Dhar, H. and Hockaday, T.D.R. (1989b). Effects of insulin on human adipose tissue metabolism *in vivo*. *Clin Sci.*, **77**, 663–70.

Coppack, S.W., Fisher, R.M., Gibbons, G.F., Humphreys, S.M., McDonough, M.J., Potts, J.L. and Frayn, K.N. (1990a). Postprandial substrate deposition in human forearm and adipose tissue *in vivo*. *Clin. Sci.*, **79**, 339–48.

Coppack, S.W., Frayn, K.W., Humphreys, S.M., Whyte, P.L. and Hockaday, T.D.R. (1990b). Arteriovenous differences across human adipose and forearm tissues after an overnight fast. *Metabolism,* **39**, 384–90.

Cornelius, P., Enerback, S., Bjursell, G., Olivecrona, T. and Pekala, P.H. (1988). Regulation of lipoprotein lipase mRNA content in 3T3-L1 cells by tumour necrosis factor. *Biochem. J.*, **249**, 765–9.

Cryer, A. (1981). Tissue lipoprotein lipase activity and its action in lipoprotein metabolism. *Int. J. Biochem.*, **13**, 525–41.

Cryer, A. and Cryer, J. (1990). Regulation of structure, growth and differentiation of adipocytes. In: *The Control of the Body Fat Content* (eds J.M. Forbes and G.R. Hervey), International Monographs in Nutrition, Metabolism and Obesity I., Smith-Gordon, London.

Cryer, A., Riley, S.E., Williamson, E.R. and Robinson, D.S. (1976). Effect of nutritional states on rat adipose tissue and post-heparin clearing factor lipase activities: their relationship to triglyceride fatty acid uptake by fat cells and to plasma insulin concentrations. *Clin. Sci. Mol. Med.*, **50**, 213–21.

Cubero, A., Ros, M., Lobato, M.F., Carcia-Ruiz, J.P. and Moreno, F.J. (1983). Co-ordination of glucose metabolism and NADPH formation in the adipose tissue and mammary gland during lactation-weaning transition. *Enzyme*, **30**, 38–47.

Cupp, M., Bensadoun, A. and Melford, K. (1987). Heparin decreases the degradation rate of lipoprotein lipase in adipocytes. *J. Biol. Chem.*, **262**, 6383–8.

Deeb, S.S. and Peng, R. (1989). Structure of the human lipoprotein lipase gene. *Biochemistry*, **28**, 4131–5.

De Gasquet, P., Péquignot, E., Lemonnier, D. and Alexia, A. (1973). Adipose-tissue lipoprotein lipase activity and cellularity in the genetically obese Zucker rat (fa/fa). *Biochem. J.*, **132**, 633–5.

Denton, R.M., Brownsey, R.W. and Belsham, G.J. (1981). A partial view of the mechanism of insulin action. *Diabetologia*, **21**, 347–62.

Denton, R.M., Midgeley, P.J.W., Rutter, G.A., Thomas, A.P. and McCormack, J.G. (1989). Studies into the mechanism whereby insulin activates pyruvate dehydrogenase complex in adipose tissue. *Ann. N.Y. Acad. Sci.*, **573**, 285–96.

Dinarello, C.A. (1988). Biology of interleukin-1. *FASEB J.*, **2**, 108–15.

Doolittle, M.H., Ben-Zeev, O., Elovson, J., Martin, D. and Kirchgessner, T.G. (1990). The response of lipoprotein lipase to feeding and fasting. Evidence for posttranslational regulation. *J. Biol. Chem.*, **265**, 4570–7.

Eckel, R.H. (1987). Adipose tissue lipoprotein lipase. In: *Lipoprotein Lipase* (ed. J. Borensztajn), Evener, Chicago, pp. 79–132.

Eckel, R.H. (1989). Lipoprotein lipase. A multifunctional enzyme relevant to common metabolic diseases. *N. Engl. J. Med.*, **320**, 1060–8.

Elia, M., Zed, C., Neale, G. and Livesey, G. (1987). The energy cost of triglyceride–fatty acid recycling in non-obese subjects after an overnight fast and four days of starvation. *Metabolism*, **36**, 252–5.

Elkeles, R.S. (1974). Lipoprotein lipase in human adipose tissue. *Clin. Sci. Mol. Med.*, **46**, 753–62.
Evans, R.D. and Williamson, D.H. (1988). Tissue-specific effect of rapid tumour growth on lipid metabolism in the rat during lactation and on litter removal. *Biochem. J.*, **252**, 65–72.
Evans, R.D., Argilés, J.M. and Williamson, D.H. (1989). Metabolic effects of tumour necrosis factor-α (cachectin) and interleukin-1. *Clin. Sci.*, **77**, 357–64.
Flatt, J.P. (1972). Role of increased adipose tissue mass in apparent insulin insensitivity of obesity. *Am. J. Clin. Nutr.*, **25**, 1189–92.
Flint, D.J., Clegg, R.A. and Vernon, R.G. (1981). Prolactin and the regulation of adipose-tissue metabolism during lactation in rats. *Mol. Cell Endocrinol.*, **22**, 265–75.
Flint, D.J., Sinnett-Smith, P.A., Clegg, R.A. and Vernon, R.G. (1979). Role of insulin receptors in the changing metabolism of adipose tissue during pregnancy and lactation in the rat. *Biochem. J.*, **182**, 421–7.
Fraker, D.L., Merino, M.J. and Norton, J.A. (1989). Reversal of the toxic effects of cachectin by concurrent insulin administration. *Am. J. Physiol.*, **256**, E725–31.
Frayn, K.N., Coppack, S.W., Humphreys, S.M. and Whyte, P.L. (1989). Metabolic characteristics of human adipose tissue *in vivo*. *Clin. Sci.*, **76**, 509–16.
Fried, S.K. and Zechner, R. (1989). Cachectin/tumor necrosis factor decreases human adipose tissue lipoprotein lipase mRNA levels, synthesis and activity. *J. Lipid Res.*, **30**, 1917–23.
Galton, D.J. (1971). In: *The Human Adipose Cell: A Model for Errors in Metabolic Regulation*, London, Butterworth.
Gandemer, G., Pascal, G. and Durand, G. (1982). *In vivo* changes in the rates of total lipid and fatty acid synthesis in liver and white adipose tissue of male rats during post-weaning growth. *Int. J. Biochem.*, **14**, 797–804.
Garton, A.J., Campbell, D.G., Carling, D., Hardie, D.G., Colbran, R.J. and Yeaman, S.J. (1989). Phosphorylation of bovine-sensitive lipase by the AMP-activated protein kinase. A possible antilipolytic mechanism. *Eur. J. Biochem.*, **179**, 244–54.
Gibbins, J.M., Denton, R.M. and McCormack, J.G. (1985). Evidence that noradrenaline increases pyruvate dehydrogenase activity and increases acetyl-CoA carboxylase activity in rat interscapular brown adipose tissue *in vivo*. *Biochem. J.*, **228**, 751–5.
Gibbons, G.F. (1990). Assembly and secretion of hepatic very-low-density lipoprotein. *Biochem. J.*, **268**, 1–13.
Godbole, V. and York, D.A. (1978). Lipogenesis *in situ* in the genetically obese Zucker fatty rat (fa/fa). Role of hyperphagia and hyperinsulinaemia. *Diabetologia*, **14**, 191–7.
Golay, A., Felber, J.P., Meyer, H.U., Curchod, B., Maeder, E. and Jéquier, E. (1984). Studies on lipid metabolism on obesity diabetes. *Metabolism*, **33**, 111–6.
Gough, N.M., Gearing, D.P., King, J.A., Willson, T.A., Hilton, D.J., Nicola, N.A. and Metcalf, D. (1988). Molecular cloning and expression of the human homologue of the murine gene encoding myeloid leukemia-inhibitory factor. *Proc. Natl Acad. Sci. USA*, **85**, 2623–7.
Greenwood, M.R.C. (1985). The relationship of enzyme activity to feeding behaviour in rats: lipoprotein lipase as the metabolic gatekeeper. *Int. J. Obesity*, **9**, 67–70.
Gruen, R.E., Hietanen, E. and Greenwood, M.R.C. (1978). Increased adipose tissue lipoprotein lipase activity during development of the genetically obese rat (fa/fa). *Metabolism,* **27**, 1955–66.

Grundy, S.M. and Mok, H.Y.I. (1976). Chylomicron clearance in normal and hyperlipidemic man. *Metabolism,* **25**, 1225–39.

Haggarty, P., Reeds, P.J., Fletcher, J.M. and Wahle, K.W.J. (1986). The fate of ^{14}C derived from radioactively labelled dietary precursors in young rats of the Zucker strain. (Fa/- and fa/fa). *Biochem. J.*, **235**, 323–7.

Hagström, E., Arner, P., Ungerstedt, U. and Bolinder, J. (1990). Subcutaneous adipose tissue: a source of lactate production after glucose ingestion in humans. *Am. J. Physiol.*, **258**, E888–93.

Hales, C.N., Luzio, J.P. and Siddle, K. (1978). Hormonal control of adipose-tissue lipolysis. *Biochem. Soc. Symp.*, **43**, 97–135.

Halestrap, A.P. and Denton, R.M. (1974). Hormonal regulation of adipose-tissue acetyl coenzyme A carboxylase by changes in the polymeric state of the enzyme. The role of long-chain fatty acyl-coenzyme A thioesters and citrate. *Biochem. J.*, **142**, 368–77.

Halperin, M.L. and Denton, R.M. (1969). Regulation of glycolysis and L-glycerol-3-phosphate concentration in rat epididymal adipose tissue *in vitro. Biochem. J.*, **113**, 207–14.

Hansson, P., Newsholme, E.A. and Williamson, D.H. (1987). Effects of lactation and removal of pups on the rate of triacyl-glycerol/fatty acid substrate cycling in white adipose tissue of the rat. *Biochem. J.*, **243**, 267–71.

Hardie, D.G. (1989). Regulation of fatty acid synthesis via phosphorylation of acetyl-CoA carboxylase. *Prog. Lipid Res.*, **28**, 117–46.

Hardie, D.G., Carling, D. and Sim, A.T.R. (1989). The AMP-activated protein kinase: a multisubstrate regulator of lipid metabolism. *Trends Biochem. Sci.*, **14**, 20–3.

Harper, R.D. and Saggerson, E.D. (1976). Factors affecting fatty acid oxidation in fat cells isolated from rat white adipose tissue. *J. Lipid Res.*, **17**, 516–26.

Harris, R.B.S. and Martin, R.J. (1986). Metabolic response to a specific lipid-depleting factor in parabiotic rats. *Am. J. Physiol.*, **250**, R276–86.

Harris, R.B.S., Bruch, R.C. and Martin, R.J. (1989). *In vitro* evidence for an inhibitor of lipogenesis in serum from overfed obese rats. *Am. J. Physiol.*, **257**, R326–36.

Hartman, A.D. (1977). Lipoprotein lipase distribution in rat adipose tissues: effect on chylomicron uptake. *Am. J. Physiol.*, **232**, E316–23.

Haystead, T.A., Moore, F., Cohen, P. and Hardie, D.G. (1990). Roles of the AMP-activated and cyclic-AMP-dependent protein kinase in the adrenaline-induced inactivation of acetyl-CoA carboxylase in rat adipocytes. *Eur. J. Biochem.*, **187**, 199–205.

Hervey, G.R. and Duggan, J.P. (1990). The concept of feedback control of body fat content and the evidence for it. In: *The Control of Body Fat Content* (eds J.M. Forbes and G.R. Hervey) International Monographs in Nutrition, Metabolism and Obesity I, Smith-Gordon, London, pp. 1–18.

Holm, C., Belfrage, P. and Fredrickson, G. (1989). Human adipose tissue hormone-sensitive lipase: identification and comparison with other species. *Biochim. Biophys. Acta*, **1006**, 193–7.

Horwitz, B.A., Inokuchi, T., Wickler, S.J. and Stern, J.S. (1984). Lipoprotein lipase activity and cellularity in brown and white adipose tissue in Zucker obese rats. *Metabolism*, **33**, 354–7.

Howanitz, P.J. and Levy, H.R. (1965). Acetyl-CoA carboxylase and citrate cleavage enzyme in rat mammary gland. *Biochim. Biophys. Acta*, **106**, 430–3.

Jamdar, S.C. and Osborne, L.J. (1982). Glycerolipid biosynthesis in rat adipose tissue. 10. Changes during a starvation and refeeding cycle. *Biochim. Biophys. Acta*, **713**, 647–56.

Jansson, P.-A., Smith, U. and Lönnroth, P. (1990). Evidence for lactate production by human adipose tissue *in vivo*. *Diabetologia*, **33**, 253–6.

Jeanrenaud, B. (1979). Insulin and obesity. *Diabetologia*, **17**, 133–8.

Jones, R.G., Ilic, V. and Williamson, D.H. (1984a). Physiological significance of altered insulin metabolism in the conscious rat during lactation. *Biochem. J.*, **220**, 455–60.

Jones, R.G., Ilic, V. and Williamson, D.H. (1984b). Regulation of lactating-rat mammary-gland lipogenesis by insulin and glucagon *in vivo*. The role and site of action of insulin in the transition to the starved state. *Biochem. J.*, **223**, 345–51.

Jungas, R.L. (1968). Fatty acid synthesis in adipose tissue incubated in tritiated water. *Biochemistry*, **7**, 3708–17.

Kahn, B.B., Cushman, S.W. and Flier, J.S. (1989). Regulation of glucose transporter-specific mRNA levels in rat adipose cells with fasting and refeeding. Implications for *in vivo* control of glucose transporter number. *J. Clin. Invest.*, **83**, 199–204.

Karnieli, E., Hissin, P.J., Simpson, I.A., Salans, L.B. and Cushman, S.W. (1981a). A possible mechanism of insulin resistance in the rat adipose cell in streptozotocin-induced diabetes mellitus. *J.Clin. Invest.*, **68**, 811–4.

Karnieli, E., Zarnowski, M.J., Hissin, P.J., Simpson, I.A., Salans, L.B. and Cushman, S.W. (1981b). Insulin-stimulated translocation of glucose transport systems in the isolated rat adipose cell. *J. Biol. Chem.*, **256**, 4772–7.

Katz, J. and Wals, P.A. (1974). Lipogenesis from lactate in rat adipose tissue. *Biochim. Biophys. Acta*, **348**, 344–56.

Kern, P.A. (1988). Recombinant human tumor necrosis factor does not inhibit lipoprotein lipase in primary cultures of isolated human adipocytes. *J. Lipid Res.*, **29**, 1379–85.

Kern, K.A. and Norton, J.A. (1988). Cancer cachexia. *J. PEN.*, **12**, 286–98.

Kern, P.A., Marshall, S. and Eckel, R.H. (1985). Regulation of lipoprotein lipase in primary cultures of isolated human adipocytes. *J. Clin. Invest.*, **75**, 199–208.

Kern, P.A., Ong, J.M., Saffari, B. and Carty, J. (1990). The effects of weight loss on the activity and expression of adipose-tissue lipoprotein lipase in very obese humans.

Kilgour, E. and Vernon, R.G. (1987). Tissue-specific changes in the ability of insulin and noradrenaline to activate pyruvate dehydrogenase *in vivo* during lactation in the rat. *Biochem. J.*, **243**, 69–74.

Kirchgessner, T.G., Chuat, J.-C., Heinzmann, C., Etienne, J., Guilhot, S., Svenson, K., Ameis, D., Pilon, C., D'Auriol, L., Andalibi, A., Schotz, M.C., Galibert, F. and Lusis, A.J. (1989). Organization of the human lipoprotein lipase gene and evolution of the lipase gene family. *Proc. Natl Acad. Sci. USA*, **86**, 9647–51.

Kitada, S., Hays, E.F. and Mead, J.F. (1980). A lipid mobilising factor in serum of tumor-bearing mice. *Lipids*, **15**, 168–74.

Lanza-Jacoby, S., Lansey, S.C., Miller, E.E. and Cleary, M.P. (1984). Sequential changes in the activities of lipoprotein lipase and lipogenic enzymes during tumor growth in rats. *Cancer Res.*, **44**, 5062–7.

Lawson, N., Pollard, A.D., Jennings, R.J., Gurr, M.I. and Brindley, D.N. (1981). The activities of lipoprotein lipase and of enzymes involved in triacylglycerol synthesis in rat adipose tissue. Effects of starvation, dietary modification and of corticotropin injection. *Biochem. J.*, **200**, 285–94.

Lent, B.A., Lee, K.-H. and Kim, K.H. (1978). Regulation of rat liver acetyl-CoA carboxylase. Stimulation of phosphorylation and subsequent inactivation of liver acetyl-CoA carboxylase by cyclic 3′,5′-monophosphate and effect on structure of the enzyme. *J. Biol. Chem.*, **253**, 8149–56.

Londos, C., Honnor, R.C. and Dhillon, G.S. (1985). cAMP-dependent protein kinase and lipolysis in rat adipocytes. III. Multiple modes of insulin regulation of lipolysis and regulation of insulin response by adenylate cyclase regulators. *J. Biol. Chem.*, **260**, 15139–45.

Lönnroth, P., Janson, P.-A. and Smith, U. (1988). Adipose tissue metabolism measured by microdialysis. In: *Obesity in Europe 88* (eds P. Björntorp and S. Rössner), John Libbey, London, pp. 209–214.

Mabrouk, G.M., Helmy, I.M., Thompy, K.G. and Wakil, S.J. (1990). Acute hormonal control of acetyl-CoA carboxylase. The roles of insulin, glucagon and epinephrine. *J. Biol. Chem.*, **265**, 6330–8.

Mackay, A.G., Oliver, J.D. and Rogers, M.P. (1990). Regulation of lipoprotein lipase activity and mRNA content in rat epididymal adipose tissue *in vitro* by recombinant tumour necrosis factor. *Biochem. J.*, **269**, 123–6.

Maggio, C. and Greenwood, M.R.C. (1982). Triglyceride clearance and adipose tissue lipoprotein lipase activity and triglyceride uptake in Zucker rats. *Physiol. Behav.*, **29**, 1147–52.

Marin, P., Rebuffé-Scrive, M., Smith, U. and Björntorp, P. (1987). Glucose uptake in human adipose tissue. *Metabolism*, **36**, 1154–60.

Martin, R.J., Harris, R.B.S. and Jones, D.D. (1986). Evidence for a central mechanism of obesity in the Zucker fatty rat (fa/fa). *Proc. Soc. Exp. Biol. Med.*, **183**, 1–10.

Mayes, P.A. (1976). Control of hepatic triacylglycerol metabolism. *Biochem. Soc. Trans.*, **4**, 575–80.

Miyoshi, H., Shulman, G.I., Peters, E.J., Wolfe, M.H. and Wolfe,R.R. (1988). Hormonal control of substrate cycling in humans. *J. Clin. Invest.*, **81**, 1545–55.

Moller, F., Green, P. and Harkness, E.J. (1977). Soluble rat adipocyte phosphatidate phosphatase activity: characterization and effects of fasting and various lipids. *Biochim. Biophys. Acta.*, **486**, 359–68.

Mori, M., Yamaguchi, K. and Abe, K. (1989). Purification of a lipoprotein lipase-inhibiting protein produced by a melanoma cell line associated with cancer cachexia. *Biochem. Biophys. Res.Commun.*, **160**, 1085–92.

Newsholme, E.A. (1980). Reflections on the mechanism of action of hormones. *FEBS Lett.*, **117**, K121–34.

Nicolls, A. and Lewis, B. (1980). Evaluation of the roles of lipoprotein lipase in lipoprotein metabolism: *in vivo* and *in vitro* studies in man. *Eur. J. Clin. Invest.*, **10**, 487–95.

Nimmo, H.G. and Houston, B. (1978). Rat adipose tissue glycerol-phosphate acyltransferase can be inactivated by cyclic AMP dependent protein kinase. *Biochem. J.*, **176**, 607–10.

Oliff, A., Defeo-Jones, D., Boyer, M., Martinez, D., Kiefer, D., Vuocolo, G., Wolfe, A. and Socher, S.H. (1987). Tumors secreting human TNF/cachectin induce cachexia in mice. *Cell*, **50**, 555–63.

Olivecrona, T. and Bengtsson-Olivecrona, G. (1987). Lipoprotein lipase from milk—the model enzyme in lipoprotein lipase research. In: *Lipoprotein lipase* (ed. J. Borensztajn), Evener, Chicago, pp. 15–58.

Olivecrona, T., Chernick, S.S., Bengtsson-Olivecrona, G., Paterniti, J.R., Brown, W.V. and Scow, R.O. (1985). Combined lipase deficiency (cld/cld) in mice. Demonstration that an inactive form of lipoprotein lipase is synthesized. *J.Biol. Chem.*, **260**, 2552–7.

Olivecrona, T., Chernick, S.S., Bengtsson-Olivecrona, G., Garrison, M. and Scow, R.O. (1987). Synthesis and secretion of lipoprotein lipase in 3T3-L1 adipocytes. Demonstration of inactive forms of lipase in cells. *J. Biol. Chem.*, **262**, 10748–59.

Oller do Nascimento, C.M. and Williamson, D.H. (1986). Evidence for the conservation of dietary lipid in the rat during lactation and the immediate period after removal of the litter. Decreased oxidation of oral [1-^{14}C]triolein. *Biochem. J.*, **239**, 233–6.

Oller do Nascimento, C.M. and Williamson, D.H. (1988).Tissue-specific effects of starvation and refeeding on the disposal of oral [1-^{14}C]triolein in the rat during lactation and on removal of litter. *Biochem. J.*, **254**, 539–46.

Oller do Nascimento, C.M., Ilic, V. and Williamson, D.H. (1989). Re-examination of the putative roles of insulin and prolactin in the regulation of lipid deposition and lipogenesis *in vivo* in mammary gland and white and brown adipose tissue of lactating rats and litter-removed rats. *Biochem. J.*, **258**, 273–8.

Olsson, H. and Belfrage, P. (1987). The regulatory and basal phosphorylation sites of hormone-sensitive lipase are dephosphorylated by protein phosphatase-1, 2A and 2C, but not by protein phosphatase-2B. *Eur. J. Biochem.*, **168**, 399–405.

Ong, J.M. and Kern, P.A. (1989). Effect of feeding and obesity on lipoprotein lipase activity, immunoreactive protein, and messenger RNA levels in human adipose tissue. *J. Clin. Invest.*, **84**, 305–11.

Ong, J.M., Kirchgessner, T.G., Schotz, M.C. and Kern, P.A. (1988). Insulin increases the synthetic rate and messenger RNA level of lipoprotein lipase in isolated rat adipocytes. *J. Biol. Chem.*, **263**, 12933–8.

Ookhtens, M., Montisano, D., Lyon, I. and Baker, N. (1986). Inhibition of fatty acid incorporation into adipose tissue triglycerides in Ehrlich Ascites tumour-bearing mice. *Cancer Res.*, **46**, 633–8.

Opie, L.H. and Walfisch, P.G. (1963). Plasma free fatty acid concentrations in obesity. *N. Engl. J. Med.*, **268**, 757–60.

Palacin, M., Lasuncion, M.A. and Herrera, E. (1988). Utilization of glucose, alanine, lactate and glycerol as lipogenic substrates by periuterine adipose tissue *in situ* in fed and starved rats. *J. Lipid Res.*, **29**, 26–32.

Pallardo, F.V. and Williamson, D.H. (1989). Comparison of the flux of carbon to hepatic glycogen deposition and fatty acid and cholesterol synthesis on refeeding rats fed *ad libitum* or meal-fed rats with a chow-diet meal. *Biochem. J.*, **257**, 607–10.

Randle, P.J. (1986). Fuel selection in animals. *Biochem. Soc. Trans.*, **14**, 799–806.

Randle, P.J., Garland, P.B., Hales, C.N. and Newsholme, E.A. (1963). The glucose-fatty acid cycle. Its role in insulin sensitivity and the metabolic disturbances of diabetes mellitus. *Lancet*, **i**, 785–9.

Rider, M.H. and Saggerson, E.D. (1983). Regulation by noradrenaline of the mitochondrial and microsomal forms of glycerol phosphate acyltransferase in rat adipocytes. *Biochem. J.*, **214**, 235–46.

Robins, S.J., Fasulo, J.M., Robins, V.F. and Patton, G.M. (1987). Response of serum triglycerides of endogenous origin to the administration of triglyceride-rich particles. *Am. J. Physiol.*, **257**, E860–5.

Robinson, D.S. (1970). The function of the plasma triglycerides in fatty acid transport. In: *Comprehensive Biochemistry*, Vol. 18 (ed. M. Florkin and E.H. Stotz), Elsevier, Amsterdam, pp. 51–116.

Robinson, D.S. and Speake, B.K. (1989). Role of insulin in the control of lipoprotein lipase activity. *Biochem. Soc. Trans.*, **17**, 40–6.

Robinson, A.M., Girard, J.R. and Williamson, D.H. (1978). Evidence for a role of insulin in the regulation of lipogenesis in lactating rat mammary gland. Measurements of lipogenesis *in vivo* and plasma hormone concentrations in response to starvation and refeeding. *Biochem. J.*, **176**, 343–6.

Robson, N.A., Clegg, R.A. and Zammit, V.A. (1984). Regulation of peripheral lipogenesis by glucagon. Inability of the hormone to inhibit lipogenesis in rat mammary acini *in vitro* in the presence or absence of agents which alter its effects on adipocytes. *Biochem. J.*, **217**, 743–9.

Rosell, S. and Belfrage, E. (1979). Blood circulation in adipose tissue. *Physiol. Rev.*, **59**, 1078–104.

Rothwell, N.J., Stock, M.J. and Trayhurn, P. (1983). Reduced lipogenesis in cafeteria-fed rats exhibiting diet-induced thermogenesis. *Biosci. Rep.*, **3**, 217–24.

Sadur, C.N. and Eckel, R.H. (1982). Insulin stimulation of adipose tissue lipoprotein lipase: use of euglycemic clamp technique. *J. Clin. Invest.*, **69**, 1119–25.

Saggerson, E.D. (1985). Hormonal regulation of biosynthetic activities in white adipose tissue. In: *New perspectives in adipose tissue: structure, function and development* (ed. A. Cryer and R.L.R. Van), Butterworths, London.

Scharrer, E. and Langhans, W. (1990). Mechanisms for the effect of body fat on food intake. In: *The Control of Body Fat Content* (eds J.M. Forbes and P.G.R. Hervey), Smith-Gordon, London, pp. 63–99.

Schonfeld, G., Felski, C. and Howald, M.A. (1974). Characterization of the plasma lipoproteins in the genetically obese hyperlipoproteinaemic Zucker fatty rat. *J. Lipid Res.*, **15**, 457–64.

Scow, R.O. (1977). Metabolism of chylomicrons in perfused adipose and mammary tissue of the rat. *Fed. Proc.*, **36**, 182–5.

Scow, R.O. and Chernick, S.S. (1987). Role of lipoprotein lipase during lactation. In: *Lipoprotein Lipase* (ed. J. Borensztajn), Evener, Chicago, pp. 149–85.

Scow, R.O., Chernick, S.S. and Fleck, T.R. (1977). Lipoprotein lipase and uptake of triacylglycerol, cholesterol and phosphatidyl choline from chylomicrons by mammary and adipose tissue of lactating rats *in vivo*. *Biochim. Biophys. Acta*, **487**, 297–306.

Seals, J.R. and Jarett, L. (1980). Activation of pyruvate dehydrogenase by direct addition of insulin to an isolated plasma membrane mitochondria mixture. Evidence for generation of insulin's second messenger in a subcellular system. *Proc. Natl Acad. Sci. USA*, **77**, 77–81.

Semb, H. and Olivecrona, T. (1987). Mechanisms for the turnover of lipoprotein lipase in guinea pig adipocytes. *Biochim. Biophys. Acta*, **921**, 104–15.

Semenkovich, C.F., Wims, M., Noe, L., Etienne, J. and Chan, L. (1989). Insulin regulation of lipoprotein lipase activity in 3T3-L1 adipocytes is mediated at post-transcriptional and post-translational levels. *J. Biol.Chem.*, **264**, 9030–8.

Sherry, B.A., Gelin, J., Fong, Y., Marano, M., Wei, H., Cerami, A., Lowry, S.F., Lundholm, K.G. and Moldawer, L.L. (1989). Anticachectin/tumor necrosis factor-α antibodies attenuate development of cachexia in tumor models. *FASEB J.*, **3**, 1956–62.

Simpson, I.A. and Cushman, S.W. (1986). Hormonal regulation of mammalian glucose transport. *Ann. Rev. Biochem.*, **55**, 1059–89.

Sinnett-Smith, P.A., Vernon, R.G. and Mayer, R.J. (1980). Lipogenic enzymes in rat maternal adipose tissue in the perinatal period. *Biochem. J.*, **186**, 937–44.

Sivitz, W.I., DeSoutel, S.L., Kayano, T., Bell, G.I. and Pressin, J.E. (1989). Regulation of glucose transporter messenger RNA in insulin-deficient states. *Nature*, **340**, 72–4.

Smith, R.W. and Walsh, A. (1976). Effect of lactation on lipolysis in rat adipose tissue. *Lipids*, **11**, 418–20.

Smith, V., Kuroda, M. and Simpson, I.A. (1984). Counter-regulation of insulin-stimulated glucose transport by catecholamines in the isolated rat adipose cell. *J. Biol. Chem.*, **259**, 8758–63.

Sooranna, S.R. and Saggerson, E.D. (1978). A stable decrease in long-chain fatty acyl-CoA synthetase activity after treatment of rat adipocytes with adrenaline. *FEBS Lett.*, **92**, 241–4.

Speake, B.K., Parkin, S.M. and Robinson, D.S. (1985a). Degradation of lipoprotein lipase in rat adipose tissue. *Biochim. Biophys. Acta*, **840**, 419–22.

Speake, B.K., Parkinson, C. and Robinson, D.S. (1985b). The effect of insulin on the synthesis of lipoprotein lipase in rat adipose tissue. *Horm. Metab. Res.*, **12**, 637–40.

Spooner, P.M., Chernick, S.S., Garrison, M. and Scow, R.O. (1979). Insulin regulation of lipoprotein lipase activity and release in 3T3-L1. Separation and dependence of hormonal effects on hexose metabolism and synthesis of RNA and protein. *J. Biol. Chem.*, **254**, 10023–9.

Stansbie, D., Brownsey, R.W., Crettaz, M. and Denton, R.M. (1976). Acute effects of anti-insulin serum on rates of fatty acid synthesis and activities of acetyl-coenzyme A carboxylase and pyruvate dehydrogenase in liver and epididymal adipose tissue of fed rats. *Biochem. J.*, **160**, 413–6.

Steinberg, D. (1963). Fatty acid mobilization. Mechanisms of regulation and metabolic consequences. *Biochem. Soc. Symp.*, **24**, 111–38.

Steingrimsdottir, L., Greenwood, M.R.C. and Brasel, J.A. (1980). Effects of pregnancy, lactation and a high fat diet on adipose tissue in Osborne–Mendel rats. *J. Nutr.*, **110**, 600–9.

Stovroff, M.C., Fraker, D.L. and Norton, J.A. (1989). Cachectin activity in the serum of cachectic, tumor-bearing rats. *Arch. Surg.*, **124**, 94–9.

Strälfors, P. and Belfrage, P. (1982). Properties and purification of the catalytic sub-unit of cyclic AMP-dependent protein kinase of adipose tissue. *Biochim. Biophys. Acta*, **721**, 434–40.

Strälfors, P. and Belfrage, P. (1983). Phosphorylation of hormone-sensitive lipase by cyclic AMP-dependent protein kinase. *J. Biol. Chem.*, **258**, 15146–52.

Strälfors, P., Björgell, P. and Belfrage, P. (1984). Hormonal regulation of hormone-sensitive lipase in intact adipocytes: identification of phosphorylated sites and effects on the phosphorylation by lipolytic hormones and insulin. *Proc. Natl Acad. Sci. USA*, **81**, 3317–21.

Suzuki, H., Kawarabayasi, Y., Kundo, J., Tataaki, A., Nishikawa, K., Kimura, S., Hashimoto, T. and Yamamoto, T. (1990). Structure and regulation of rat long-chain acyl-CoA synthetase. *J. Biol. Chem.*, **265**, 8681–5.

Tedstone, A.E., Ilic, V. and Williamson, D.H. (1991). The effect of previous feeding patterns on the oxidation and tissue [^{14}]lipid accumulation after a meal containing [1-^{14}C]-triolein in the rat. *Proc. Nutr. Soc.*, **50**, 39A.

Thiede, M.A. and Rodan, G.A. (1988). Expression of a calcium-mobilizing parathyroid hormone-like peptide in lactating mammary tissue. *Science*, **242**, 278–80.

Thompson, M.P., Koons, J.E., Tan, E.T.H. and Grigor, M.R. (1981). Modified lipoprotein lipase activities, rates of lipogenesis and lipolysis as factors leading to lipid depletion in C57BL mice bearing the preputial tumor (ESR-586). *Cancer Res.*, **41**, 3228–32.

Torbay, N., Bracco, E.F., Geliebter, A., Stewart, I.M. and Hashim, S.A. (1985). Insulin increases body fat despite control of food intake and physical activity. *Am. J. Physiol.*, **248**, R120–4.

Vannier, C. and Ailhaud, G. (1989). Biosynthesis of lipoprotein lipase in cultured mouse adipocytes. II. Processing, subunit assembly, and intracellular transport. *J. Biol. Chem.*, **264**, 13206–16.

Vannier, C., Amri, E.-Z., Etienne, J., Negrel, R. and Ailhaud, G. (1985). Maturation and secretion of lipoprotein lipase in cultured adipose cells. I. Intracellular activation of the enzyme. *J. Biol. Chem.*, **260**, 4424–31.

Vannier, C., Deslex, S., Pradines-Figuères, A. and Ailhaud, G.A. (1989). Biosynthesis of lipoprotein lipase in cultured mouse adipocytes. I. Characterization of a specific antibody and relationships between the intracellular and secreted pools of the enzyme. *J. Biol. Chem.*, **264**, 13199–205.

Vernon, R.G. and Clegg, R.A. (1985). The metabolism of white adipose tissue *in vivo* and *in vitro*. In: *New Perspectives in Adipose Tissue: Structure, Function and Development* (eds A. Cryer and R.L.R. Van), Butterworths, London, pp. 65–86.

Vernon, R.G. and Flint, D.J. (1983). Control of fatty acid synthesis in lactation. *Proc. Nutr. Soc.*, **42**, 315–31.

Vlassara, H., Spiegel, R.J., San Doval, D. and Cerami, A. (1986). Reduced plasma lipoprotein lipase activity in patients with malignancy-associated weight loss. *Horm. Metab. Res.*, **18**, 698–703.

Vydelingum, N., Drake, R.L., Etienne, J. and Kissebah, A.M. (1983). Insulin regulation of fat cell ribosomes, protein synthesis and lipoprotein lipase. *Am. J. Physiol.*, **245**, E121–31.

Wardzala, L.J., Hirshman, M., Pofcher, E., Horton, E.D., Mead, P.M., Cashman, S.W. and Horton, E.S. (1985). Regulation of glucose utilization in adipose tissue cells and muscle after long-term experimental hyperinsulinaemia in rats. *J. Clin. Invest.*, **76**, 460–9.

Williamson, D.H. (1980). Integration of metabolism in tissues of the lactating rat. *FEBS Lett.*, **117**, K93–105.

Williamson, D.H. (1989). Role of insulin in the integration of lipid metabolism in mammalian tissues. *Biochem. Soc. Trans.*, **17**, 37–40.

Williamson, D.H. and Robinson, A.M. (1977). Control of glucose metabolism in the lactating-rat mammary-gland: Effects of other substrates, insulin and starvation. *Biochem. Soc. Trans.*, **5**, 829–34.

Witters, L.A., Watts, T.D., Daniels, D.L. and Evans, J.L. (1988). Insulin stimulates the dephosphorylation and activation of acetyl-CoA carboxylase. *Proc. Natl Acad. Sci. USA*, **85**, 5473–7.

Witzum, J.L. and Schonfeld, G. (1979). Lipoproteins in the plasma and hepatic perfusates of the Zucker fatty rat. *Diabetes*, **28**, 509–16.

Wolfe, R.R. and Durkot, M.J. (1985). Role of very low density lipoproteins in the energy metabolism of the rat. *J. Lipid Res.*, **26**, 210–7.

Wolfe, R.R. and Peters, E.J. (1987). Lipolytic response to glucose infusion in human subjects. *Am. J. Physiol.*, **252**, E218–23.

Wolfe, R.R., Herndon, D.N., Jahoor, F., Miyoshi, H. and Wolfe, M.H. (1987). Effect of severe burn injury on substrate cycling by glucose and fatty acids. *N. Engl. J. Med.*, **317**, 403–8.

Wolfe, R.R., Klein, S., Carraro, F. and Weber, J.M. (1990). Role of triglyceride-fatty acid cycle in controlling fat metabolism in humans during and after exercise. *Am. J. Physiol.*, **258**, E382–9.

York, D.A. and Bray, G.A. (1973). Genetic obesity in rats. II. The effect of food

restriction on the metabolism of adipose tissue. *Metabolism*, **22**, 443–54.
York, D.A., Steinke, J. and Bray, G.A. (1972). Hyperinsulinaemia and insulin resistance in genetically obese rats. *Metabolism*, **21**, 277–84.
Zammit, V.A. (1985). Regulation of lipogenesis in rat tissues during pregnancy and lactation. *Biochem. Soc. Trans.*, **13**, 831–3.
Zammit, V.A. and Corstophine, G.G. (1982). Inhibition of acetyl-CoA carboxylase activity in isolated rat adipocytes incubated with glucagon. Interactions with effects of insulin, adrenaline and adenosine deaminase. *Biochem. J.*, **208**, 783–8.
Zechner, R., Newman, T.C., Sherry, B., Cerami, A. and Breslow, J.L. (1988). Recombinant human cachectin/tumor necrosis factor but not interleukin-1α down-regulates lipoprotein lipase gene expression at the transcriptional level in mouse 3T3-L1 adipocytes. *Mol. Cell. Biol.*, **8**, 2394–401.
Zucker, L.M., (1965). Hereditary obesity in the rat associated with hyperlipaemia. *Ann. NY Acad. Sci.*, **131**, 447–58.

Chapter Six

Neuroendocrine and Metabolic Derangements in Animal Models of Obesity

David A. York
Pennington Biomedical Research Center, 6400 Perkins Road, Baton Rouge, LA 70808, USA

Obesity is the major nutritional disorder confronting the western world. A wide variety of animal models have been utilized to study the metabolic consequences of the obese state and the possible biochemical cause or causes of obesity. These models are listed in Table 6.1 in which the obesities are subdivided upon the basis of their primary causative factor. Thus the obesity may arise as the result of either a genetic defect, from damage to or stimulation of specific hypothalamic centers, from manipulation of diet, its composition or content, or from endocrine-related manipulations. A comprehensive review of all these syndromes is beyond the scope of this paper, but readers requiring further detailed information are referred to previous reviews (Bray and York, 1979; Bray, 1984; Bray *et al.*, 1989, 1990). In this paper I shall describe the 'autonomic hypothesis' which provides an understanding of the basic mechanisms which appear to underlie all the experimental obesities listed in Table 6.1 and subsequently suggest mechanisms through which adrenal glucocorticoids may act to mediate their essential requirement for the development of all obesity. While the chromosomal location of several of the gene defects causing obesity is known, neither the genes nor their products have yet been identified. Although a strong genetic component is now recognized in human obesity (Bouchard, 1989) the relevance of these single gene animal models to our understanding of human obesity will not be clear until the proteins encoded by these defective genes are identified. However, it is possible that the

Obesity and Cachexia. Edited by N.J. Rothwell and M.J. Stock

Table 6.1 Classification of experimental models of obesity in rodents

I.	*Hypothalamic*	
	a. Electrical or thermal lesions (VMH, PVN, amygdala).	
	b. Knife cuts (parasagittal or hypothalamic islands).	
	c. Chemical damage (GTG, MSG, BPM, antagonists to 5HT and DA).	
	d. Viral infection	
	e. Chemical infusions (NA into VMH, NPY into PVN).	
	f. Electrical stimulation of LH.	
II.	*Genetic*	
	a. Single gene dominant	$A^{y}a$ yellow mouse; chromosome 2
		Adipose mouse; chromosome 7
	b. Single gene recessive	obob obese mouse; chromosome 6
		dbdb diabetes mouse; chromosome 4
		tubby mouse; chromosome 7
		fat mouse; chromosome 8
		PAS mouse
		fafa fatty rat; chromosome 5
		$fa^{cp}fa^{cp}$ or $fa^{k}fa^{k}$ corpulent or Koletsky rat
	c. Polygenic	NZO New Zealand mouse
		KK Japanese mouse
		PBB/Ld Paul Bailey Black mouse
		Wellesley mouse
		Osborne–Mendel rats—high fat diet
		Sand rat—lab chow diet
		Psammys obesus—lab chow diet
III.	*Dietary*	
	a. Altered diet	High fat diets—species specific
		Cafeteria diets—species and age dependent
		High sucrose/high fat diets
		Sucrose drinking solutions
	b. Meal pattern change	Meal feeding
		Tube feeding
IV.	*Endocrine/Other*	Ovariectomy
		Chronic peripheral insulin administration
		Viral—Scrapie or canine distemper

VMH, ventromedial hypothalamus; PVN, paraventricular nucleus; LH, lateral hypothalamus; GTG, gold thioglucose; MSG, monosodium glutamate; BPM, bipiperidyl mustard; DA, dopamine; 5HT, 5-hydroxytryptamine (serotonin); NA, noradrenaline; NPY, neuropeptide Y. Readers are referred to Bray and York (1979) or Bray *et al.* (1989) for further details of these models.

apparent heterogeneity of the human obesities may reflect the variety of background genomes rather than a multiplicity of biochemical abnormalities since studies with the animal models has shown that the phenotypic expression of an obese gene is altered greatly by the species of mouse in which it is expressed (Coleman, 1978). Nevertheless, the recognition that the degree of obesity induced by dietary manipulations varies with animal species may provide examples of gene–nutrient interactions which are particularly pertinent to human obesity.

THE AUTONOMIC HYPOTHESIS

The autonomic hypothesis was originally proposed to explain the obesity which resulted from destructive lesions to the ventromedial hypothalamus (Inoue *et al.*, 1977; Bray and York, 1979; Bray, 1984), but subsequent work has shown that this hypothesis may explain all forms of experimental animal obesity and may be relevant to human obesity (see Bray *et al.*, 1989, 1990). The autonomic hypothesis (Figure 6.1) proposes that obesity results from an imbalance in the autonomic nervous system whereby sympathetic drive to 'thermogenic' tissues is impaired, particularly in response to nutrient related signals (Seydoux *et al.*, 1981; Hogan *et al.*, 1982; Holt *et al.*, 1983; Rohner-Jeanrenaud *et al.*, 1983a; Triandafillou and Himms-Hagen, 1983;

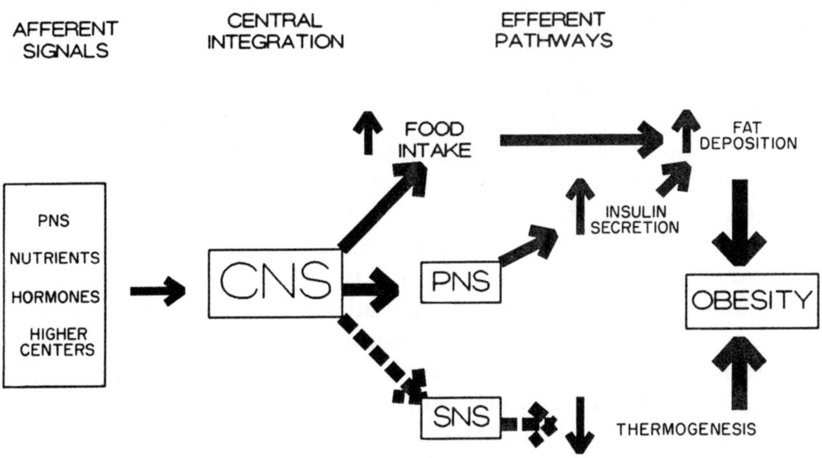

Figure 6.1 The autonomic hypothesis: Imbalance in the efferent responses from hypothalamic centers results from either an abnormal afferent signal, failure to recognize a signal or abnormal central control of the coupling of afferent information to efferent pathways. Imbalance in the autonomic pathways, coupled to the increase in food intake which is often present, leads to excessive fat synthesis and impaired thermogenesis, the net result of which is the excess fat deposition of obesity

York *et al.*, 1985; Romsos, 1985; Bray *et al.*, 1990), while afferent parasympathetic activity from the gastrointestinal tract (Ionescu *et al.*, 1988) and efferent parasympathetic drive to the endocrine pancreas is enhanced (Rohner-Jeanrenaud *et al.*, 1983a,b; Rohner-Jeanrenaud and Jeanrenaud, 1985b). This imbalance in activity of the autonomic nervous system leads to a reduction in the thermogenic responses to feeding (Holt *et al.*, 1983, 1987) and an increase in insulin secretion (Rohner-Jeanrenaud and Jeanrenaud, 1985) which itself drives excess lipogenesis and fat deposition (Godbole and York, 1978). The combination of these effects leads to obesity. Not all of these autonomic changes are found in all obesities, but some loss of sympathetic activity or sympathetic response appears to be universally present with the possible exception of obesity in the paraventricular-lesioned rat (Bray *et al.*, 1989, 1990; York, 1987).

The hypothalamic centers which control sympathetic responses to feeding have been identified and the pathways from them through the brain stem to peripheral tissues have been mapped (Luiten *et al.*, 1987). Using a variety of methods for measuring sympathetic activity, either a low basal sympathetic drive to brown adipose tissue and/or the absence of the normal increase in sympathetic firing rate after feeding has been reported in all the forms of experimental obesity that have been studied (Seydoux *et al.*, 1984; York *et al.*, 1985; Romsos, 1985; Sakaguchi *et al.*, 1988a; Holt and York, 1989a; Bray *et al.*, 1989, 1990). This observation can be illustrated by reference to two examples. After electrolytic or chemical lesions to the ventromedial nucleus, which result in obesity, the activity of sympathetic nerves to the interscapular brown adipose tissue depot was decreased (Rohner-Jeanrenaud *et al.*, 1983c; Sakaguchi *et al.*, 1988a) whereas after stimulation of the VMN (Saito *et al.*, 1989) or lesions to the lateral hypothalamus (Holt and York, 1988; Sakaguchi *et al.*, 1988b) which lead to loss of body weight and body fat, sympathetic activity is enhanced. Likewise in the genetically obese fafa rat, the activity of the sympathetic innervation to brown adipose tissue is attenuated and fails to respond to a variety of dietary manipulations (Holt and York, 1989a,c; Holt *et al.*, 1983). Since the thermogenic activity of brown adipose tissue is controlled by the activity of the sympathetic innervation, it has been possible to substantiate that the absence or attenuation of brown adipose tissue thermogenesis in response to dietary signals results from the abnormal control of the sympathetic system rather than a defect in the tissue itself (Holt *et al.*, 1988).

Both afferent and efferent parasympathetic vagal activity are enhanced in obesity. Brooks *et al.* (1946) were the first to suggest the link between the ventromedial hypothalamus (VMH) and vagal nerve after showing that VMH lesions enhanced gastric acid secretions. Powley and coworkers (see Powley, 1977) showed that the vagally mediated cephalic responses to feeding were enhanced in VMH obesity. After damage to the VMH there is a rapid increase in both basal and stimulated insulin secretion that can be prevented with atropine or subdiaphragmatic vagotomy (Powley and Opsahl, 1974; Powley, 1977; Berthoud and Jeanrenaud, 1979; Jeanrenaud, 1985; Rohner-Jeanrenaud *et al.*, 1983b) and is not observed when pancreatic islets are transplanted below the renal capsule away from their vagal innervation (Inoue *et al.*, 1978). In contrast to lesions of the VMH, lesions to the paraventricular nucleus (PVN) primarily increase parasympathetic activation of pancreatic β-cells and glucagon release (Tokunaga *et al.*, 1986), the increase in insulin being secondary to the hyperphagia.

Food intake is increased in nearly all forms of experimental obesity, the exceptions being the obesity which follows monosodium glutamate lesions and VMH lesions in weanling rats (Bray and York, 1979; Bray *et al.*, 1989, 1990). While the increase in food intake associated with the change in diet is a necessary stimulus for the development of the dietary type of obesity, hyperphagia is not an essential requirement for the development of the genetic or VMH obesities (Bernardis, 1973; Bray *et al.*, 1973). This indicates that the obesity in the models must result from a reduction in energy expenditure, any increase in food intake exacerbating the impaired energy expenditure in response to food intake and exaggerating the resultant obesity.

The mechanism underlying the hyperphagia of the genetic and dietary obesities is not known. Numerous hypothalamic centers influence food intake, the paraventricular nucleus (PVN), the ventromedial nucleus(VMN) and lateral hypothalamus (LH) being particularly recognized. The PVN is the sole of an α_2-noradrenergic system which acutely regulates carbohydrate feeding (Leibowitz *et al.*, 1984; Leibowitz, 1986). Of the many neuropeptides that influence food intake (Table 6.2) neuropeptide Y injected into the PVN is known to be particularly effective in stimulating food intake and repeated administration of this peptide leads to hyperphagia and obesity (Stanley *et al.*, 1985, 1986).

In contrast CRF injected acutely into the third ventricle suppresses food intake and enhances the activity of sympathetic nerves innervating BAT. Chronically infused into the third ventricle, CRF stops or reverses the development of obesity, this effect being associated with the reduction in food intake, an increase in BAT function (increase in sympathetic activity) and a reduction in basal and stimulated insulin secretion (a decrease in parasympathetic function) (Arase *et al.*, 1988, 1989a, b; Rohner-Jeanrenaud *et al.*, 1989; Holt and York, 1989; Stubbs and York, 1989 and unpublished observations).

There are many reports on the concentration of neuropeptides in the hypothalamus and other brain regions of obese rats and more recent ones have used micropunch techniques to study specific nuclei. The information reported has provided much data but, as yet, little insight into the potential

Table 6.2 Some neuropeptides that alter food intake

Increase food intake	Decrease food intake
Dynorphin	Anorectin
β-Endorphin	Calcitonin
Galanin	Cholecystokinin-8
Growth hormone releasing hormone (GHRH)	Corticotropin releasing hormone (CRH)
	CycloHisPro
Neuropeptide Y	Enterostatin
	Interleukin-1β
	Glucagon
	Neurotensin
	Procolipase activation peptide (VPDPR)
	Satietin
	Thyrotropin releasing hormone (TRH)
	Tumor necrosis factor

role of abnormal neuropeptide secretion in the development of obesity, since changes in neuropeptide concentration alone are difficult to interpret. Two such studies, however, are worthy of mention. First, the observation that NPY mRNA concentration in the PVN of obese fafa rats is elevated (White *et al.*, 1988) might be important in the light of the potency of this peptide in its effects on food intake and sympathetic activity (Stanley *et al.*, 1986; Egawa *et al.*, 1990). Second, the observation that CRF mRNA levels are reduced in a few specific hypothalamic nuclei of obese rats (Nakaishi *et al.*, 1988) might also be consistent with the development of obesity (but see later for further discussion). An alternative possibility to a primary neuropeptide role in the hyperphagic behavior arises from the observation that sympathetic activity and food intake are always reciprocally related (Table 6.3). It is possible that the sympathetic drive to BAT or some other tissue stimulates the release of a circulating afferent signal which is monitored by the hypothalamus to reduce feeding. In situations of low sympathetic activity the absence of such a circulating signal would lead to feeding. The existence of such a putative feeding signal has been suggested previously from studies of parabiotic unions of lean and obese rats and mice (Hervey, 1959; Coleman and Hummel, 1969; Nishizawa and Bray, 1980), but its identity is unknown. A further extension of this hypothesis would be that a reduction in sympathetic drive to BAT or other tissues might be the primary cause of the hyperphagia as it is of the defective thermogenesis. Developmental studies indicate that the attenuated sympathetic activity and the exaggerated parasympathetic function may be the earliest detectable abnormalities in the genetic obesities and thus be closely related to the gene defect. Lavau and colleagues (Bazin *et al.*, 1984) have presented evidence

Table 6.3 The reciprocal control of food intake and sympathetic drive to interscapular brown adipose tissue

Condition		Food intake	Sympathetic activity
A.	Substrate ICV/IP		
	Glucose	↓	↑
	2-Deoxy-D-glucose	↑	↓
	β-Hydroxybutyrate	↓	↑
	High fat diets	↑	↓
B.	Hormones/Peptides ICV		
	Insulin	↓	↓
	CRH	↓	↑
	NPY	↑	↓
	Interleukin-1β	↓	↑
C.	Hypothalamic controls		
	VMH lesion	↑	↓
	LH lesion	↓	↑
	VMH stimulation	↓	↑
	LH stimulation	↑	↓
	PVN lesion	↑	–
D.	Other		
	Night time	↑	↓
	Daytime	↓	↑
	DL-Fenfluramine	↓	↑
	Genetic obesity	↓	↑

↑ represents an increase, ↓ a decrease, and – indicates no change.

to suggest the BAT function (presumably a reflection of sympathetic activity) is reduced in the first few days of life in the Zucker fafa rat as is energy expenditure (Planche *et al.*, 1983); assay of plasma and pancreatic insulin in fetuses of litters containing or not containing the recessive obese (fafa) genotype suggests that the parasympathetic and insulin secretory response defects observed in suckling fafa rats (Rohner-Jeanrenaud and Jeanrenaud 1983a, 1985a) may be expressed even before birth (Turkenkopf *et al.*, 1983).

Insulin

Insulin plays an important regulatory role in the development of obesity. Hyperinsulinemia is a common feature of most obesities (Bray and York, 1979), the most prominent exception to this being in the obesity that results from high fat diets (Inoue *et al.*, 1977; Fisler and Bray, 1985). Regulation of insulin secretory capacity with drugs such as streptozotocin severely or completely restricts the excessive tissue lipogenesis and fat deposition of obese animals (Godbole and York, 1978; Godbole *et al.*, 1978). The hypersecretion of insulin itself, leads to the development of insulin resistance in a differential manner between tissues and between individual metabolic pathways within a specific target tissue that itself leads to significant changes in energy and nutrient partitioning between muscle and adipose tissue (Cusin *et al.*, 1990). This is illustrated in Figure 6.2

Glucocorticoids

Adrenal glucocorticoids are essential for the development of all forms of obesity irrespective of the type. This universal characteristic leads to the hypothesis that abnormal glucocorticoid regulation of autonomic responses, food intake and other aspects of metabolism may predispose an animal to the development of obesity. The basis for this proposal is briefly outlined below.

Adrenalectomy prevents the further development of all forms of obesity and restores the majority of the described metabolic abnormalities back or towards normal (Allars and York, 1987; Bray *et al.*, 1989, 1990; York, 1989). Corticosterone replacement reverses the effects of adrenalectomy (Freedman *et al.*, 1985, 1986). The reciprocal increase in CRF secretion from the PVN that results from removal of corticosteroids may play an important role in this response. In contrast, the rise in ACTH secretion does not appear to play a role in the prevention of obesity by adrenalectomy (Holt *et al.*, 1987).

Obese animals are more responsive to the effects of adrenal glucocorticoids (Freedman *et al.*, 1985, 1986; Fletcher and MacKenzie, 1988). Adrenalectomy has small effects on lean rodents and replacement doses of corticosterone or other glucocorticoids that cause gross obesity in adrenalectomized obese rats have only small effects on weight gain, food intake and fat deposition in adrenalectomized lean rats. Thus both the sensitivity and responsiveness to adrenal glucocorticoids appears to be increased in obesity.

Corticosterone binds to two types of glucocorticoid receptors in the central nervous system; the MR (type I) receptor is similar to the peripheral mineralocorticoid receptor and the GR (type II) receptor is similar to the peripheral glucocorticoid receptor (Reul and DeKloet, 1985). By using an antagonist (RU-38486) which is specific to both GR and progesterone receptors (Philibert, 1984), we have been able to show that GR activity is an essential requirement for the obesity of Zucker fafa rats (Langley and York, 1990a). Additionally, we have shown that RU-38486 also blocks the high fat diet-induced obesity of Osborne–Mendel rats (Okada *et al.*, 1990), a form of obesity that had previously appeared to be partially resistant to

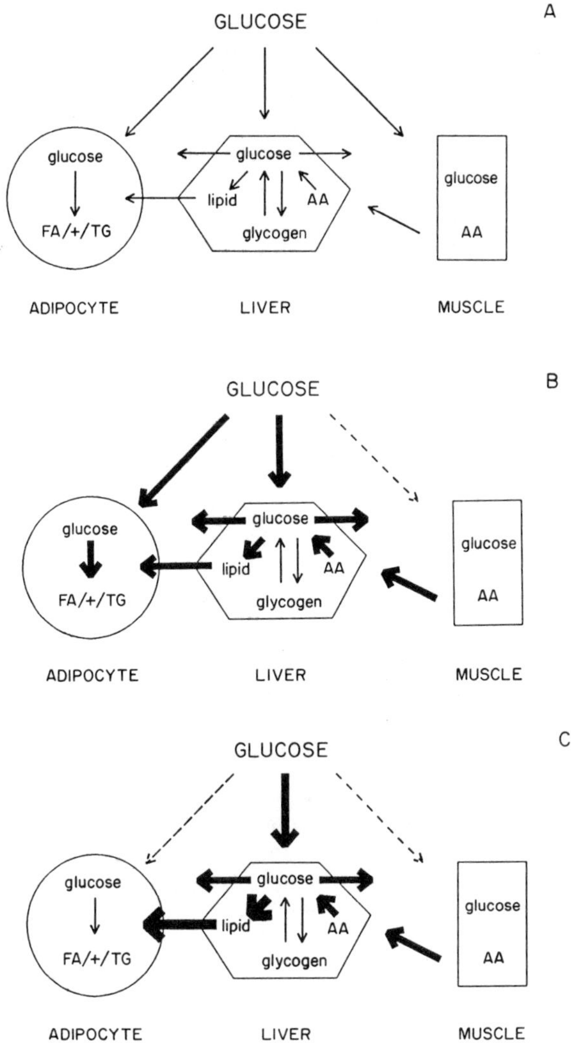

Figure 6.2 Insulin-sensitive pathways of carbohydrate metabolism at different stages of the development of obesity. The figure shows pathways in lean rats (A), rats in the early stages of obesity (B) and in gross obesity (C). Thick arrows represent enhanced flux through pathways from excess insulin stimulation or impaired insulin inhibition (gluconeogenesis) and dotted arrows indicate reduced flux as the result of insulin resistance. In the early stages of the development of obesity (B) insulin enhances glucose uptake into adipocytes and the lipogenic pathways of both liver and adipocytes are under maximal insulin stimulation while glucose uptake into muscle and gluconeogenesis are both resistant to the effects of insulin. In later stages (C) the glucose uptake and lipogenic pathways of adipocytes lose insulin sensitivity and the liver becomes the major site of lipid synthesis while adipocyte lipoprotein lipase remains at high activity to clear both dietary and hepatic lipoprotein triglyceride

the effects of adrenalectomy (Smith and Romsos, 1985; Fisler *et al.*, 1989).

Using Scatchard analysis of ligand binding, we have recently shown that maximum binding to the GR receptor (B_{max}) is increased in both the hippocampus and hypothalamus of obese fafa rats although the affinity (K_d) of the receptors was consistently reduced (Table 6.4) (Langley and York, 1990b). Tsai and Romsos (1990) have recently observed a similar reduction in GR receptor affinity in the brains of obese obob mice. At this stage, such changes in receptor affinity are difficult to understand. They could reflect a change in the structure of the GR receptor itself or more likely perhaps, an altered interaction of the receptor with its associated heat shock protein (HSP90) molecules (Mendel and Orti, 1988). Further studies are underway to investigate these possibilities. However, it is important to recognize that the increase in receptor number occurs despite the presence of normal or slightly elevated corticosterone in the brain tissues of the non-adrenalectomized obese rat (York *et al.*, 1988). An increase in receptor number in the presence of normal endogenous corticosterone concentrations suggests that either the regulation of GR synthesis or turnover of the GR receptor itself are defective. This suggestion gains credence from recent unpublished experimental observations which show that the normal upregulation of GR ligand binding that begins 12–24 h after adrenalectomy is absent in the obese genotype.

While the evidence above is consistent with a misregulation of GR receptor function in obese fafa rats, the regulation of gene transcription by the receptor–glucocorticoid complex appears also to be abnormal in these rats. In recent work, George Bray, Marcel Lavau and I (unpublished observations) investigated the regulation by glucocorticoids of the level of mRNA for malic enzyme and glyceraldehyde-3-phosphate dehydrogenase (GAPDH) in liver and adipose tissue of lean and obese fafa rats. Two significant observations were made. First, malic enzyme mRNA, an enzyme whose

Table 6.4 Scatchard analysis of [^{3}H]corticosterone binding to glucocorticoid type II receptors (GR) in lean and obese rats

	B_{max} (fmol/mg protein)	K_d (nM)	*n*
Hippocampus			
lean	124 ± 7	0.32 ± 0.18	3
obese	222 ± 5	1.05 ± 0.08	3
Hypothalamus			
lean	56	0.51	2
obese	79	2.23	2

Figures are means ± SEM for hippocampus or mean values for hypothalamus. The number of observations is given by *n*.

activity is normally increased by glucocorticoids, was more sensitive and responsive to change in corticosterone levels in the obese rat. Second, the constitutive enzyme, GAPDH, which is not affected by changes in glucocorticoid status in lean rats, was induced in a dose and time-dependent manner in obese rats. These data need to be interpreted in the light of the apparent failure of glucocorticoids to down regulate its own receptor in the brain. Assuming that only a single defect exists in the glucocorticoid response system in fafa rats, there are a number of possible defects which might explain the observations. These include a mutation in the GR receptor which would alter its binding to positive and negative glucocorticoid response elements on glucocorticoid target genes or the absence of a transcriptional factor which normally inhibits or prevents glucocorticoid induced transcription.

Alterations in circulating glucocorticoids effect numerous central peptides, hormone and neurotransmitter systems in addition to CRF. Of particular importance to the obesity of the yellow ($A^{y}a$) obese mouse are the peptides derived from the propiomelanocortin (POMC) parent molecule. In the yellow mouse, the degree of obesity is closely correlated to coat color (Woolf, 1965) and the gene defect has been located close to the Aqouti-locus on chromosome 2. Injection of α-melanocyte stimulating hormone (α-MSH) restored the normal coat color of these mice (Geschwind *et al.*, 1972) suggesting that the coat color, and its associated obesity, might result from lack of α-MSH activity. MSH occurs in two forms differing in acetylation of the N-terminal amino acid. The acetyl-MSH is the biologically active form of the hormone for regulation of coat pigmentation. We have shown that the ratio of desacetyl- to acetyl-MSH is increased in the pituitary of yellow mice (Bray *et al.*, 1988). This observation was particularly pertinent since subsequent studies by Bray and colleagues (Shimizu *et al.*, 1989a,b) showed that the desacetyl-MSH increased food intake, weight gain and serum corticosterone but was without effect on coat color. Abnormal acetylation of MSH may thus provide the basis for the association of coat color with obesity in these mice. Whether the defect in acetylation is restricted to MSH is not known but the recognition that acetylation of β-endorphin abolishes its effect in stimulating food intake (Bray *et al.*, 1990), opens the possibility that acetylation of neuropeptides may have an important regulatory role in the control of energy intake and expenditure and that defects in this regulation may be responsible for obesity observed in the yellow ($A^{y}a$) obese mouse. The effect of adrenalectomy on *N*-acetylation of desacetyl-MSH is not known. Since adrenalectomy prevents the obesity of yellow mice, it would be anticipated that glucocorticoids may regulate, directly or indirectly, the acetylation of peptides in addition to regulating the post-translational processing of POMC.

With the increasing use of molecular biological approaches, we can look forward to the identification of some of the primary protein defects that will cause obesity in the next few years and the clarification of some of the observations presented in this paper.

REFERENCES

Allars, J. and York, D.A. (1987). Energetic efficiency and brown adipose tissue uncoupling protein of obese Zucker rats fed high carbohydrate and high fat diets: effect of adrenalectomy. *Int. J. Obesity*, **11**, 591–602.

Arase, K., Shargill, N.S. and Bray, G.A. (1989a). Effect of corticotropin releasing factor on genetically obese (fatty) rats. *Physiol. Behav.*, **45** 1–6.

Arase, K., Shargill, N.S. and Bray, G.A. (1989b). Effects of intraventricular infusion of corticotropin releasing factor on VMH-lesioned obese rats. *Am. J. Physiol.*, **256**, R751–6.

Arase, K., York, D.A., Shimizu, H., Shargill, N.S. and Bray, G.A. (1988). Effect of corticotropin releasing factor on food intake and brown adipose tissue thermogenesis in rats. *Am. J. Physiol.*, **255**, E255–9.

Bazin, R., Eteve, D. and Lavau, M. (1984). Evidence for decreased GDP binding to brown adipose tissue mitochondria of obese Zucker rats in the very first few days of life. *Biochem. J.*, **221**, 241–5.

Bernardis, L. (1973). Disruption of diurnal feeding and weight gain cycles in weanling rats by ventromedial and dorsomedial hypothalamic lesions. *Physiol. Behav.*, **10**, 855–61.

Berthoud, H.R. and Jeanrenaud, B. (1979). Acute hyperinsulinaemia and its reversal by vagotomy after lesions of the ventromedial hypothalamus in anaesthetized rats. *Endocrinology*, **105**, 146–51.

Bouchard, C. (1989). Genetic factors in obesity. *Med. Clin. North Am.*, **73**, 67–81.

Bray, G.A. (1984). Hypothalamic and genetic obesity: an appraisal of the autonomic and endocrine hypotheses. *Int. J. Obesity*, **8**, 119–37.

Bray, G.A., Fisler, J. and York, D.A. (1990). Neuroendocrine control of the development of obesity: Understanding gained from studies of experimental animal models. *Front. Neuroendocrinol.*, **11**, 128–81.

Bray, G.A., Shimizu, H., Retsias, A.D., Shargill, N.S. and York, D.A. (1988). Reduced acetylation of MSH: a biochemical explanation for the yellow obese mouse. In: *Obesity in Europe 88* (eds P. Bjorntorp and S. Rossner), John Libbey, London, pp. 259–70.

Bray, G.A. and York, D.A. (1979). Hypothalamic and genetic obesity in experimental animals: an autonomic and endocrine hypothesis. *Physiol. Rev.*, **59**, 719–809.

Bray, G.A., York, D.A. and Fisler, J. (1989). Experimental obesity: a homeostatic failure due to defective nutrient stimulation of the sympathetic nervous system. *Vitam. Horm.*, **45**, 1–124.

Bray, G.A., York, D.A. and Swerdloff, R. (1973). Genetic obesity in rats. The effects of food restriction on body composition and hypothalamic function. *Metab. Clin. Exp.*, **2**, 435–42.

Brooks, C.M., Lockwood, R.A. and Wiggins, M.I. (1946). A study of the effect of hypothalamic lesions on the eating habits of the albino rat. *Am J. Physiol.*, **147**, 735–41.

Coleman, D.L. (1978). Obesity and diabetes: two mutant genes causing diabetes-obesity syndromes in mice. *Diabetologia*, **14**, 141–8.

Coleman, D.L. and Hummel, K.P. (1969). The effects of parabiosis of normal with genetically diabetic mice. *Am. J. Physiol.* , **217**, 1298–306.

Cusin, I., Terretaz, J., Rohner-Jeanrenaud, F. and Jeanrenaud, B. (1990). Metabolic consequences of hyperinsulinaemia imposed on normal rats on glucose handling by white adipose tissue, muscles and liver. *Biochem. J.*, **267**, 99–103.

Egawa, M., Yoshimatsu, H. and Bray, G.A. (1990). Effects of CRH and NPY on electrophysiological activity of sympathetic nerves to interscapular brown adipose tissue. *Neuroscience*, **34**, 771–5.

Fisler, J. and Bray, G.A. (1985). Dietary obesity: effects of drugs on food intake in S5B/P1 and Osborne–Mendel rats. *Physiol. Behav.*, **34**, 225–31.

Fisler, J., York, D.A. and Bray, G.A. (1989). Adrenalectomy in Osborne–Mendel rats. *FASEB. J.*, **3**, A763.

Fletcher, J. and MacKenzie, J. (1988). The parasympathetic nervous system and glucocorticoid mediated hyperinsulinaemia in the genetically obese Zucker fafa rat. *J. Endocrinol.*, **118**, 87–92.

Freedman, M.R., Castonguay, T.W. and Stern, J.S. (1985). Effect of adrenalectomy and corticosterone replacement on meal patterns of Zucker rats. *Am. J. Physiol.*, **249**, R584–94.

Freedman, M.R., Horwitz, B.A. and Stern, J.S. (1986). Effect of adrenalectomy and glucocorticoid replacement on development of obesity. *Am. J. Physiol.*, **250**, R595–607.

Geschwind, H., Huseby, R.A. and Nishioka, R. (1972). The effect of melanocyte stimulating hormone on coat color in the mouse. *Rec. Prog. Horm. Res.*, **28**, 91–130.

Godbole, V. and York, D.A. (1978). Lipogenesis *in situ* in genetically obese Zucker fatty rat: role of hyperphagia and hyperinsulinaemia. *Diabetologia*, **14**, 191–7.

Godbole, V., York, D.A. and Bloxham, D.B. (1978). Development changes in the fatty rat: evidence for defective thermogenesis preceding hyperlipogenesis and hyperinsulinaemia. *Diabetologia*, **15**, 41–4.

Hervey, G.R. (1959). The effects of lesions in the hypothalamus in parabiotic rats. *J. Physiol. (Lond).*, **145**, 336–45.

Hogan, S., Coscina, D.V. and Himms-Hagen, J. (1982). Brown adipose tissue of rats with obesity induced with medial hypothalamic lesions. *Am. J. Physiol.*, **243**, E338–43.

Holt, S.J., Rothwell, N.S., Stock, M.J. and York, D.A. (1987). Changes in energy balance, thermogenesis and brown adipose tissue activity following hypophysectomy in obese Zucker rats. *Am. J. Physiol.*, **254**, E162–7.

Holt, S.J., Wheal, H.V. and York, D.A. (1988). Hypothalamic control of brown adipose tissue in Zucker lean and obese rats: effects of electrical stimulation of the ventromedial nucleus and other hypothalamic centers. *Brain Res.*, **405**, 227–33.

Holt, S.J. and York, D.A. (1988). Effects of lateral hypothalamic lesions on brown adipose tissue of Zucker lean and obese rats. *Physiol. Behav.*, **43**, 293–9.

Holt, S.J. and York, D.A. (1989a). Studies on the sympathetic efferents of brown adipose tissue of lean and obese Zucker rats. *Brain Res.*, **491**, 106–12.

Holt, S.J. and York, D.A. (1989b). The effects of adrenalectomy, corticotropin releasing factor and vasopressin on the sympathetic firing rate of nerves to interscapular brown adipose tissue in Zucker obese rats. *Physiol. Behav.*, **45**, 1123–9.

Holt, S.J. and York, D.A. (1989c). Interaction of intra-cerebroventricular insulin and glucose in the regulation of the activity of sympathetic efferent nerves to brown adipose tissue in lean and obese Zucker rats. *Brain Res.*, **500**, 384–8.

Holt, S.J., York, D.A. and Fitzsimmons, J.T.R. (1983). The effects of corticosterone, cold exposure and overfeeding sucrose on brown adipose tissue of obese Zucker rats. *Biochem. J.*, **214**, 215–23.

Inoue, S., Bray, G.A. and Mullen, Y. (1978). Transplantation of pancreatic B-cells prevents the development of hypothalamic obesity in rats. *Am. J. Physiol.*, **235**, E266–71.

Inoue, S., Campfield, L.A. and Bray, G.A. (1977). Comparison of metabolic alterations in hypothalamic and high fat induced obesity. *Am. J. Physiol.*, **223**, R162–8.

Ionescu, E., Rohner-Jeanrenaud, F., Proietto, J., Rivest, R.W. and Jeanrenaud, B. (1988). Taste induced changes in plasma insulin and glucose turnover in lean and genetically obese rats. *Diabetes*, **37**, 773–9.

Jeanrenaud, B. (1985). An hypothesis on the aetiology of obesity: dysfunction of the central nervous system as a primary cause. *Diabetologia*, **28**, 502–13.

Langley, S. and York, D.A. (1990a). Increased type II glucocorticoid receptor numbers and glucocorticoid-sensitive enzyme activities in the brain of the obese Zucker rat. *Brain Res.*, **533**, 268–74.

Langley, S. and York, D.A. (1990b). The effects of the antiglucocorticoid RU-486 on the development of obesity in the obese fafa rat. *Am. J. Physiol.*, **259**, R539–44.

Leibowitz, S.F. (1986). Brain neurochemistry and eating behaviour. *Adv. Biosci.*, **60**, 65–72.

Leibowitz, S.F., Roland, C.R., Hor, L. and Squillari, V. (1984). Noradrenergic feeding elicited via the paraventricular nucleus is dependent upon circulating corticosterone. *Physiol. Behav.*, **32**, 857–64.

Luiten, P.G.M., ter Horst, J.G. and Steffens, A.B. (1987). The hypothalamus. Intrinsic connections and outflow pathways to the endocrine system in relation to the control of feeding and metabolism. *Prog. Neurobiol.*, **28**, 1–54.

Mendel, D.B. and Orti, E. (1988). Isoform composition and stoichiometry of the approximately 90kDa heat shock protein associated with glucocorticoid receptors. *J. Biol. Chem.*, **263**, 6695–702.

Nakaishi, S., Nakai, Y., Naitoh, Y., Fukata, J., Tominaga, T., Murakami, N., Tsukada, T., Usui, T., Ikeda, H. and Matsuo, T. (1988). Hypothalamo-pituitary-adrenal function in lean and obese Zucker rats. *J. Cell. Biochem.* (Suppl.) **13E**, 240 (Abstr.).

Nishizawa, Y. and Bray, G.A. (1980). Evidence of a circulating ergostatic factor: studies on parabiotic rats. *Am. J. Physiol.*, **239**, R344–51.

Okada, S., York, D.A. and Bray, G.A. (1990). Blockade of glucocorticoid receptors with RU-486 prevents dietary obesity. *FASEB J.*, **4**, A918.

Philibert, D. (1984). RU-38486: An original multifaceted antihormone *in vivo*. In: *Adrenal Steroid Antagonism* (ed. M.K. Agarwal), deGruyter & Co., Berlin, pp. 77–101.

Planche, E., Joliff, M., deGasquet, P. and LeLiepvre, X. (1983). Evidence of a defect in energy expenditure in the 7 day old Zucker rat (fafa). *Am. J. Physiol.*, **245**, E107–13.

Powley, T.L. (1977). The ventromedial hypothalamic syndrome, satiety and a cephalic phase hypothesis. *Psychol. Rev.*, **84**, 89–126.

Powley, T.L. and Opsahl, C.A. (1974). Ventromedial hypothalamic obesity abolished by subdiaphragmatic vagotomy. *Am. J. Physiol.*, **226**, 25–33.

Reul, J.M.H.M. and DeKloet, E.R. (1985). Two receptor systems for corticosterone in rat brain: microdissection and differential occupation. *Endocrinology*, **117**, 2505–11.

Rohner-Jeanrenaud, F., Hochstrasser, A-C. and Jeanrenaud, B. (1983a). Hyperinsulinaemia of preobese and obese fafa rats is partly vagus nerve mediated. *Am. J. Physiol.*, **244**, E317–22.

Rohner-Jeanrenaud, F., Ionescu, E. and Jeanrenaud, B. (1983b). The origins and role of efferent vagal nuclei in hyperinsulinaemia in hypothalamic and genetically obese rodents. *J. Auton. Nerv. Syst.*, **9**, 173–84.

Rohner-Jeanrenaud, F. and Jeanrenaud, B. (1985a). Involvement of the cholinergic system in insulin and glucagon oversecretion in preobesity. *Endocrinology*, **116**, 830–4.

Rohner-Jeanrenaud, F. and Jeanrenaud, B. (1985b). A role for the vagus in the etiology and maintenance of the hyperinsulinaemia of genetically obese fafa rats. *Int. J. Obesity*, **9**, 71–5.

Rohner-Jeanrenaud, F., Seydoux, J., Chinet, A., Bas, S., Giacobino, J-B., Assimocopoulos-Jeannet, F., Jeanrenaud, B. and Girardier, L. (1983c). Defective diet induced but normal cold induced brown adipose tissue adaptation in hypothalamic obesity. *J. Physiol. (Paris)*, **78**, 833–7.

Rohner-Jeanrenaud, F., Walker, C-D., Greco-Perotto, R. and Jeanrenaud, B. (1989). Central corticotropin releasing factor administration prevents the excessive body weight gain of genetically obese fafa rats. *Endocrinology*, **124**, 733–9.

Romsos, D.R. (1985). Norepinephrine turnover in obese mice and rats. *Int. J. Obesity*, **9**, 55–62.

Saito, M., Minokoshi, Y. and Shimazu, T. (1989). Accelerated norepinephrine turnover in peripheral tissues after ventromedial hypothalamic stimulation in rats. *Brain Res.*, **481**, 298–303.

Sakaguchi, T., Arase, K. and Bray, G.A. (1988a). Sympathetic activity and food intake of rats with ventromedial hypothalamic lesions. *Int. J. Obesity*, **12**, 43–9.

Sakaguchi, T., Takahashi, M. and Bray, G.A. (1988b). The lateral hypothalamus and sympathetic firing rate. *Am. J. Physiol.*, **255**, R507–12.

Seydoux, J., Rohner-Jeanrenaud, F., Assimacopoulos-Jeannet, F., Jeanrenaud, B. and Girardier, L. (1981). Functional disconnection of brown adipose tissue in hypothalamic obesity. *Pflugers Arch.*, **390**, 1–4.

Shimizu, H., Shargill, N.S., Bray, G.A., Yen, T.T. and Gesellchen, P.D. (1989a). Effects of MSH on food intake, body weight and coat color of the yellow obese mouse. *Life Sci.*, **45**, 543–52.

Shimizu, H., Shargill, N.S. and Bray, G.A. (1989b). Adrenalectomy and the response to corticosterone and MSH in the genetically obese yellow mouse. *Am. J. Physiol.*, **256**, R494–500.

Smith, C.K. and Romsos, D.R. (1985). Effects of adrenalectomy on energy balance of obese mice are diet dependent. *Am. J. Physiol.*, **249**, R13–22.

Stanley, B.G., Daniel, D.R., Chin, A.S. and Leibowitz, S.F. (1985). Paraventricular nucleus injections of peptide YY and neuropeptide Y preferentially enhance carbohydrate ingestion. *Peptides*, **6**, 1205–11.

Stanley, B.G., Krykouli, S.E., Lampert, S. and Leibowitz, S.F. (1986). Neuropeptide Y chronically injected into the hypothalamus: a powerful neurochemical inducer of hyperphagia and obesity. *Peptides*, **7**, 1189–92.

Stubbs, M. and York, D.A. (1989). Effect of CRF infusion on glucose turnover in obese Zucker rats. *Int. J. Obesity*, **13**, (suppl. 1), 53.

Tokunaga, K., Fukushima, M., Kemnitz, J. and Bray, G.A. (1986). Effect of vagotomy on serum insulin in rats with paraventricular or ventromedial hypothalamic lesions. *Endocrinology*, **119**, 1708–11.

Triandafillou, J. and Himms-Hagen, J. (1983). Brown adipose tissue in genetically obese fafa rats: response to cold and diet. *Am. J. Physiol.*, **244**, E145–50.

Tsai, H.J. and Romsos, D. (1990). Cytosolic corticosterone receptors in obob mice. *FASEB J.*, **4**, A918, (Abstr.).

Turkenkopf, I.J., Johnson, P.R. and Greenwood, M.R.C. (1983). Development of pancreatic and plasma insulin in prenatal and suckling Zucker rats. *Am. J. Physiol.*, **24**, E220–5.

White, J.D., Sanacora, G. and Kershaw, M. (1988). Increased hypothalamic neuropeptide Y mRNA in obese Zucker fatty rats. *J. Cell. Biochem. Suppl.*, **13E**, 243.

Wolff, G.L. (1965). Body composition and coat color correlation in different phenotypes of 'viable yellow' mice. *Science*, **147**, 1145–7.

York, D.A. (1987). Neural activity in hypothalamic and genetic obesity. *Proc. Nutr. Soc.*, **46**, 105–17.

York, D.A. (1989). Corticosterone inhibition of thermogenesis in obese animals. *Proc. Nutr. Soc.*, **48**, 275–9.

York, D.A., Holt, S.J., Allars, J. and Payne, J. (1988). Glucocorticoids and the central control of sympathetic activity in the obese fafa rat. In: *Obesity in Europe 88*, (eds P. Bjorntorp and S. Rossner), John Libbey, London, pp. 219–24.

York, D.A., Marchington, D., Holt, S.J. and Allars, J. (1985). Regulation of sympathetic activity in lean and obese Zucker fafa rats. *Am. J. Physiol.*, **249**, E299–305.

Chapter Seven

Insulin Resistance in Obesity and in Wasting Disorders

Keith N. Frayn
Sheikh Rashid Diabetes Unit, Radcliffe Infirmary, Oxford OX2 6HE, UK

INTRODUCTION

The description by Banting and Best in 1922 (Banting and Best, 1922) of the isolation of insulin marked the beginning of an era of intensive study of this hormone. Insulin has been a leader amongst hormones in many respects: it was the first hormone to be used for the treatment of a deficiency disease, the first protein whose primary structure was determined (Ryle *et al.*, 1955) and the first protein produced by recombinant-DNA techniques to be used therapeutically.

Although the introduction of insulin into clinical use was of direct life-saving benefit to many sufferers from diabetes, it soon became clear that the disease of diabetes does not always reflect simply a deficiency of insulin secretion. In the mid-1930s Himsworth and Kerr showed, by injection of insulin, that the response of an individual was related both to the dose injected, and to a variable factor which they described as the 'sensitivity' of the individual to insulin (Himsworth and Kerr, 1939). In prescient studies on patients with diabetes, they described the existence of two types of patient: the thin, young ketosis-prone patient who showed a normal or high sensitivity to injected insulin, and the overweight, older patient showing resistance to the effect of insulin. These two categories would now be referred to as the 'insulin-dependent' or type 1, and the non-insulin dependent or type 2 patient respectively, and it is still a matter of controversy

Obesity and Cachexia. Edited by N.J. Rothwell and M.J. Stock

how much of the aetiology of the latter is due to peripheral resistance to the action of insulin rather than to impaired insulin secretion (see Cahill, 1988).

In the intervening years, changes in the sensitivity of tissues, or of the whole body, to insulin have been described in many conditions. In many cases we are still ignorant as to the mechanism of these well-documented changes, and in most, if not all, cases we are ignorant as to their significance. This review will focus on two conditions: that of 'nutrient excess', or obesity, and the wasting condition associated with sepsis or major injury. Although the latter has certain features in common with the cachexia of cancer (covered in Chapters 8 and 13), it also has distinctive features and probably, at least in part, a different aetiology.

ACTION OF INSULIN AND MECHANISM OF CHANGES IN SENSITIVITY

Insulin is synthesized in the beta-cells of the pancreatic islets of Langerhans, in the form of a precursor protein, pre-proinsulin; after cleavage of a signal peptide the precursor consists of a single polypeptide chain of 86 amino acids, proinsulin (Figure 7.1). Part of this polypeptide is removed by proteolytic cleavage whilst in the storage granules of the beta-cell (Hutton, 1989); the products are 'mature' insulin, and the C- (connecting) peptide. Insulin is thus a peptide hormone consisting of two separate chains linked by inter-chain disulphide bridges. Insulin and C-peptide are secreted from the beta-cell in equimolar amounts, into the pancreatico-duodenal vein, a tributary of the hepatic portal vein. Insulin travels in the plasma, not bound to any larger carrier. A large fraction (around 50%; Waldhäusl *et al.*, 1982) of the insulin is removed during its first passage through the liver. Insulin then acts on peripheral tissues, via binding to specific cell-surface receptors. It is thought to be degraded after, or in the course of, exerting its effects on these 'target tissues'; degradation appears to follow receptor-binding (Gammeltoft, 1984). The idea that insulin acted on the cell membrane was first put forward by Levine, who, in 1949, showed by experiments on dogs that insulin increased the permeability of cells to galactose (Levine *et al.*, 1949). This idea was confirmed by the work of Cuatrecasas (1969) who showed that insulin bound to large agarose beads (too large to enter cells) exerted its expected effects on the cell. (These experiments have since been criticized for technical reasons, but the results have been confirmed by several authors; Cuatrecasas, 1973.) Nevertheless, the actions of insulin are not confined to changes in transport across the cell membrane; the activities of several intracellular enzymes are affected by insulin, showing that the signal resulting from insulin reaching the cell membrane must be transmitted

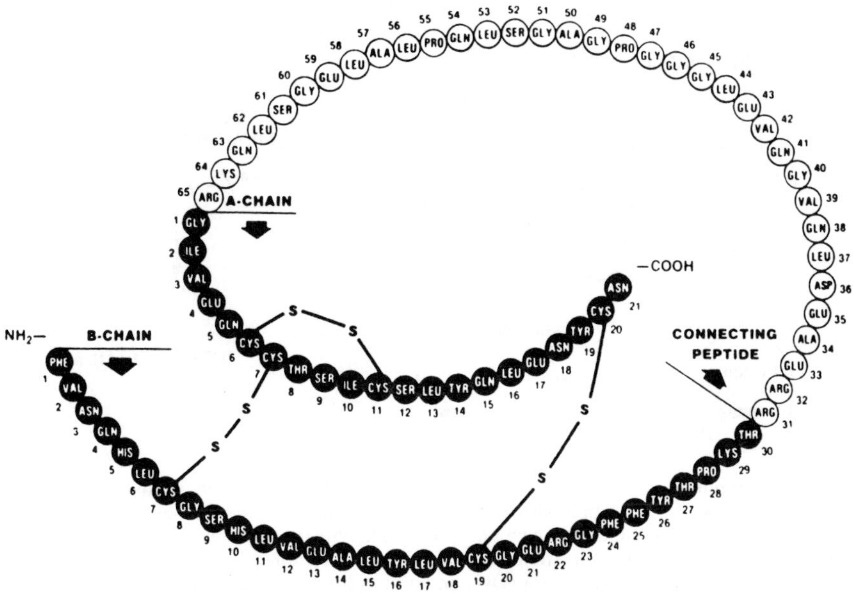

Figure 7.1 The primary sequence of human insulin, C-peptide and proinsulin. The proinsulin molecule is synthesized as a single peptide chain. Before release from the pancreatic β-cells, the connecting (C-) peptide (white circles) is removed by proteolytic cleavage, leaving the A and B chains (black circles) linked by disulphide bridges. C-peptide and insulin are then secreted in equimolar amounts. Reproduced from Johnson (1982). Copyright 1982 by the American Diabetes Association. Reprinted with permission

through the cell by a 'second messenger' system. The structure of the cell-surface insulin receptor is now known, and some of the intracellular events resulting from the binding of insulin have been clarified (Figure 7.2), but the details of the means of transmission of the signal within the cell remain obscure in most cases.

With such a complex pathway, from the release of insulin from the pancreatic beta-cells to the final effect on target pathways, there are, of course, many steps at which the normal efficiency of the process may be altered (Table 7.1). These might conveniently be grouped into four: changes in the structure of insulin or its transport through the plasma; changes in the insulin receptor of the target tissue; changes in the coupling between receptor-binding and metabolic effect; and changes in the target metabolic pathway making it more 'resistant' to modulation by insulin. Apart from a few rare conditions, such as the production of a mutant insulin or the severe insulin resistance produced by the presence of auto-antibodies against the insulin receptor (reviewed by Olefsky, Kolterman and Scarlett, 1982), we are ignorant of the details of the mechanism of changes in sensitivity to

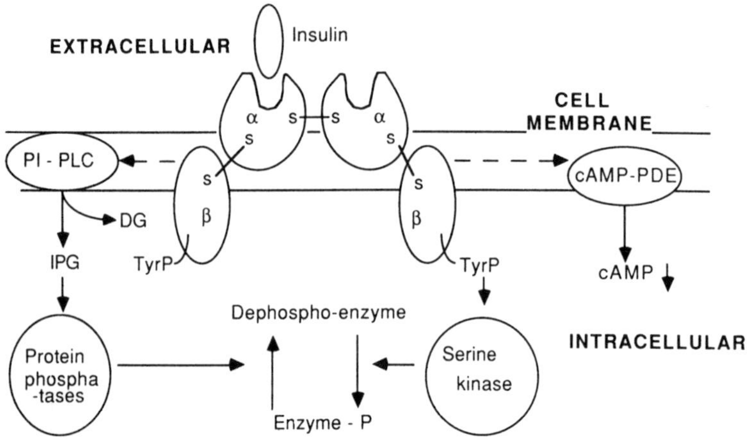

Figure 7.2 Simplified representation of the possible coupling between insulin receptor and intracellular events. Insulin binding to the α-subunit of the receptor causes activation of the tyrosine kinase in the β-subunit, which phosphorylates tyrosine residues in the β-subunit itself. Other effects may include activation of a phosphatidylinositol-specific phospholipase C (PI-PLC) with release of a diacylglycerol (DG) and an inositol phosphate-glycan (IPG) which may be responsible for activation of intracellular protein phosphatases. There may be activation of cyclic-AMP phosphodiesterase (cAMP-PDE) and possibly inhibition of adenylate cyclase, either via IPG or through other mechanisms, resulting in a lowering of the intracellular cyclic-AMP level. Most enzymes activated by insulin are dephosphorylated to their active form, but some are phosphorylated (e.g. acetyl-CoA carboxylase), perhaps through activation of a serine kinase. Based on Espinal (1987); reproduced from Frayn (1988) with permission

Table 7.1 Stages at which sensitivity to insulin may be regulated

Synthesis and secretion
Abnormal gene product
Incomplete conversion of proinsulin to insulin
Circulating insulin antagonists
Anti-insulin antibodies
Anti-insulin receptor antibodies
Elevated levels of counter-regulatory hormones
Target tissue defects
Reduced receptor number
Reduced receptor affinity
Impaired coupling between receptor and intracellular events
Direct impairment of metabolic pathways

Based on Olefsky (1981).

insulin in most clinical conditions, including such common ones as obesity and trauma.

Much of the information we do have about the mechanism of changes in sensitivity to insulin has come from an analysis of the dose–response relationships for the effects of insulin, as described by Kahn (1978) (Figure 7.3). He suggested that a distinction be made between 'diminished responsiveness' to insulin (line c) and 'diminished sensitivity' (line b, a change in the insulin concentration needed for half-maximal effect on a process). This simple distinction can give information about mechanisms, because of the well-documented phenomenon of 'spare receptors'; all insulin-target tissues have many more insulin receptors than need to be occupied to give a maximal response. It can be argued that the purpose of this apparent superfluity is to give a means for changing sensitivity to insulin; the binding of insulin to receptors is a reversible, equilibrium process, and the number of receptors occupied will be a function both of the concentration of insulin, and of the concentration of receptors (Freychet, 1976). In practice, because the proportion of receptors 'spare' is far greater than any decrease in insulin receptor number seen in insulin-resistant states, we can say that any change in responsiveness to insulin caused purely by a decrease in receptor number will be of the 'decreased sensitivity' type (line b, Figure 7.3);

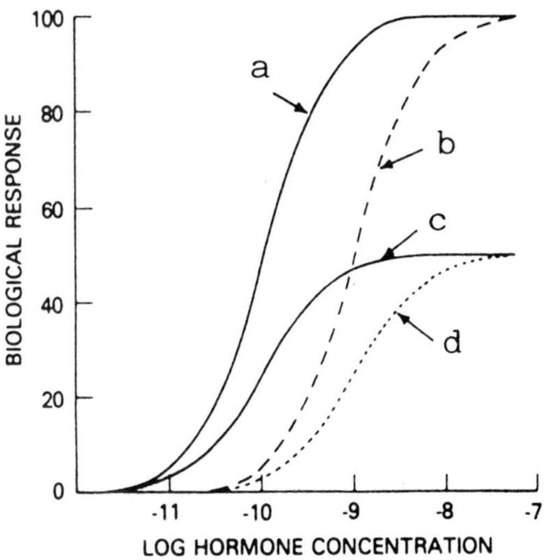

Figure 7.3 Possible variations in the effectiveness of insulin. The response of some process (*in vitro* or *in vivo*) to insulin is plotted against the logarithm of the insulin concentration (as in a standard pharmacological dose–response curve). Line (a) shows a normal response; (b) decreased sensitivity (implying decreased receptor-binding of insulin); (c) decreased responsiveness (implying a post-receptor defect); (d) combined decreases in sensitivity and in responsiveness. From Kahn (1978), reproduced, by permission of W.B. Saunders Co

a decrease in maximal responsiveness (line c) must represent an intracellular impairment of the action of insulin (either an impairment of the transmission of the insulin message, or of the target pathway).

The analysis of insulin dose–response relationships is complemented by direct studies of insulin binding to cells or membranes from target tissues, allowing the identification of changes in receptor density, or in the affinity of the receptor for insulin. The methodology for such studies was first described by Freychet *et al.* (1971), and these techniques have been widely employed in the study of insulin-resistant states such as obesity (Freychet, 1976; Olefsky, 1976a).

When a post-receptor defect is believed to exist instead of or in addition to a change in insulin binding, then this requires examination by biochemical techniques. Post-receptor defects, as defined according to Kahn's scheme (Figure 7.3 lines c and d), may include impairment of the coupling mechanisms between receptor and target pathway. There has been little study of such mechanisms in insulin-resistant states, although they are of potential importance (Farese and Cooper, 1989). The study of intracellular pathways (for instance, of glucose metabolism) has usually involved the removal of tissues and their metabolic investigation *in vitro*; the details of such studies will be discussed, where relevant, below.

One major mechanism for the regulation of intracellular glucose metabolism, and of its apparent sensitivity to insulin, has, however, become of such general importance that it will be reviewed briefly here. This is the limitation of the rate of glucose metabolism which occurs when the rate of fatty acid oxidization is increased. This phenomenon was first documented in a series of papers by Randle and colleagues in the late 1950s, and integrated into a theory of metabolic regulation in the whole organism by the same authors in 1963 (Randle *et al.*, 1963). The proposed reciprocal relationship between the metabolism of fatty acids and of glucose, and the mechanisms by which it occurred in adipose tissue and in muscle, was referred to as the glucose–fatty acid cycle. (It is not a cycle in the same sense as, for instance, the tricarboxylic acid or Krebs' cycle. The point is mentioned here, because it has now become common to refer to the 'muscle end' of this system of metabolic regulation as, for instance, the 'Randle cycle' (Bevilacqua *et al.*, 1990), a misleading description for a non-cyclical aspect of metabolic regulation.) In essence, the glucose–fatty acid cycle suggests that both the availability and the utilization of fatty acids and glucose are coordinated. Increased availability of carbohydrate (for instance, ingestion of a meal) suppresses the liberation of fatty acids from adipose tissue (through increases in the plasma glucose and insulin concentrations). The increased insulin concentration also promotes the utilization of glucose by muscle, and the body's fuel economy is thereby switched to carbohydrate consumption and sparing of fat stores. In times of lack of carbohydrate, then low plasma insulin and glucose concentrations lead to increased release of fatty acids from adipose tissue triacylglycerol stores; increased concentrations of fatty acids lead to their increased usage by muscle; and, within muscle, the oxidation of fatty acids leads, through the mechanisms elucidated by Randle and colleagues, to a restriction on the rate of glucose utilization. The body's fuel economy is thus switched to the use of fatty acids and a sparing of carbohydrate.

The suggestion then arises that an increased delivery of fatty acids in pathological states such as diabetes mellitus (arising from the lack of insulin-restraint of lipolysis) may inappropriately inhibit glucose utilization in a way which will cause its apparent 'resistance' to stimulation by insulin. Although this theory has received occasional study over the intervening decades, it has in recent years undergone a resurgence of interest, and studies on the role of such a mechanism for reduction of sensitivity to insulin will be reviewed below. The renewal of interest in this regulatory mechanism has come about for a number of reasons, not least of which is the ability to analyse substrate oxidation in the whole body (by indirect calorimetric techniques) and simultaneously to analyse whole-body glucose utilization by a 'glucose clamp' technique. For instance, in normal subjects in whom the plasma non-esterified fatty acid level was raised by infusion of a triacylglycerol emulsion, glucose disposal (both 'storage' and oxidation) was impaired during a period of hyperinsulinaemia, and the degree of impairment was related to the plasma fatty acid concentration (Thiébaud *et al.*, 1982).

ROLE OF INSULIN IN REGULATION OF BODY WEIGHT

Why is a chapter on insulin appearing in a book on obesity and cachexia? The answer is that insulin has a fundamental role in the regulation of body weight, and that disturbances in the actions of insulin are both a cause and a result of disturbances in the regulation of body weight. Insulin is often described as the 'key anabolic hormone'. It is certainly the hormone with the most general integrating role in the storage of excess nutrients. Insulin stimulates the uptake of glucose into some of the major glucose-consuming tissues, particularly skeletal muscle; it stimulates the storage of glucose in the form of glycogen, and also the pathway of synthesis of lipid from glucose. Insulin restrains the mobilization of adipose tissue fat stores, and increases the esterification—and thus storage—of fatty acids, as opposed to their oxidation. Insulin has a generally anabolic role in protein metabolism; it stimulates amino acid uptake and protein synthesis in muscle and other tissues, and may have a restraining role on muscle protein breakdown. In evolutionary terms, all these actions must have been 'good', enabling an animal to store excess nutrients when available and thus better survive times of nutrient shortage.

The importance of insulin in body weight regulation is clearly shown by the wasting of fat and muscle which used to occur in those unfortunate sufferers from diabetes mellitus before the advent of insulin therapy, and by the quite remarkable changes which occurred in patients treated with the early preparations of insulin (Bliss, 1983). It is quite reasonable to suppose, from that situation, that a deficiency of insulin might be the cause of the wasting seen in sepsis and trauma, and that an excess of insulin would lead to excess accumulation of body fat as in obesity. In practice, as will be described, the aetiology of these conditions is far from that simple, but this background will at least help to set the scene for examining the role of insulin, and of changes in the body's sensitivity to that hormone, in wasting conditions and in obesity.

WASTING DUE TO TRAUMA AND SEPSIS

The loss of body mass associated with major injury or sepsis has long been recognized. The loss of nitrogenous substances in the urine during fever was studied 100 years ago (Sassetzky, 1883). Both lean and fat body tissues are lost. The former make a smaller contribution to 'calories expended' in the post-surgical patient, but, because of the low energy density of lean tissue, actual loss of lean body mass may exceed that of fat about two-fold (Kinney *et al.*, 1968). These changes are deleterious in terms of both increased susceptibility to infection, and impaired mobility and rehabilitation. Although aggressive nutritional support can help such patients, it is not always successful even in overcoming negative nitrogen balance. What part do changes in sensitivity to insulin play in this metabolic picture?

Changes in sensitivity to insulin in sepsis and trauma

The idea of 'insulin resistance' in injured patients was perhaps first prompted by the studies of Evans and Butterfield (1951) on adrenocortical responses in burned patients. They described, in several of their severely burned subjects, a state of 'pseudodiabetes' with blood sugar concentrations of up to 70 mmol/L, which were not controlled even with the administration, in one case, of over 300 Units of insulin. (The average daily requirement of a completely insulin-dependent diabetic patient is about 50 Units; Taylor and Agius, 1988.) The authors attributed these findings to the provision of excessive dietary carbohydrate in the presence of an 'impaired, delayed action' of insulin as was known to occur in Cushing's disease. Later in that decade, Howard (1955) reported studies on American soldiers injured in the Korean war; he showed resistance—in terms of a lesser hypoglycaemic response—to the action of injected insulin, the degree of insulin resistance being broadly related to the severity of injury. The insulin resistance following injury was again highlighted by the studies of Hinton, Allison and colleagues (1971), who tried to overcome negative nitrogen balance in burned patients by giving hypertonic glucose and insulin; some patients required as much as 700 Units of insulin per day to prevent excessive hyperglycaemia.

In terms of quantitative insulin resistance, there appears to be a special case, which has been much studied, after abdominal operations of moderate severity. During the first few days after operation patients show an exaggerated (sometimes enormously so) insulin response to a glucose challenge, whilst glucose tolerance or utilization is impaired (Ross *et al.*, 1966; Giddings *et al.*, 1977; Soerjodibroto *et al.*, 1977) (Figure 7.4). In other conditions—accidental injury or sepsis—the usual finding is that prevailing insulin concentrations, and the insulin response to a glucose challenge, are around normal or slightly elevated relative to the degree of hyperglycaemia, which is also usually increased (Wolfe *et al.*, 1979; Frayn *et al.*, 1984a,b; White *et al.*, 1987). Again, the rate of glucose utilization is not as high as

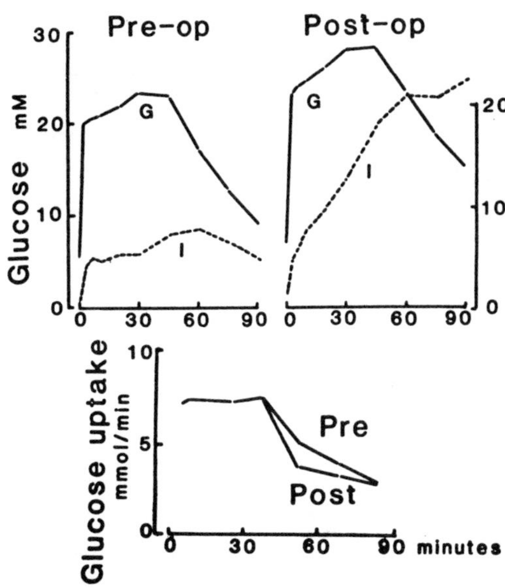

Figure 7.4 Insulin resistance after elective abdominal surgery. Eighteen patients were studied on the day before ('Pre-op') and on the day after ('Post-op') operation. Glucose was infused over 45 min and the plasma glucose (G) and insulin (I) concentrations measured. Whole-body glucose uptake was calculated from the apparent distribution space for glucose, the rate of change of concentration and the rate of infusion. In the post-operative test, plasma glucose concentrations were higher throughout; insulin concentrations were very much higher and rose throughout; and yet rates of glucose utilization were almost identical to those in the pre-operative test. Reproduced from Giddings *et al.* (1977) by permission of J.B. Lippincott Co

would be expected for the prevailing insulin and glucose concentrations (Figure 7.5). A possible explanation for the elevation of insulin concentrations, when present, is the synergistic effect which amino acids and glucose have on insulin secretion (Fajans *et al.*, 1967). Amino acid concentrations, including those of the potent insulin secretagogue arginine, are elevated during the response to injury (references in Frayn *et al.*, 1984a). The explanation for the failure of these normal or elevated insulin levels to maintain normal glucose homeostasis will be considered below.

During the last two decades many studies have been carried out with

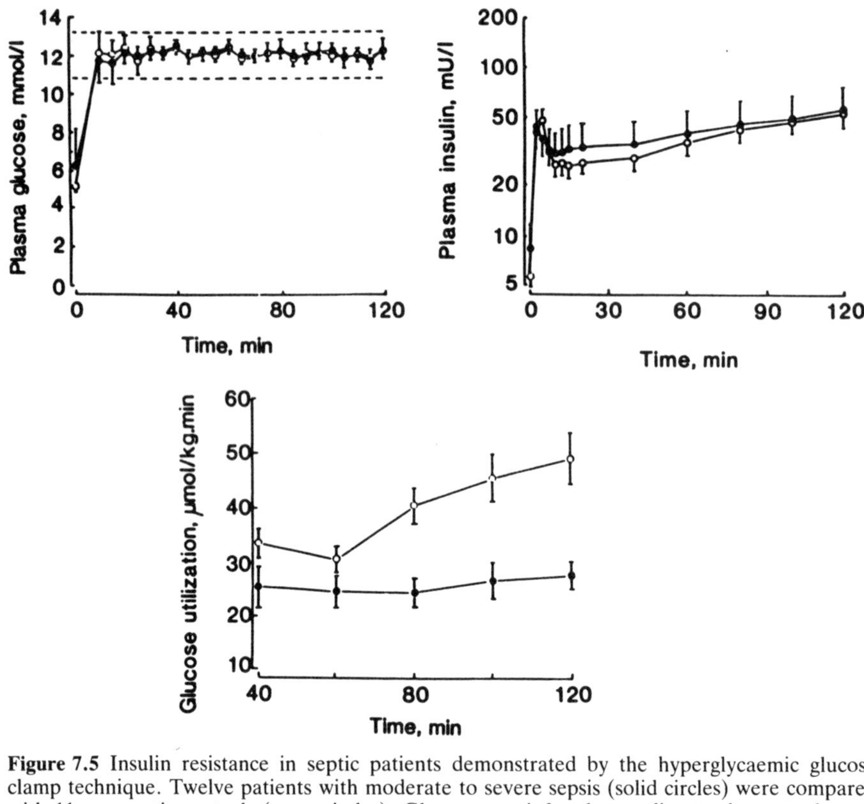

Figure 7.5 Insulin resistance in septic patients demonstrated by the hyperglycaemic glucose clamp technique. Twelve patients with moderate to severe sepsis (solid circles) were compared with 11 non-septic controls (open circles). Glucose was infused according to the procedure of DeFronzo *et al.* (1979), and the rate of infusion adjusted as necessary to maintain a plasma glucose concentration of 12 mmol/L for 120 min (top left). (The plasma glucose concentration was measured every 5 min and the rate of infusion adjusted accordingly.) The response of endogenous plasma insulin was similar in controls and patients (top right). The rate of whole-body glucose utilization (lower panel) was calculated as described by DeFronzo *et al.* (1979); despite identical glucose and insulin concentrations, the rate of glucose utilization in the patients failed to rise as it did in the controls. Reproduced from White *et al.*, (1987) by permission of Williams and Wilkins

tracer methodology and with catheterization techniques, and have led to a clearer picture of the several facets of insulin resistance in these conditions.

One component of the lack of effectiveness of insulin in reducing hyperglycaemia in such patients is an overproduction of glucose by the liver, which is not suppressed as normal by glucose or insulin infusion. This sustained glucose production must reflect gluconeogenesis, and the substrates for this process appear to be lactate, 'recycled' from peripheral tissues including sites of sepsis and injured tissue (Wilmore, 1986), amino acids liberated in the net protein breakdown in skeletal muscle, and glycerol liberated in adipose tissue lipolysis. The mean basal rate of 'total glucose production' in burned patients, for instance, was 4.25 mg/kg min, compared with a mean of 2.53 in normal subjects; the difference between rates of 'non-recycled' glucose production in the patients and controls was less marked (3.35 in patients versus 2.28 in controls), showing the elevated rate of glucose recycling (Wolfe *et al.*, 1979). Conversion of a labelled alanine tracer into glucose in injured and septic patients continued at high rates even during glucose administration at rates sufficient to completely suppress it in normal subjects (Long *et al.*, 1976; Elwyn *et al.*, 1979); hepatic vein catheterization studies show that in septic patients the liver continues to release glucose at high blood glucose concentrations, which cause a switch to hepatic glucose uptake in normal subjects (Gump *et al.*, 1974).

Another component of the insensitivity to insulin is an impairment of glucose disposal; this has been demonstrated in the whole-body by the 'glucose clamp' (DeFronzo *et al.*, 1979) technique (Black *et al.*, 1982; White *et al.*, 1987), and in forearm muscle by measurement of arteriovenous differences and their responsiveness to insulin (Brooks *et al.*, 1984; White *et al.*, 1987; Henderson *et al.*, 1991). It should be noted that, whereas injured tissue may contribute to the accelerated glucose turnover and glucose production described above by using glucose at a high rate and returning lactate to the liver, the impairment of glucose uptake and its responsiveness to insulin affects non-injured tissue. Dose–response curves have been produced for both whole-body and forearm glucose utilization by infusing insulin at different rates using a 'glucose clamp' procedure (Black *et al.*, 1982; White *et al.*, 1987; Brooks *et al.*, 1984; Henderson *et al.*, 1991), and show a clear impairment of maximal responsiveness to insulin, accompanied by a rightward shift of the curve; i.e. both diminished sensitivity and diminished maximal responsiveness appear to be involved (Figure 7.6).

The responses seem to be very similar in septic, non-injured and in injured, non-septic patients, lending support to the idea that the wasting seen in both these conditions has at least features in common. The degree of insulin resistance has not been clearly linked to the severity of the disease (sepsis or injury); White and colleagues (1987) found that septic patients of a variety of severities behaved as a reasonably homogeneous group, but all

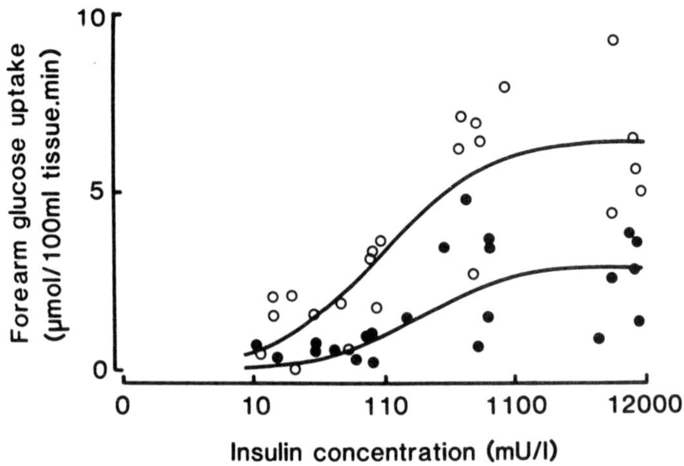

Figure 7.6 Dose–response curves for effect of insulin on glucose uptake in the human forearm. Twenty-two patients were studied about 1 week after sustaining injuries all involving fracture of long-bones (solid circles); they were compared with 24 studies on non-injured controls (open circles). The plasma insulin concentration was elevated for 120 min by infusing insulin whilst maintaining a constant plasma glucose concentration of 5 mmol/L by the euglycaemic-glucose clamp procedure (DeFronzo *et al.*, 1979). Each patient was studied once only; different rates of insulin infusion were used to produce a range of insulin concentrations. Forearm glucose uptake was measured as the product of the arteriovenous difference for glucose across the forearm, and the forearm blood flow. The dose–response curve in the injured patients shows reduced sensitivity and reduced maximal responsiveness compared with the controls. Reproduced from Henderson *et al.* (1991) by permission of the Biochemical Society and Portland Press

behaved differently from non-septic control subjects. Despite this finding, the elevation of 'stress hormones', thought by some to be the primary cause of this insulin resistance (as discussed below), was strongly related to the severity of the sepsis (White *et al.*, 1987).

Mechanisms of changes in sensitivity to insulin in sepsis and trauma

Unfortunately, remarkably little is known about the mechanism of the observed changes in sensitivity to insulin. From the dose–response relationships of peripheral glucose uptake with insulin described above (e.g. Figure 7.6), it appears that 'post-receptor' changes must be involved although with a receptor defect possibly also contributing.

We still lack detailed studies of cellular metabolism and responsiveness to insulin in these conditions. There are many reasons why such studies are difficult to perform. Many would argue that it is difficult to establish a satisfactory animal model of these human conditions, although recent work on models of sepsis may be pointing us in the right direction (e.g. de Vasconcelos *et al.*, 1987). Even in an animal model, one is faced with the problem that the response to injury or sepsis is undoubtedly a whole-body phenomenon, and changes observed in tissues after isolation may not be representative of the situation *in vivo*. A good example of this is seen in the acute 'insulin resistance' which occurs within a few hours of severe injury; some studies show that skeletal muscles isolated from small animals during this phase show normal sensitivity to insulin *in vitro* (e.g. Frayn *et al.*, 1978), and the insulin resistance seen *in vivo* presumably reflects the effects of circulating factors (counter-regulatory hormones and/or fatty acids), which are not reproduced *in vitro*.

The evidence from the studies *in vivo*, however, points strongly to the existence in the septic or injured patient of two major changes in glucose metabolism: a sustained high rate of hepatic glucose production, resistant to the normal suppression by glucose and/or insulin, and an intracellular impairment of glucose disposal. Although the former may result in part from an increased provision of substrate, as discussed earlier, it may also represent hormonal stimulation of the pathway of gluconeogenesis. Considerable attention has been devoted recently to the role of the 'counter-regulatory' or 'stress' hormones (catecholamines, glucagon, cortisol) in such catabolic conditions. Studies in which these hormones have been infused, in combination, into normal subjects show that many of the changes characteristic of sepsis or injury are produced (Bessey *et al.*, 1984; Gelfand *et al.*, 1984). However, it should be noted that the plasma levels produced during those studies are generally much greater than would be seen in patients displaying marked 'insulin resistance' (Frayn, 1986; White *et al.*, 1987).

Another factor which may provide a common explanation for both aspects of glucose metabolism is an increased rate of fat oxidation. Patients in these conditions show increased oxidation of fat as assessed either by indirect calorimetry or by production of $^{14}CO_2$ from labelled fatty acids, and this high rate of fat oxidation is resistant to the normal suppressive effect of glucose or insulin administration (Nordenström *et al.*, 1983; White *et al.*, 1987; Henderson *et al.*, 1991). This drive to mobilize fat may come from the sustained sympathetic drive, since there is a relationship between plasma noradrenaline concentrations (an index of sympathetic activity) and glycerol levels (an index of fat mobilization) (White *et al.*, 1987). In addition, Kinney's group have shown in a number of studies that administration of glucose to injured and septic patients (as is usually the case clinically) will cause an elevation in noradrenaline excretion, an index of sympathetic activity (summarised by Nordenström *et al.*, 1981).

The relevance of the increased rate of fat oxidation is two-fold. First, there is growing evidence that increased provision of fatty acids to the liver, and their consequent oxidation (which will predominate in the hormonal environment present in the 'catabolic' patient), will increase hepatic glucose production (reviewed by Reaven, 1988; Groop *et al.*, 1989). Fatty acids are not themselves substrates for gluconeogenesis, but the increased provision of acetyl-CoA and ATP from fat oxidation appears both to inhibit pyruvate dehydrogenase and to activate pyruvate carboxylase, as well as providing the energy required for gluconeogenesis (Reaven, 1988). In peripheral glucose-consuming tissues (mainly muscle), increased fat oxidation will impair glucose utilization by the mechanisms described earlier as part of the glucose–fatty acid cycle.

Another contributing factor to impaired glucose uptake may be immobility, which impairs glucose tolerance and sensitivity to insulin (Dolkas and Greenleaf, 1977; Vernikos, 1986). The means by which this occurs are uncertain (reviewed by Vernikos, 1986).

Significance of the changes in sensitivity to insulin

The significance of these changes in sensitivity to insulin, in clinical terms, concerns not so much the difficulty of controlling the plasma glucose concentration, as the resistance of other aspects of metabolism. The whole response at this stage can be seen as a picture of unresponsiveness to insulin (Frayn, 1986), at a time when plasma insulin concentrations are usually elevated. Thus, there is high rate of turnover of fatty acids and of fat oxidation, despite hyperglycaemia which would normally suppress these processes. There is net breakdown of body protein, actually reaching a peak at a time when plasma insulin levels may also peak and insulin resistance is apparently maximal (Frayn *et al.*, 1984b; Frayn, 1986); insulin would normally act to promote net deposition of body protein. This description suggests superficially a generalized change in insulin action, which might conveniently be explained, for instance, by a decrease in insulin receptor number. However, as described above, the dose–response relationships with insulin do not support such a mechanism. Otherwise, the various metabolic pathways involved seem to have little in common, and it is difficult to see a common explanation for the changes in sensitivity to insulin.

Again in clinical terms, these are important questions. We need to understand why it is difficult to reverse the protein catabolism of injury and sepsis. We need to be able to diminish the elevation in metabolic rate and loss of body mass. Pharmacological approaches are at present not very helpful (Fellows and Woolfson, 1985; Kehlet and Schulze, 1986) although growth hormone has a beneficial anabolic effect if given with suitable nutritional support (Ponting *et al.*, 1988), and β-adrenergic blockers have been shown to reduce the elevated metabolic rate (Wilmore *et al.*, 1974). The development of more effective and specific centrally or peripherally acting drugs to counteract these responses, which one hopes will follow once we understand more clearly the mechanisms involved, will provide genuinely life-saving benefits.

OBESITY

Changes in sensitivity to insulin in obesity

The essential role of insulin in the maintenance of normal body weight was described earlier. It is not, therefore, too implausible to suggest that a chronic excess of insulin might lead to excessive deposition of fat, especially since insulin stimulates so many of the metabolic pathways involved in energy storage. An early suggestion that obese subjects were demonstrating an 'adaptive hyperinsulinism' was made by Rabinowitz and Zierler (1961) on the basis of metabolic changes which they observed when studying the metabolism of forearm muscle, by arteriovenous difference techniques. They also found that when insulin was administered, in small doses, into the brachial artery supplying the forearm under study, the responses were less in the obese than in normal-weight controls. Although plasma insulin concentrations could not at that time be measured, these authors suggested that 'tissues which are chronically under the influence of excessive amounts of insulin become less responsive to further challenge by insulin', a concept common to the action of several hormones, now recognized to occur through the 'down-regulation' of receptors.

The finding of high plasma insulin concentrations in obese subjects was first made by Karam and colleagues (Figure 7.7) soon after the development of radioimmunological techniques for measurement of insulin (Karam *et al.*, 1963). They showed elevated fasting insulin levels, and an exaggerated response to intravenous glucose administration, compared with non-obese controls. The elevation of fasting plasma insulin concentration was shown to be related to the degree of obesity (Bagdade *et al.*, 1967). A high plasma insulin concentration by itself is not, of course, necessarily a sign of insulin resistance; it occurs in all of us after a meal, but is then accompanied by a high rate of glucose turnover. That obese subjects display true insulin resistance has since been confirmed in many studies, by tracer techniques or 'glucose clamp' techniques, which show that glucose turnover in the obese is not elevated in accordance with the elevation in plasma insulin (DeFronzo *et al.*, 1978; Olefsky *et al.*, 1982). The finding of a high plasma insulin concentration, and insulin resistance, in the obese does not, of course, mean that these are causal factors in the pathogenesis of obesity; this point will be discussed below.

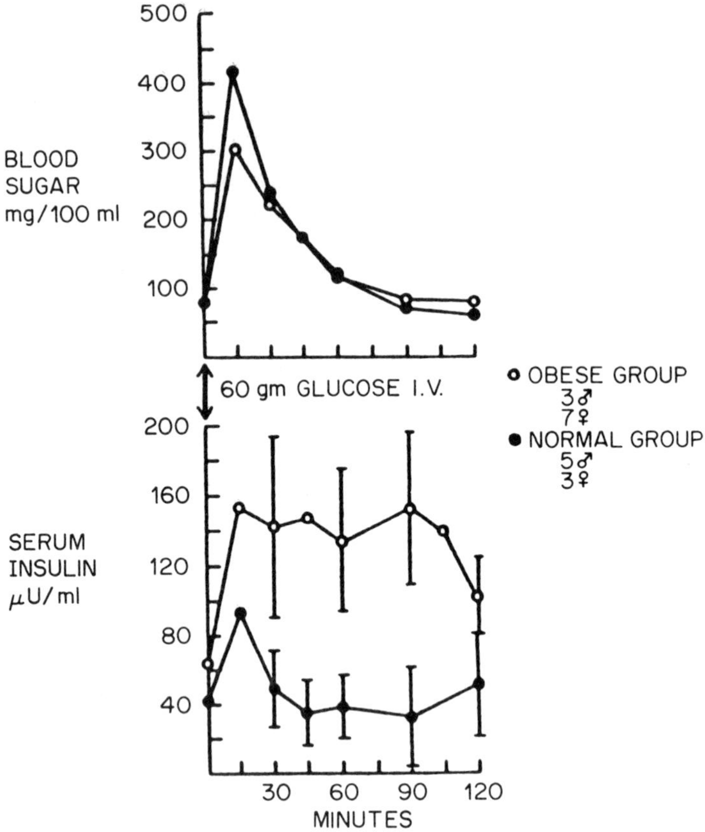

Figure 7.7 Exaggerated insulin response to intravenous glucose administration in obese subjects: an early demonstration. Reproduced from Karam *et al.* (1963). Copyright 1963 by the American Diabetes Association. Reprinted with permission

Mechanism of changes in sensitivity to insulin in obesity

There have been many studies of the mechanism of changes in sensitivity to insulin in obesity, although not universal agreement. Dose–response curves for the effects of insulin on glucose disposal *in vivo* show, according to some authors, a heterogeneous pattern, with some subjects displaying an impairment of sensitivity with a normal maximal response, others showing diminution of both sensitivity and maximal responsiveness (Olefsky *et al.*, 1982; Garvey and Kolterman, 1988). Most studies of tissues or cells from obese rodents *in vitro* find a diminished maximal responsiveness to insulin in various parameters of glucose metabolism (Olefsky, 1976b; Le Marchand-Brustel *et al.*, 1978; Crettaz *et al.*, 1980) (Figure 7.8).

The insulin resistance of obesity attracted the attention of investigators soon after the description of specific insulin receptors in the early 1970s. In several studies, it was shown that insulin binding to liver plasma membranes was reduced in various animal models of obesity, although the binding of insulin to adipocyte membranes was not so clearly diminished in the large

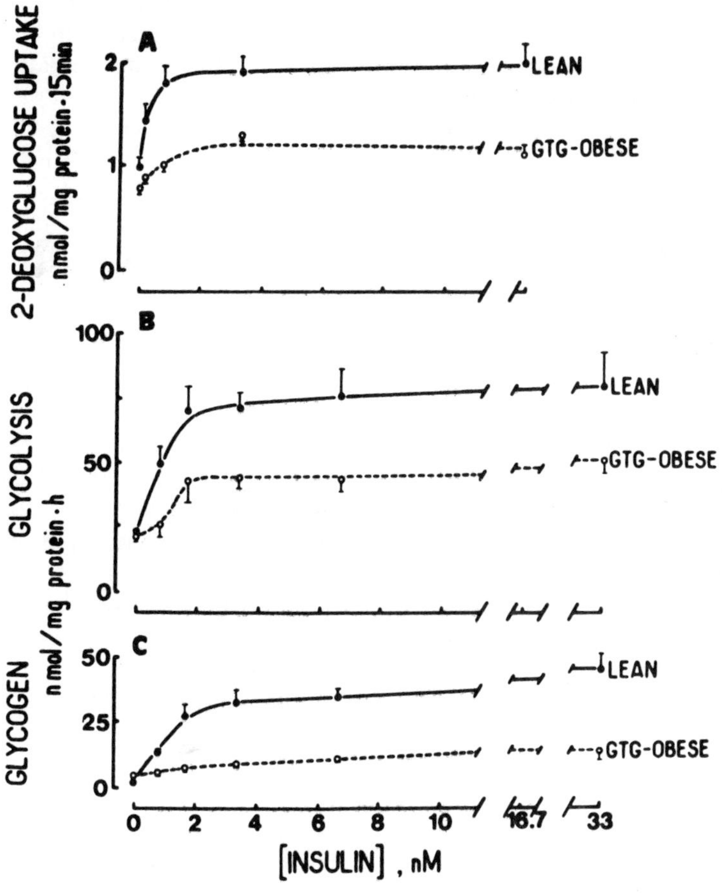

Figure 7.8 Effects of insulin on metabolism of skeletal (soleus) muscles isolated from normal, lean mice (solid points) and obese mice (open circles). The obesity was induced by injection of gold thioglucose ('GTG-obese'), which produces a hypothalamic lesion with effects similar to those described in the text for electrolytic lesions of the hypothalamus. Muscles were incubated *in vitro* with different concentrations of insulin and the following parameters measured: Top panel, 2-deoxyglucose uptake (a measure of glucose transport into the cell and phosphorylation); middle panel, glycolysis as measured by detritiation of ^{3}H-glucose; lower panel, incorporation of labelled glucose into glycogen (a measure of the rate of glycogen synthesis). All three parameters show marked impairment of maximal responsiveness in the obese mice compared with the controls. Reproduced from Le Marchand-Brustel *et al.* (1978) by permission of the American Physiological Society

fat cells of obese rodents or humans (Lockwood *et al.*, 1975; Freychet, 1976; Olefsky, 1976b; Amatruda *et al.*, 1985). Olefsky has suggested (1976a) that even the latter studies can be interpreted as showing a decrease in insulin receptor number per unit area of cell surface, and thus a relative deficiency of insulin receptors. Reduced insulin binding has also been shown in skeletal muscles and in muscle cell membranes and in monocytes from obese rodents or humans (Bar *et al.*, 1976; Freychet, 1976; Olefsky *et al.*, 1976; DeFronzo *et al.*, 1978; Le Marchand-Brustel *et al.*, 1978). The reduction in binding is found, in most studies, to represent a loss of receptor number rather than a diminution in affinity for insulin.

There has been some, albeit rather little, study of the coupling of insulin receptor-binding to intracellular events. Thus, Le Marchand-Brustel *et al.* (1985) showed a diminished activity of the intrinsic insulin receptor tyrosine kinase activity in muscle from obese mice; Jeanrenaud (1988) and Farese and Cooper (1989) have summarized evidence of defects in phospholipid-signalling mechanisms in obesity.

Some studies of post-receptor events in obesity suggest an impairment of the glucose transport process itself (Ciaraldi *et al.*, 1981; Cuendet *et al.*, 1976), but it also seems likely that an impairment of intracellular glucose metabolism is involved. For instance, in a detailed study of 'basal' (non-insulin stimulated) metabolism of soleus muscles from ob/ob genetically obese mice, Cuendet *et al.* (1976) found markedly impaired rates of glycolysis and glucose oxidation, suggesting a fundamental defect in intracellular glucose metabolism. Maximal insulin-stimulated rates of glycolysis, glycogen synthesis and glucose oxidation are reduced, and insulin-suppression of lipid utilization impaired, in muscles from obese animals (Le Marchand-Brustel *et al.*, 1978; Crettaz *et al.*, 1980).

Again, various lines of evidence point to a role for increased fat oxidation in the aetiology of the insulin resistance of obesity. Plasma non-esterified fatty acid (NEFA) concentrations have long been known to be elevated in obese subjects (reviewed by Flatt, 1972). This elevation represents in part simply the increased mass of adipose tissue (their source) (Flatt, 1972), but also a resistance in obesity to the normal suppressive effect of glucose and/or insulin on lipolysis (Felber *et al.*, 1987; Jensen *et al.*, 1989). Increased provision of fatty acids to skeletal muscle and other tissues will have the effects described earlier: increased hepatic glucose production, diminished sensitivity of glucose uptake to insulin, and also—an important aspect which will be discussed further below—increased hepatic secretion of very-low-density lipoprotein triacylglycerol (VLDL-TAG). The rate of fat oxidation, measured in the whole body by indirect calorimetry or in rodent muscles *in vitro*, is high in obesity (Crettaz *et al.*, 1980; Felber *et al.*, 1987), and is inversely correlated with the whole-body rate of glucose oxidation (Felber *et al.*, 1987).

Role of insulin resistance in the development of obesity

The finding of insulin resistance in humans or rodents with developed obesity may imply that insulin resistance causes or predisposes to obesity, but it may equally mean that insulin resistance develops as a consequence of obesity, or that both obesity and insulin resistance develop secondarily to some other underlying change. This question is difficult to answer in man, since it inevitably requires longitudinal studies of subjects during the development of obesity, and, hence, presupposes that 'pre-obese' subjects can be identified. This is more feasible in animal models. Even in rodents, some of the best studied models of obesity—the fa/fa Zucker rat and the ob/ob mouse, for instance—show no characteristic identifying features before the onset of hyperphagia and a gain in body weight. Much of our information on the longitudinal development of obesity has come from the intensive study of a model of obesity in the rat initiated by the production of a lesion in the ventromedial hypothalamus (VMH). Since this procedure is carried out in previously normal animals, the development of obesity and its associated metabolic events may be followed. The findings in this model agree well with observations made in the genetic models of obesity (fa/fa rats, ob/ob mice) (Jeanrenaud, 1985). Much of the evidence from these models points to the initial abnormality being an oversecretion of insulin from the pancreas (Goldman *et al.*, 1972; Le Magnen, 1983; Jeanrenaud, 1985), which Jeanrenaud believes reflects an alteration at the hypothalamic level in the neural control of the endocrine pancreas (Jeanrenaud, 1985; Jeanrenaud *et al.*, 1985; Jeanrenaud, 1988). The tissues of these animals, interestingly, show an initial phase of normal or increased sensitivity to insulin (Frohman *et al.*, 1972; Pénicaud *et al.*, 1986; Debant *et al.*, 1987; Jeanrenaud, 1988), which only changes to diminished sensitivity with the development of the obesity.

It is presumed that hypersecretion of insulin will lead to eventual 'down-regulation' of insulin receptors and loss of sensitivity to insulin. At the same time, other changes probably accentuate the development of both the obesity and the insulin resistance. Several studies indicate that glucose metabolism in adipocytes remains insulin-sensitive, or even hypersensitive, at a stage when other tissues (muscle in particular) are becoming insulin-resistant (Carnie *et al.*, 1979; Debant *et al.*, 1987; Pénicaud *et al.*, 1987). Lipogenesis in adipose tissue is a process which even in normal animals shows considerable sensitivity to insulin, but it appears to be particularly insulin-sensitive in adipose tissue of obese rodents (Carnie *et al.*, 1979; Jeanrenaud, 1985). Hepatic lipogenesis is also increased (Jeanrenaud, 1985), and adipose tissue lipoprotein lipase activity is high (Jeanrenaud, 1985; Jeanrenaud *et al.*, 1985). The store of body fat is thus increased, and as the triacylglycerol content of muscle, in particular, increases, so its insulin sensitivity will decrease (Crettaz *et al.*, 1980; Falholt *et al.*, 1988). The consequent impairment of glucose utilization would lead to further increases in the circulating concentrations of glucose and insulin.

The metabolic picture may be accentuated by changes in other tissues. If glucose uptake by the hypothalamic glucose-sensing centre is reduced by the insulin resistance, or by diversion of glucose into pathways such as lipogenesis, then hyperphagia will be accentuated (Cuendet *et al.*, 1976; Le Magnen, 1983). There is evidence for a considerable diminution in insulin sensitivity of brown adipose tissue (BAT) in obese rodents (Ferré *et al.*, 1986; Tanti *et al.*, 1986; Pénicaud *et al.*, 1987; Cawthorne, 1989). This tissue may be responsible, in rodents, for a significant proportion of glucose utilization, particularly when stimulated post-prandially by insulin (Ferré *et al.*, 1986); impairment of its response may then lead to further diminished glucose utilization, and also to impaired diet-induced thermogenesis, with consequent further increase in body weight. Other evidence also links the post-prandial rise in insulin with diet-induced thermogenesis (Rothwell and Stock, 1981; Felig, 1984), and if this process, whatever its metabolic basis, becomes insulin-resistant, then energy storage will become yet more efficient in the obese subject. The combination of excessive insulin secretion, obesity and insulin resistance is thus a potentially powerful self-perpetuating cycle.

Relevance of insulin resistance in obesity; body fat distribution and the adverse medical consequences of obesity

The animal models discussed above tend to show that impairment of sensitivity to insulin is not the primary change leading to obesity, although it may, once it occurs, accentuate the tendency to excess fat deposition. Why, then, is so much attention focused on insulin resistance in obesity? One answer is that obesity is a condition which predisposes to a number of adverse medical conditions, including particularly non-insulin dependent diabetes mellitus (NIDDM), cardiovascular disease (atherosclerosis) and hypertension (van Itallie, 1985). Insulin resistance is increasingly being seen as the focal point, or even the common link in the aetiology, of these pathological conditions (Reaven, 1988; Kaplan, 1989).

To appreciate this further, it is necessary to realize that obesity is not a uniform condition of excess body weight deposition. The pattern of distribution of the excess fat stores is now recognized to be at least as important as their amount, in terms of the excess cardiovascular mortality associated with obesity. This observation is most clearly seen in the comparison of typical male and female fat distribution, and from the fact that cardiovascular mortality is so much higher in men than in women of similar age. Males tend to accumulate what is often referred to as 'upper body fat', especially at the abdominal level, whilst women tend to have a 'lower body' distribution. These characteristic fat distributions were first studied systematically by Vague more than 40 years ago; he termed them the 'android' (masculine, upper body) and 'gynoid' distributions, and showed that susceptibility to conditions including diabetes, atherosclerosis and gout were associated with the masculine distribution of fat (Vague, 1956). The male pattern is determined by the action of androgens (in conjunction with excess calorie deposition), and the spread of body fat distributions seen amonst both men and women is related to their relative androgen to oestrogen balance; thus, atypical upper-body (android) fat distributions are seen in women with conditions such as idiopathic hirsutism and polycystic ovary disease which are characterized by a relative androgen excess (Kissebah and Peiris, 1989).

What has now become clear is that insulin resistance is associated particularly with 'upper body' obesity. The waist/hip ratio is a good measure of the risks associated with upper body obesity (Krotkiewski *et al.*, 1983; Kissebah and Peiris, 1989). Insulin resistance during an oral glucose tolerance test or as measured by a glucose clamp technique is greater in subjects with high waist–hip ratios than in those with low ratios (Krotkiewski *et al.*, 1983; Haffner *et al.*, 1988; Kissebah and Peiris, 1989) (Figure 7.9).

What is the relevance of the distribution of adipose tissue to the action of insulin, and to the development of adverse medical conditions? Most authors agree that the important feature in upper-body obesity is the accumulation of intra-abdominal fat (which is measured by the waist/hip ratio). The relationship between intra-abdominal, or 'visceral', adipose tissue and waist–hip ratio has been investigated by computed tomography (CT); the two techniques show general agreement (Ashwell *et al.*, 1985; Fujioka *et al.*, 1987; Kissebah and Peiris, 1989), although the CT measurement of visceral/subcutaneous fat correlates even more highly with metabolic complications (Fujioka *et al.*, 1989) than does the simpler anthropometric measure. The metabolic properties of intra-abdominal adipocytes are quantitatively different from those of subcutaneous fat cells; basal and catecholamine-stimulated rates of lipolysis, in particular, are higher in the former, especially in cells from obese subjects (reviewed by Arner, 1989; Kaplan, 1989), perhaps because of a greater density of β-adrenergic receptors on the abdominal cells (Wahrenberg *et al.*, 1989). In addition, some studies suggest that insulin-suppression of lipolysis is less in abdominal than in subcutaneous adipocytes (Bolinder *et al.*, 1983; Björntorp, 1988). Excessive fatty acid liberation into the portal vein from intra-abdominal adipocytes may have a number of effects: increased delivery of fatty acids into the systemic circulation, with the consequences of diminished glucose utilization as described earlier; increased hepatic VLDL-TAG secretion; and decreased insulin extraction. The often-observed link between insulin resistance and hypertriglyceridaemia (Tobey *et al.*, 1981; Haffner *et al.*, 1988; Kaplan,

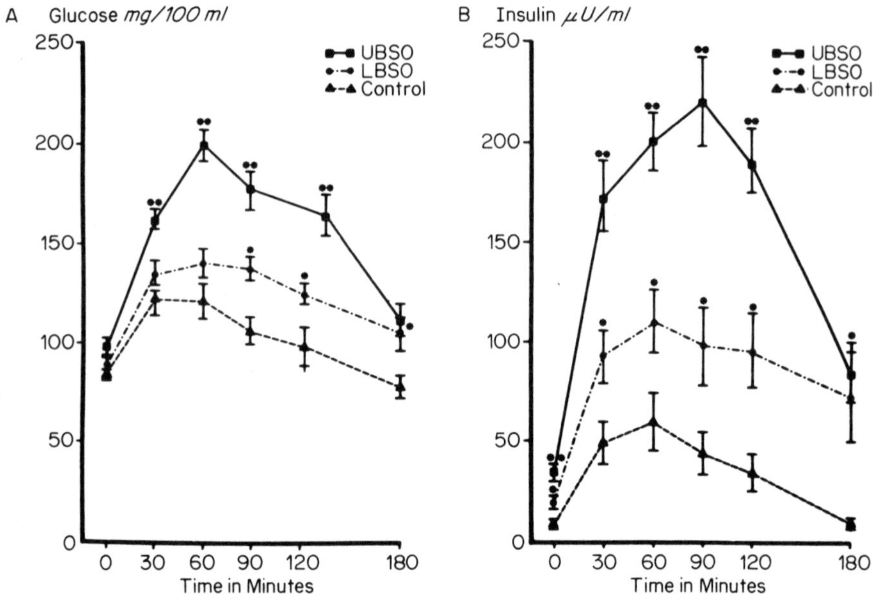

Figure 7.9 The insulin resistance of obesity is related to the distribution of body fat: exaggerated insulin response and impaired glucose tolerance in 'upper body segment obese' (UBSO) subjects compared with 'lower body segment obese' (LBSO) and normal-weight subjects. Reproduced from Kissebah and Peiris (1989) by permission of John Wiley & Sons Inc

1989) is thus explained. The mechanism of the decrease in hepatic insulin extraction is unclear, but it is a consistent finding in upper-body obesity (Björntorp, 1988; Kissebah and Peiris, 1989; Kaplan, 1989), and may be one explanation for the high insulin levels in this condition, although increased insulin secretion (as assessed by plasma C-peptide concentrations) is also seen with increased intra-abdominal fat (Björntorp, 1988; Bergstrom *et al.*, 1990).

Other work, outside the scope of this review, suggests that other components of this picture are the development of hypertension (reviewed by Reaven, 1988; Haffner *et al.*, 1988; Kaplan, 1989), and the development of other changes in lipoprotein metabolism of significance for the pathogenesis of atherosclerosis, particularly a decrease in the protective high-density lipoprotein cholesterol fraction (Haffner *et al.*, 1988; Reaven, 1988) (Figure 7.10). The excess mortality associated with obesity is thus seen as part of a syndrome of upper-body obesity, insulin resistance, hypertriglyceridaemia and hypertension, termed by Reaven 'Syndrome X' (Table 7.2), although it has been pointed out (Lancet, 1989) that it might be more appropriately referred to as 'Reaven's syndrome'.

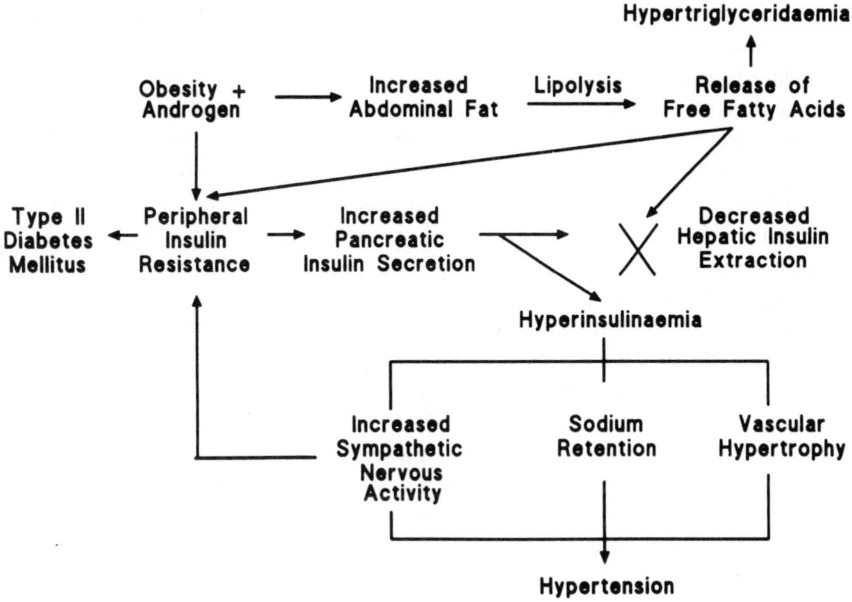

Figure 7.10 Possible aetiology of adverse medical effects of upper-body obesity. The diagram is explained in the text. Based on Kaplan (1989) *Arch. Intern. Med.*, **149**, 1514–20. Copyright 1989 American Medical Association

CONCLUSION

Insulin plays a key role in the maintenance of normal body weight. Impairment of sensitivity to insulin is seen in two major disorders of body weight regulation, the wasting associated with trauma and sepsis, and obesity. It seems important to conclude by asking how it can be that two such dissimilar conditions involve insulin resistance as a common theme. The answer may be that the term 'insulin resistance' is far too vague, and hides completely different underlying causes in the two conditions.

Table 7.2 Characteristics of 'Syndrome X' or Reaven's syndrome

Resistance to insulin-stimulated glucose uptake
Glucose intolerance
Hyperinsulinaemia
Increased very-low-density lipoprotein triacylglycerol
Decreased high-density lipoprotein cholesterol
Hypertension

From Reaven (1988).

In trauma and sepsis, we see a state of hypermetabolism, driven probably by increased sympatho-adrenal activity. All routes of metabolism are accelerated; the liberation of fatty acids, because of increased sympatho-adrenal drive, becomes unresponsive to the anti-lipolytic effect of insulin, and, because fat oxidation tends to 'win' in competition with carbohydrate oxidation—through the operation of part of the glucose–fatty acid cycle—the ability of insulin to stimulate glucose utilization is impaired. Other hormonal changes, and possibly other humoral factors not yet clearly identified such as cytokines, drive other aspects of metabolism such as the net protein breakdown.

In obesity, on the other hand, we see the effects of an increased fat mass dominating the metabolic picture. (It should not be forgotten that, as Flatt (1972) has pointed out, the percentage increase in fat mass when someone becomes obese is far higher than the percentage increase in total body mass.) Whole-body metabolic rate is increased in obese subjects compared with non-obese, but the metabolic activity of adipose tissue per unit weight of tissue is low, so measures such as glucose utilization per unit body weight show a reduction relative to normal subjects. In addition, the increased mass of adipose tissue is liberating increased quantities of fatty acids into the circulation, and these are producing further resistance to the action of insulin. Persistent high insulin concentrations lead to down-regulation of insulin sensitivity in target tissues. In contrast to the situation in sepsis or trauma, sympathetic activity may well be reduced, increasing metabolic efficiency and leading to further tendency to fat storage. Insulin resistance is seen as a response to the accumulation of body fat and is a feature of (in some respects) a 'sluggish' but energy-efficient metabolism, unlike the very active, energy-inefficient metabolic picture in sepsis and trauma.

REFERENCES

Amatruda, J.M., Livingston, J.N. and Lockwood, D.H. (1985). Cellular mechanisms in selected states of insulin resistance: human obesity, glucocorticoid excess, and chronic renal failure. *Diabet./Metab. Rev.*, **1**, 293–317.

Arner, P. (1989). Regulation of lipolysis in different regions of human adipose tissue. In: *Obesity in Europe 88* (eds P.R. Björntorp and S. Rössner) John Libbey, London, pp. 201–7.

Ashwell, M., Cole, T.J. and Dixon, A.K. (1985). Obesity: new insight into the anthropometric classification of fat distribution shown by computed tomography. *Br. Med. J.*, **290**, 1692–4.

Bagdade, J.D., Bierman, E.L. and Porte, D. Jr. (1967). The significance of basal insulin levels in the evaluation of the insulin response to glucose in diabetic and nondiabetic subjects. *J. Clin. Invest.*, **46**, 1549–57.

Banting, F.G. and Best, C.H. (1922). The internal secretion of the pancreas. *J. Lab. Clin. Med.*, **7**, 251–66.

Bar, R.S., Gorden, P., Roth, J., Kahn, C.R. and De Meyts, P. (1976). Fluctuations in the affinity and concentration of insulin receptors on circulating monocytes of obese patients. Effects of starvation, re-feeding, and dieting. *J. Clin. Invest.*, **58**, 1123–35.

Bergstrom, R.W., Newell-Morris, L.L., Leonetti, D.L., Shuman, W.P., Wahl, P.W. and Fujimoto, W.Y. (1990). Association of elevated fasting C-peptide level and increased intra-abdominal fat distribution with development of NIDDM in Japanese-American men. *Diabetes*, **39**, 104–11.

Bessey, P.Q., Watters, J.M., Aoki, T.T. and Wilmore, D.W. (1984). Combined hormonal infusion simulates the metabolic response to injury. *Ann. Surg.*, **200**, 264–81.

Bevilacqua, S., Buzzigoli, G., Bonadonna, R., Brandi, L.S., Oleggini, M., Boni, C., Geloni, M. and Ferrannini, E. (1990). Operation of Randle's cycle in patients with NIDDM. *Diabetes*, **39**, 383–9.

Björntorp, P. (1988). Abdominal obesity and the development of noninsulin-dependent diabetes mellitus. *Diabet./Metab. Rev.*, **4**, 615–22.

Black, P.R., Brooks, D.C., Bessey, P.Q., Wolfe, R.R. and Wilmore, D.W. (1982). Mechanisms of insulin resistance following injury. *Ann. Surg.*, **196**, 420–35.

Bliss, M. (1983). *The Discovery of Insulin*, Paul Harris, Edinburgh.

Bolinder, J., Kager, L., Östman, J. and Arner, P. (1983). Differences at the receptor and postreceptor levels between human omental and subcutaneous adipose tissue in the action of insulin on lipolysis. *Diabetes*, **32**, 117–9.

Brooks, D.C., Bessey, P.Q., Black, P.R., Aoki, T.T. and Wilmore, D.W. (1984). Post-traumatic insulin resistance in uninjured forearm tissue. *J. Surg. Res.*, **37**, 100–7.

Cahill, G.F. (1988). Beta-cell deficiency, insulin resistance, or both? *N. Engl. J. Med.*, **318**, 1268–70.

Carnie, J.A., Smith, D.G. and Mavris-Vavayannis, M. (1979). Effects of insulin on lipolysis and lipogenesis in adipocytes from genetically obese (ob/ob) mice. *Biochem. J.*, **184**, 107–12.

Cawthorne, M.A. (1989). Does brown adipose tissue have a role to play in glucose homeostasis? *Proc. Nutr. Soc.*, **48**, 207–14.

Ciaraldi, T.P., Kolterman, O.G. and Olefsky, J.M. (1981). Mechanism of the postreceptor defect in insulin action in human obesity. Decrease in glucose transport system activity. *J. Clin. Invest.*, **68**, 875–80.

Crettaz, M., Prentki, M., Zaninetti, D. and Jeanrenaud, B. (1980). Insulin resistance in soleus muscle from obese Zucker rats. Involvement of several defective sites. *Biochem. J.*, **186**, 525–34.

Cuatrecasas, P. (1969). Interaction of insulin with the cell membrane: the primary action of insulin. *Proc. Natl Acad. Sci. USA*, **63**, 450–7.

Cuatrecasas, P. (1973). Insulin receptor of liver and fat cell membranes. *Fed. Proc.*, **32**, 1838–46.

Cuendet, G.S., Loten, E.G., Jeanrenaud, B. and Renold, A.E. (1976). Decreased basal, noninsulin-stimulated glucose uptake and metabolism by skeletal soleus muscle isolated from obese-hyperglycemic (ob/ob) mice. *J. Clin. Invest.*, **58**, 1078–88.

Debant, A., Guerre-Millo, M., Le Marchand-Brustel, Y., Freychet, P., Lavau, M., and van Obberghen, E. (1987). Insulin receptor kinase is hyperresponsive in adipocytes of young obese Zucker rats. *Am. J. Physiol.*, **252**, E273–8.

DeFronzo, R.A., Soman, V., Sherwin, R.S., Hendler, R. and Felig, P. (1978). Insulin binding to monocytes and insulin action in human obesity, starvation and refeeding. *J. Clin. Invest.*, **62**, 204–13.

DeFronzo, R.A., Tobin, J. and Andres, R. (1979). Glucose clamp technique: a method for quantifying insulin secretion and resistance. *Am. J. Physiol.*, **237**, E214–23.

Dolkas, C.B. and Greenleaf, J.E. (1977). Insulin and glucose responses during bed rest with isotonic and isometric exercise. *J. Appl. Physiol.*, **43**, 1033–8.

Elwyn, D.H., Kinney, J.M., Jeevanandam, M., Gump, F.E. and Broell, J.R. (1979). Influence of increasing carbohydrate intake on glucose kinetics in injured patients. *Ann. Surg.*, **190**, 117–27.

Espinal, J. (1987). Mechanism of insulin action. *Nature*, **328**, 574–5.

Evans, E.I. and Butterfield, W.J.H. (1951). The stress response in the severely burned. *Ann. Surg.*, **134**, 588–613.

Fajans, S.S., Floyd, J.C., Knopf, R.F. and Conn, J.E. (1967). Effect of amino acids and proteins on insulin secretion in man. *Rec. Prog. Horm. Res.*, **23**, 617–62.

Falholt, K., Jensen, I., Lindkaer Jensen, S., Mortensen, H., Vølund, Aa., Heding, L.G., Petersen, P.N. and Falholt, W. (1988). Carbohydrate and lipid metabolism of skeletal muscle in type 2 diabetic patients. *Diab. Med.*, **5**, 27–31.

Farese, R.V. and Cooper, D.R. (1989). Potential role of phospholipid-signaling systems in insulin action and states of clinical insulin resistance. *Diabet./Metab. Rev.*, **5**, 455–74.

Felber, J.-P., Ferrannini, E., Golay, A., Meyer, H.U., Theibaud, D., Curchod, B., Maeder, E., Jequier, E. and DeFronzo, R.A. (1987). Role of lipid oxidation in pathogenesis of insulin resistance of obesity and type II diabetes. *Diabetes*, **36**, 1341–50.

Felig, P. (1984). Insulin is the mediator of feeding-related thermogenesis: insulin resistance and/or deficiency results in a thermogenic defect which contributes to the pathogenesis of obesity. *Clin. Physiol.*, **4**, 267–73.

Fellows, I.W. and Woolfson, A.M.J. (1985). Effects of therapeutic intervention on the metabolic responses to injury. *Br. Med. Bull.*, **41**, 287–94.

Ferré, P., Burnol, A.-F., Leturque, A., Terrataz, J., Penicaud, L., Jeanrenaud, B. and Girard, J. (1986). Glucose utilization *in vivo* and insulin-sensitivity of rat brown adipose tissue in various physiological and pathological conditions. *Biochem. J.*, **233**, 249–52.

Flatt, J.-P. (1972). Role of the increased adipose tissue mass in the apparent insulin insensitivity of obesity. *Am. J. Clin. Nutr.*, **25**, 1189–92.

Frayn, K.N. (1986). Hormonal control of metabolism in trauma and sepsis. *Clin. Endocrinol.*, **24**, 577–99.

Frayn, K.N. (1988). L'insulino-résistance en nutrition artificielle. *Nutr. Clin. Métabol.*, **2**, 76–82.

Frayn, K.N., Le Marchand-Brustel, Y. and Freychet, P. (1978). Studies on the mechanism of insulin resistance after injury in the mouse. *Diabetologia*, **14**, 337–41.

Frayn, K.N., Little, R.A., Stoner, H.B. and Galasko, C.S.B. (1984a). Metabolic control in non-septic patients with musculoskeletal injuries. *Injury*, **16**, 73–9.

Frayn, K.N., Price, D.A., Maycock, P.F. and Carroll, S.M. (1984b). Plasma somatomedin activity after injury in man and its relationship to other hormonal and metabolic changes. *Clin. Endocrinol.*, **20**, 179–87.

Freychet, P. (1976). Interactions of polypeptide hormones with cell membrane specific receptors: studies with insulin and glucagon. *Diabetologia*, **12**, 83–100.

Freychet, P., Roth, J. and Neville, D.M. (1971). Insulin receptors in the liver: specific binding of [^{125}I]insulin to the plasma membrane and its relation to insulin bioactivity. *Proc. Natl Acad. Sci. USA*, **68**, 1833–7.

Frohman, L.A., Goldman, J.K. and Bernardis, L.L. (1972). Studies of insulin sensitivity *in vivo* in weanling rats with hypothalamic obesity. *Metabolism*, **21**, 1133–42.

Fujioka, S., Matsuzawa, Y., Tokunaga, K., Kawamoto, T., Kobatake, T., Keno, Y. and Tarui, S. (1989). Comparison of a novel classification of obesity (visceral fat obesity and subcutaneous fat obesity) with previous classifications of obesity concerning body features or adipose tissue cellularity. In: *Obesity in Europe 88* (eds P.R. Björntorp and S. Rössner), John Libbey, London, pp. 85–9.

Fujioka, S., Matsuzawa, Y., Tokunaga, K. and Tarui, S. (1987). Contribution of intra-abdominal fat accumulation to the impairment of glucose and lipid metabolism in human obesity. *Metabolism*, **36**, 54–9.

Gammeltoft, S. (1984). Insulin receptors: binding kinetics and structure–function relationship of insulin. *Physiol. Rev.*, **64**, 1321–78.

Garvey, W.T. and Kolterman, O.G. (1988). Correlation of *in vivo* and *in vitro* actions of insulin in obesity and noninsulin-dependent diabetes mellitus: role of the glucose transport system. *Diabet./Metab. Rev.*, **4**, 543–69.

Gelfand, R.A., Matthews, D.E., Bier, D.M. and Sherwin, R.S. (1984). Role of counterregulatory hormones in the catabolic response to stress. *J. Clin. Invest.*, **74**, 2238–48.

Giddings, A.E.B., Mangnall, D., Rowlands, B.J. and Clark, R.G. (1977). Plasma insulin and surgery 1. Early changes due to operation in the insulin response to glucose. *Ann. Surg.*, **186**, 681–6.

Goldman, J.K., Bernardis, L.L. and Frohman, L.A. (1972). Insulin responsiveness *in vitro* of diaphragm and adipose tissue from weanling rats with hypothalamic obesity. *Horm. Metab. Res.*, **4**, 328–31.

Groop, L.C., Bonadonna, R.C., DelPrato, S., Ratheiser, K., Zyck, K., Ferrannini, E. and DeFronzo, R.A. (1989). Glucose and free fatty acid metabolism in non-insulin-dependent diabetes mellitus. *J. Clin. Invest.*, **84**, 205–13.

Gump, F.E., Long, C., Killian, P. and Kinney, J.M. (1974). Studies of glucose intolerance in septic injured patients. *J. Trauma*, **14**, 378–88.

Haffner, S.M., Fong, D., Hazuda, H.P., Pugh, J.A. and Patterson, J.K. (1988). Hyperinsulinemia, upper body adiposity, and cardiovascular risk factors in non-diabetics. *Metabolism*, **37**, 338–45.

Henderson, A.A., Frayn, K.N., Little, R.A. and Galasko, C.S.B. (1991). Dose-response relationships for the effects of insulin on glucose and fat metabolism in injured patients and control subjects. *Clin. Sci.*, **80**, 25–32.

Himsworth, H.P. and Kerr, R.B. (1939). Insulin-sensitive and insulin-insensitive types of diabetes mellitus. *Clin. Sci.*, **4**, 119–52.

Hinton, P., Allison, S.P., Littlejohn, S. and Lloyd, J. (1971). Insulin and glucose to reduce catabolic response to injury in burned patients. *Lancet*, **1**, 767–9.

Howard, J.M. (1955). Studies of the absorption and metabolism of glucose following injury. The systemic response to injury. *Ann. Surg.*, **141**, 321–6.

Hutton, J.C. (1989). The insulin secretory granule. *Diabetologia*, **32**, 271–81.

van Itallie, T.B. (1985). Health implications of overweight and obesity in the United States. *Ann. Intern. Med.*, **103**, 983–88.

Jeanrenaud, B. (1985). An hypothesis on the aetiology of obesity: dysfunction of the central nervous system as a primary cause. *Diabetologia*, **28**, 502–13.

Jeanrenaud, B. (1988). Neuroendocrine and metabolic basis of type II diabetes as studied in animal models. *Diabet./Metab. Rev.*, **4**, 603–14.

Jeanrenaud, B., Halimi, S. and van de Werve, G. (1985). Neuro-endocrine disorders seen as triggers of the triad: obesity-insulin resistance-abnormal glucose tolerance. *Diabet./Metab. Rev.*, **1**, 261–91.

Jensen, M.D., Haymond, M.W., Rizza, R.A., Cryer, P.E. and Miles, J.M. (1989).

Influence of body fat distribution on free fatty acid metabolism in obesity. *J. Clin. Invest.*, **83**, 1168–73.

Johnson, I.F. (1982). Authenticity and purity of human insulin (recombinant DNA). *Diabetes Care*, **5**, (Suppl. 2) 4–12.

Kahn, C.R. (1978). Insulin resistance, insulin insensitivity, and insulin unresponsiveness: a necessary distinction. *Metabolism*, **27**, (Suppl. 2), 1893–902.

Kaplan, N.M. (1989). The deadly quartet. Upper-body obesity, glucose intolerance, hypertriglyceridemia, and hypertension. *Arch. Intern. Med.*, **149**, 1514–20.

Karam, J.H., Grodsky, G.M. and Forsham, P.H. (1963). Excessive insulin response to glucose in obese subjects as measured by immunochemical assay. *Diabetes*, **12**, 197–204.

Kehlet, H. and Schulze, S. (1986). Modification of injury response. In: *The Scientific Basis for the Care of the Critically Ill* (eds R.A. Little and K.N. Frayn) Manchester University Press, Manchester, pp. 153–68.

Kinney, J.M., Long, C.L., Gump, F.E. and Duke, J.H. (1968). Tissue composition of weight loss in surgical patients. 1. Elective operation. *Ann. Surg.*, **168**, 459–74.

Kissebah, A.H. and Peiris, A.N. (1989). Biology of regional body fat distribution: relationship to non-insulin-dependent diabetes mellitus. *Diabet./Metab. Rev.*, **5**, 83–109.

Krotkiewski, M., Björntorp, P., Sjöström, L. and Smith, U. (1983). Impact of obesity on metabolism in men and women. Importance of regional adipose tissue distribution. *J. Clin. Invest.*, **72**, 1150–62.

Lancet. (1989). Type 2 diabetes or NIDDM: looking for a better name. *Lancet*, **1**, 589–91.

Le Marchand-Brustel, Y., Jeanrenaud, B. and Freychet, P. (1978). Insulin binding and effects in isolated soleus muscle of lean and obese mice. *Am. J. Physiol.*, **234**, E348–58.

Le Marchand-Brustel, Y., Grémeaux, T., Ballotti, R., and van Obberghen, E. (1985). Insulin receptor tyrosine kinase is defective in skeletal muscle of insulin-resistant obese mice. *Nature*, **315**, 676–9.

Le Magnen, J. (1983). Body energy balance and food intake: a neuroendocrine regulatory mechanism. *Physiol. Rev.*, **63**, 314–86.

Levine, R., Goldstein, M., Klein, S. and Huddlestun, B. (1949). The action of insulin on the distribution of galactose in eviscerated nephrectomized dogs. *J. Biol. Chem.*, **179**, 985–6.

Lockwood, D.H., Livingston, J.N. and Amatruda, J.M. (1975). Relation of insulin receptors to insulin resistance. *Fed. Proc.*, **34**, 1564–9.

Long, C.L., Kinney, J.M. and Geiger, J.W. (1976). Nonsuppressability of gluconeogenesis by glucose in septic patients. *Metabolism*, **25**, 193–201.

Nordenström, J., Jeevanandam, M., Elwyn, D.H., Carpentier, Y.A., Askanazi, J., Robin, A. and Kinney, J.M. (1981). Increasing glucose intake during total parenteral nutrition increases norepinephrine excretion in trauma and sepsis. *Clin. Physiol.*, **1**, 525–34.

Nordenström, J., Carpentier, Y.A., Askanazi, J., Robin, A.P., Elwyn, D.H., Hensle, T.W. and Kinney, J.M. (1983). Free fatty acid mobilization and oxidation during total parenteral nutrition in trauma and infection. *Ann. Surg.*, **198**, 725–35.

Olefsky, J.M. (1976a). The insulin receptor: its role in insulin resistance of obesity and diabetes. *Diabetes*, **25**, 1154–62.

Olefsky, J.M. (1976b). The effects of spontaneous obesity on insulin binding, glucose transport, and glucose oxidation of isolated rat adipocytes. *J. Clin. Invest.*, **57**, 842–51.

Olefsky, J.M. (1981). Insulin resistance and insulin action. An *in vitro* and *in vivo* perspective. *Diabetes*, **30**, 148–62.
Olefsky, J., Bacon, V.C. and Baur, S. (1976). Insulin receptors of skeletal muscle: specific insulin binding sites and demonstration of decreased numbers of sites in obese rats. *Metabolism*, **25**, 179–91.
Olefsky, J.M., Kolterman, O.G. and Scarlett, J.A. (1982). Insulin action and resistance in obesity and noninsulin-dependent type II diabetes mellitus. *Am. J. Physiol.*, **243**, E15–30.
Pénicaud, L., Rohner-Jeanrenaud, F. and Jeanrenaud, B. (1986). *In vivo* metabolic changes as studied longitudinally after ventromedial hypothalamic lesions. *Am. J. Physiol.*, **250**, E662–8.
Pénicaud, L., Ferré, P., Terrataz, J., Kinebanyan, M.F., Leturque, A., Doré, E., Girard, J., Jeanrenaud, B. and Picon, L. (1987). Development of obesity in Zucker rats. Early insulin resistance in muscles but normal sensitivity in white adipose tissue. *Diabetes*, **36**, 626–31.
Ponting, G.A., Halliday, D., Teale, J.D. and Sim, A.J.W. (1988). Postoperative positive nitrogen balance with intravenous hyponutrition and growth hormone. *Lancet*, **1**, 438–40.
Rabinowitz, D. and Zierler, K.L. (1961). Forearm metabolism in obesity and its response to intra-arterial insulin. Evidence for adaptive hyperinsulinism. *Lancet*, **1**, 690–2.
Randle, P.J., Garland, P.B., Hales, C.N. and Newsholme, E.A. (1963). The glucose fatty-acid cycle. Its role in insulin sensitivity and the metabolic disturbances of diabetes mellitus. *Lancet*, **1**, 785–9.
Reaven, G.M. (1988). Role of insulin resistance in human disease. *Diabetes*, **37**, 1595–607.
Ross, H., Johnston, I.D.A., Welborn, T.A. and Wright, A.D. (1966). Effect of abdominal operation on glucose tolerance and serum levels of insulin, growth hormone, and hydrocortisone. *Lancet*, **2**, 563–6.
Rothwell, N.J. and Stock, M.J. (1981). A role for insulin in the diet-induced thermogenesis of cafeteria-fed rats. *Metabolism*, **30**, 673–8.
Ryle, A.P., Sanger, F., Smith, L.F. and Kitai, R. (1955). The disulphide bonds of insulin. *Biochem. J.*, **60**, 541–56.
Sassetzky, N.A. (1883). Über den Einfluss fieberhafter Zustände und antipyretischer Behandlung auf den Umsatz der stickstoffhaltigen Substanzen und die Assimilation stickstoffhaltiger Bestandtheile der Milch. *Virchow's Arch.*, **94**, 485–541.
Soerjodibroto, W.S., Heard, C.R.C., James, W.P.T., Few, J.D. and Bloom, S.R. (1977). Metabolic and hormonal changes after surgery: hyperinsulinaemia during glucose infusion. *Eur. J. Clin. Invest.*, **7**, 579–86.
Tanti, J.-F., Grémeaux, T., Brandenburg, D., van Obberghen, E. and Le Marchand-Brustel, Y. (1986). Brown adipose tissue in lean and obese mice. Insulin-receptor binding and tyrosine kinase activity. *Diabetes*, **35**, 1243–8.
Taylor, R. and Agius, L. (1988). The biochemistry of diabetes. *Biochem. J.*, **250**, 625–40.
Thiébaud, D., DeFronzo, R.A., Jacot, E., Golay, A., Acheson, K., Maeder, E.J., Jéquier, E. and Felber, J.-P. (1982). Effect of long chain triglyceride infusion on glucose metabolism in man. *Metabolism*, **31**, 1128–36.
Tobey, T.A., Greenfield, M., Kraemer, F. and Reaven, G.M. (1981). Relationship between insulin resistance, insulin secretion, very low density lipoprotein kinetics, and plasma triglyceride levels in normotriglyceridemic man. *Metabolism*, **30**, 165–71.

Vague, J. (1956). The degree of masculine differentiation of obesities: a factor determining predisposition to diabetes, atherosclerosis, gout and uric calculous disease. *Am. J. Clin. Nutr.*, **4**, 20–34.

de Vasconcelos, P.R.L., Kettlewell, M.G.W. and Williamson, D.H. (1987). Time course of changes in hepatic metabolism in response to sepsis in the rat: impairment of gluconeogenesis and ketogenesis *in vitro*. *Clin. Sci.*, **72**, 683–91.

Vernikos, J. (1986). Metabolic and endocrine changes. In: *Inactivity: Physiological Effects* (eds H. Sandler and J. Vernikos) Academic Press, Orlando, pp. 99–121.

Wahrenberg, H., Lönnqvist, F. and Arner, P. (1989). Mechanisms underlying regional differences in lipolysis in human adipose tissue. *J. Clin. Invest.*, **84**, 458–67.

Waldhäusl, W., Bratusch-Marrain, P., Gasic, S., Korn, A. and Nowotny, P. (1982). Insulin production rate, hepatic insulin retention and splanchnic carbohydrate metabolism after oral glucose ingestion in hyperinsulinaemic type 2 (non-insulin dependent) diabetes mellitus. *Diabetologia*, **23**, 6–15.

White, R.H., Frayn, K.N., Little, R.A., Threlfall, C.J., Stoner, H.B. and Irving, M.H. (1987). Hormonal and metabolic responses to glucose infusion in sepsis studied by the hyperglycemic glucose clamp technique. *JPEN*, **11**, 345–53.

Wilmore, D.W. (1986). The wound as an organ. In: *The Scientific Basis for the Care of the Critically Ill* (eds R.A. Little and K.N. Frayn) Manchester University Press, Manchester, pp. 45–59.

Wilmore, D.W., Long, J.M., Mason, A.D., Skreen, R.W. and Pruitt, B.A. (1974). Catecholamines. Mediator of the hypermetabolic response to thermal injury. *Ann. Surg.*, **180**, 653–69.

Wolfe, R.R., Durkot, M.J., Allsop, J.R. and Burke, J.F. (1979). Glucose metabolism in severely burned patients. *Metabolism*, **28**, 1031–9.

Chapter Eight

Cancer Cachexia

M.J. Tisdale
Pharmaceutical Sciences Institute, Aston University, Birmingham B4 7ET, UK

INTRODUCTION

Weight loss is one of the major characteristics of the cancer patient and is an important cause of mortality. The cachectic syndrome, comprising anorexia and nausea, anaemia, muscle weakness, malabsorption and diarrhoea, altered host metabolism and weight loss varies with tumour type, ranging from 8 to 84%, with gastrointestinal and lung cancers having the greatest incidence (Strain, 1979). The tumour types most frequently associated with weight loss also have the greatest degree of weight loss (>10%) and these patients also have a shorter survival time (De Wys, 1985). Pretreatment weight loss is also associated with a lower frequency of response to chemotherapy. Reversal of the cachectic process might also be expected not only to improve the quality of life of the cancer patient, but also increase the overall survival.

Weight loss can develop through an imbalance between input and expenditure of calories. Although anorexia is invariably present in cancer patients, interestingly some studies have not been able to demonstrate a significant reduction in caloric intake in weight losing cancer patients. Thus the mean daily energy intake in a group of cancer patients, who were below normal body weight and body cell mass, was not significantly different from that of a group of healthy subjects within the same age range, although the intake of the cancer patients varied significantly (Warnold *et al.*, 1978). Similar observations were made in a small study comparing the energy intake in a group of cancer patients with those suffering from anorexia

Obesity and Cachexia. Edited by N.J. Rothwell and M.J. Stock

nervosa with an equivalent weight loss (Heymsfield and McManus, 1985). However, in the former study both the daily energy expenditure and the resting metabolic rate were significantly greater in the cancer patients than in the controls. Thus these cancer-bearing individuals appear to be maintaining an elevated metabolic rate despite a reduction in the size of their metabolizing tissues.

Additional studies have shown that anorexia is only a partial cause of the wasting process. In undernourished cancer patients, the measured energy intake fails to correlate with the degree of protein-calorie undernutrition (Burke *et al.*, 1980) and in both rodent and man, breakdown of both muscle and adipose tissue frequently precedes a detectable fall in food intake (Costa, 1963). Also tumour-bearing rats undergo an increased loss of both muscle and adipose mass compared with pair-fed controls (Garattini *et al.*, 1980). These results suggest that metabolic disturbances within the tumour or host tissues also contribute to the loss of body mass by increasing energy requirements. This could explain the clinical observation that cancer patients fail to gain weight with levels of nutritional therapy which would cause weight gain in cancer-free malnourished patients, and why total parenteral nutrition has failed to show a uniform survival benefit (Burt *et al.*, 1981).

MODELS OF CACHEXIA

Although a large amount of research has been conducted into the mechanism(s) of cancer cachexia there is little definitive data as to its cause. Part of the problem in determining mechanism has been the lack of appropriate animal models of the condition. Thus most animal tumours have been selected to test the effectiveness of antiproliferative agents and are rapidly growing. Weight loss is usually not observed until the tumour reaches a large proportion of total body weight and the animal is near to death, and arises solely from anorexia. Perhaps the only two models which can be considered appropriate to the human situation are both in mice. A hypernephroma from a patient with established weight loss has been established in immuno-suppressed CBA/Lac mice as a non-metastasizing transplantable xenograft (Strain *et al.*, 1980) and a dimethylhydrazine-induced colonic adenocarcinoma (MAC16) has been passaged in NMRI mice (Bibby *et al.*, 1987). In both cases, weight loss occurs when the tumour comprises less than 5% of the total body weight and is not associated with anorexia. In fact anorexia associated with the growth of two rapidly growing murine tumours was not accompanied by body weight loss, possibly due to the rapid expansion of tumour size. Xenografts from patients without weight loss grew to a much larger size than the hypernephroma, but produced no significant weight loss in identical mice. These results confirm the clinical observation that the degree of cachexia bears no simple correlation to tumour burden or anatomical site of involvement, since individual patients may have widespread metastasis with no recognizable cachexia (Costa, 1977). In fact carcass lipid depletion in tumour-bearing nude mice bearing human xenografts has been shown to be a function of tumour type rather than tumour burden (Hollander *et al.*, 1986).

The MAC16 adenocarcinoma also has a slower growth rate than other tumours of the same series which do not produce weight loss (Beck and Tisdale, 1987). However, weight loss is directly proportional to tumour mass when the tumour weight exceeds 100 mg. In this model there is a disproportionate loss of host body tissue compared with the tumour mass such that a 600 mg tumour in a 30 g male mouse causes a weight loss of 10 g, comprising loss of both adipose and muscle mass, without a change in the total body water content (Bibby *et al.*, 1987). This suggests that weight loss is not due to a direct competition between the host and the tumour for available nutrients. Excision of the tumour results in cessation of weight loss with subsequent weight gain upon recovery from surgery, showing that cachexia is a reversible process. Although hypoglycaemia is evident in cachectic animals associated with a reduced plasma insulin level, preliminary results suggest that this arises from a production of insulin-like growth factor (IGFII) (McDevitt and Tisdale, unpublished results) by the tumour, and is not related to the weight loss. Loss of muscle mass is associated with an elevated urinary nitrogen excretion, when the weight loss does not exceed about 10%, but above this level there is a conservation of nitrogen and the excretion level fell to or below that found in non-tumour bearing animals (Beck and Tisdale, 1989a). The presence of the tumour alone appears to be insufficient to account for the elevated nitrogen excretion, since animals bearing a related colon adenocarcinoma (MAC13) that did not induce weight loss had a nitrogen excretion pattern similar to that of non-tumour-bearing controls. These results all suggest that cachexia may be mediated by a humoral factor elaborated either by the tumour or by host tissues.

POTENTIAL MEDIATORS OF HOST CATABOLISM

Considerable interest has recently been directed towards a macrophage product tumour necrosis factor alpha (TNF), which has been shown to be homologous to cachectin (Beutler *et al.*, 1985) the mediator of cachexia in animals infected with trypanosomes (Rouzer and Cerami, 1980). Tumour cells transfected with the human TNF/cachectin gene have been shown to produce weight loss in recipient animals (Oliff *et al.*, 1987). Weight loss in rodents upon TNF administration is typified by a marked anorexia, which is directly proportional to the drop in body weight (Mahony *et al.*, 1988), and thus differs from that found in both the hypernephroma and MAC16 models of cachexia. Indeed in the MAC16 model no elevated serum or tumour levels of TNF were detectable, even after endotoxin administration. Anticachectin antibodies delayed, but did not fully prevent the anorexia associated with a methylcholanthrene-induced sarcoma in mice (Sherry *et al.*, 1989). While apparent reversal of loss of some body weight was seen, two important factors were not taken into consideration. First, animals administered the anti-TNF antibodies were eating more than sham-injected controls at the time body weight and body composition were measured, and, second, the anticachectin treatment reduced tumour growth, and catabolism of host body compartments was reduced in proportion to the reduction in tumour weight, since it is known that weight loss is proportional to tumour weight. In mice bearing the Lewis lung carcinoma anticachectin immunoglobulin did not alter food intake, change in body weight or final tumour weight (Sherry *et al.*, 1989). It did, however, partially reverse the depletion of body lipids and rise in serum triglycerides. These results suggest that if TNF is involved in weight loss in some experimental models, then other factors are also present to explain all of the features of cachexia.

The role of TNF in human cancer cachexia is also questionable. Thus while acute administration of TNF to cancer patients has been shown to stimulate certain of the metabolic disturbances seen in the presence of a malignant tumour, such as a rise in C-reactive protein, reduction in serum zinc and an increased efflux of total amino acids from the forearm (Warren *et al.*, 1987), chronic administration of TNF is not associated with an accelerated cachexia, although anorexia may be present during the administration (Sherman *et al.*, 1988). Most studies with cancer patients have reported undetectable serum levels of TNF (Selby *et al.*, 1987), even in patients with clinical cancer cachexia that had lost 8 to 40% of their pre-morbid weight (Socher *et al.*, 1988). However, elevated levels of TNF have been detected in children with malignancy (Saarinen *et al.*, 1990) and a study with an ELISA technique employing a single monoclonal antibody deteced an elevated 'TNF-like activity' in 50% of the serum samples from cancer patients with active disease (Balkwell *et al.*, 1987). In both cases this appeared to be related to the tumour-bearing state rather than the development of cachexia, since in neither of the studies was the level of TNF related to the weight loss.

Thus if TNF is not able to account fully for the manifestations of clinical cancer cachexia then there may be other factors which are responsible for the catabolism of host tissues. Most of the other studies have been concerned with tumour factors capable of directly catabolizing adipose tissue. Evidence that neoplastic cells are capable of elaborating a lipid mobilizing substance was first provided by the injection of non-viable preparations of Krebs-2 carcinoma into male Swiss mice, where a decrease in carcass fat similar to that produced by viable preparations of these tumour cells was observed (Costa and Holland, 1966). Kitada *et al.*, (1980) also reported that both the serum of mice bearing a thymic lymphoma, as well as extracts of the tumour, when injected into AKR mice, with ^{14}C-fatty acid labelled adipose tissue, caused an enhanced release of $^{14}CO_2$. Such activity was also obtained by the injection of samples of the culture medium from the lymphoma cell culture, suggesting that it was tumour rather than host derived. Initial characterization of the lipid mobilizing factor indicated that it was a heat stable acidic protein of molecular weight about 5000 (Kitada *et al.*, 1981), although later studies (Kitada *et al.*, 1982) suggested the active substance to be of a much higher molecular weight, and to be formed by ageing of the extracts for several days at low temperature. Unfortunately, further characterization of this material has not been presented.

A similar lipid mobilizing factor with a molecular weight of 65 000 to 75 000 and pI 4.7, has been detected in the pleural fluid of patients with malignant lymphoma and the ascites fluids from patients with ovarian tumours or hepatomas (Masuno *et al.*, 1981). In addition to the direct catabolic effect, injection of this substance, called toxohormone-L, into the lateral ventricle of rats significantly suppressed food and water intake (Masuno *et al.*, 1984). These results all suggest that tumour cells may release a lipid mobilizing substance which contributes to the loss of carcass fat in tumour-bearing animals.

Weight loss in animals bearing the MAC16 adenocarcinoma is also associated with an elevated plasma lipolytic activity, as measured by the release of free fatty acids or glycerol from murine epididymal adipocytes (Beck and Tisdale, 1987). Related colon adenocarcinomas such as the MAC13, also display some lipolytic activity, but the activity level is only about one-tenth of that found in the MAC16 tumour, and plasma levels of this material are not elevated. This suggests that mobilization of host lipids may be a quantitative phenomenon with all tumours having the capacity to secrete lipid mobilizing factors, and would explain why depletion of host lipids is a function of tumour type rather than tumour burden (Hollander *et al.*, 1986). The presence of such materials in other tumours suggests a possible requirement for host components, and in particular lipids by the tumour, since it is known that a number of tumours have a limited ability to synthesize fatty acids, and obtain a substantial amount preformed by the host (Spector and Burns, 1987). Thus, the ability of tumours to elaborate a lipolytic factor may be essential for their growth and reproduction, in which case inhibition of such activity might be expected to lead to tumour regression.

Preliminary results (Beck and Tisdale, unpublished results) suggest that the lipid mobilizing factor associated with the MAC16 colon adenocarcinoma is acidic in character as for other tumour factors, but of low molecular weight. Similar activity is also found in the serum of patients with clinical cancer cachexia, but is absent from non-tumour-bearing controls, even after acute starvation. The homology between the murine and human materials awaits the structural elucidation.

ROLE AND REVERSAL OF LIPOLYTIC ACTIVITY

The role of lipid mobilizing factors in the cachectic state can be inferred from two types of experiments. First, injection of partially purified material into tumour-bearing animals without weight loss, causes a loss of body weight (Beck *et al.*, 1990). Second, inhibitors of the lipid mobilizing factor *in vitro* should also be inhibitors of cachexia *in vivo*. The lipid mobilizing factor from the MAC16 tumour is inhibited *in vitro* by both insulin and 3-hydroxybutyrate and is thus subject to normal metabolic control (Beck and Tisdale, 1987). Animals bearing the MAC16 tumour fed an isocaloric, isonitrogenous diet in which part of the carbohydrate calories were replaced by medium chain triglycerides (MCT), a potent generator of 3-hydroxybutyrate *in vivo*, had a reduction in both the extent of weight loss and tumour weight (Tisdale *et al.*, 1987), and both the nitrogen balance and urea excretion were restored to that of non tumour-bearing animals (Beck and Tisdale, 1989a). The increase in carcass weight was attributed to an increase in both the fat and non-fat carcass mass. A small clinical trial employing five patients with malignant disease and severe weight loss (mean 32%) also showed the benefits of MCT. Thus a highly significant weight reversal was observed when they were changed to an isocaloric, isonitrogenous diet in which 70% of their calories were supplied as MCT, plus additional 3-hydroxybutyrate (Fearon *et al.*, 1988). Since weight gain in both cases was not associated with an increased caloric intake, it suggests that cancer cachexia is not due to a nutritional deficiency *per se*, but to metabolic alterations in the host produced by the tumour.

Insulin has long been proposed as a potential modulator of the cachectic state, since it increases feeding and body weight, while reducing tumour glycolysis and metabolism, and reduces gluconeogenesis and host catabolism (Chance *et al.*, 1986). Animals bearing the MAC16 tumour administered 20 U insulin kg^{-1} day^{-1} had a significantly reduced weight loss, which, however, was only achieved by an increase (50%) in tumour growth, which interestingly was counteracted by the inclusion of 3-hydroxybutyrate in the drinking water (Beck and Tisdale, 1989b). This suggests that a ketogenic diet is more advantageous than insulin in reversing weight loss.

Perhaps the most interesting and effective reversal of weight loss in animals bearing the MAC16 tumour was achieved when the carbohydrate calories were replaced by lipid in the form of fish oil. In this case complete prevention of weight loss was achieved when the fish oil comprised 50% of the total calories, and was accompanied by a significant reduction in tumour weight. Weight reversal was not associated with an alteration in total calories consumed (Tisdale and Dhesi, 1990). Since weight loss is proportional to tumour weight, the reduction in weight loss could have arisen from the reduction in tumour weight. However, the ratio of host body weight gain to tumour weight decreases as the concentration of fish oil in the diet increases, suggesting that the primary effect is on the reduction of catabolism of host tissues, which is followed by a secondary reduction in tumour growth rate.

Fish oil is composed of unsaturated fatty acids of the $(n-3)$ series of which eicosapentaenoic acid (EPA) and docosahexaenoic acid (DHA) predominate. Of these two fatty acids, only EPA was effective in inhibiting the catabolic activity of the tumour lipid mobilizing factor in an *in vitro* assay. Fatty acids of the $(n-6)$ series, such as arachidonic acid, which has the same number of carbon atoms as EPA but one less double bond, were also ineffective, showing the inhibition of tumour lipid mobilizing activity to be highly structurally specific. *In vivo* experiments with the pure fatty acids showed only EPA to be effective in preventing host weight loss in the presence of the MAC16 tumour, and only EPA to be an inhibitor of the growth of the tumour (Tisdale and Beck, 1991). This suggests that the anti-cachectic and anti-tumour effects of fish oil are due to the presence of EPA.

Again a relationship is observed between cachexia and the maintenance of tumour growth. This suggests a new avenue for the development of anti-tumour agents directed not towards replication of the tumour cells, but interference with the lipid requirements. Such agents might be expected to display more selectivity and lower host toxicity than conventional anti-tumour agents.

ACKNOWLEDGEMENTS

This work has been supported by a research grant from the Cancer Research Campaign.

REFERENCES

Balkwell, F., Osborne, K., Burke, F., Naylor, S., Talbot, D., Durbin, H., Tavernier, J. and Fiers, W. (1987). Evidence for tumor necrosis factor/cachectin production in cancer. *Lancet*, **11**, 1229–32.

Beck, S.A. and Tisdale, M.J. (1987). Production of lipolytic and proteolytic factors by a murine tumor-producing cachexia in the host. *Cancer Res.*, **47**, 5919–23.

Beck, S.A. and Tisdale, M.J. (1989a). Nitrogen excretion in cancer cachexia and its modification by a high fat diet in mice. *Cancer Res.*, **49**, 3800–4.
Beck, S.A. and Tisdale, M.J. (1989b). Effect of insulin on weight loss and tumour growth in a cachexia model. *Br. J. Cancer*, **59**, 677–81.
Beck, S.A., Mulligan, H.D. and Tisdale, M.J. (1990). Lipolytic factors associated with murine and human cancer cachexia. *J. Natl Cancer Inst.*, **82**, 1922–5.
Beutler, B., Greenwald, D., Hulmes, J.D., Chang, M., Pan, Y.-C.E., Matheson, J., Ulevitch, R. and Cerami, A. (1985). Identity of tumour necrosis factor and the macrophage-secreted factor cachectin. *Nature*, **316**, 552–4.
Bibby, M.C., Double, J.A., Ali, S.A., Fearon, K.C.H., Brennan, R.A. and Tisdale, M.J. (1987). Characterization of a transplantable adenocarcinoma of the mouse colon producing cachexia in recipient animals. *J. Natl Cancer Inst.*, **78**, 539–46.
Burke, M., Bryson, E.I. and Kark, A.E. (1980). Dietary intakes, resting metabolic rates and body composition in benign and malignant gastrointestinal disease. *Br. Med. J.*, **280**, 211–5.
Burt, M.E., Gorschboth, C.M. and Brennan, M.F. (1981). A controlled prospective randomized trial evaluating metabolic effects of enteral and parenteral nutrition in the cancer patient. *Cancer*, **49**, 1092–105.
Chance, W.T., Muggia-Sullam, M., Chen, M.-H., Murphy, R.F. and Fischer, J.E. (1986). Reversal of tumour-induced biochemical abnormalities by insulin treatment in rats. *J. Natl Cancer Inst.*, **77**, 497–503.
Costa, G. (1963). Cachexia, the metabolic component of neoplastic diseases, *Prog. Exp. Tumor Res.*, **3**, 321–69.
Costa, G. (1977). Cachexia, the metabolic component of neoplastic diseases. *Cancer Res.*, **37**, 2327–35.
Costa, G. and Holland, J.F. (1966). Effects of Krebs-2 carcinoma on the lipid metabolism of male Swiss mice. *Cancer Res.*, **22**, 1081–3.
De Wys, W.D. (1985). Management of cancer cachexia. *Semin. Oncol.*, **12**, 452–60.
Fearon, K.C.H., Borland, W., Preston, T., Tisdale, M.J., Shenkin, A. and Calman, K.C. (1988). Cancer cachexia: influence of systemic ketosis on substrate levels and nitrogen metabolism. *Am. J. Clin. Nutr.*, **47**, 42–8.
Garattini, S., Bizzi, A., Donelli, M.G., Guaitani, A., Samanin, R. and Spreafico, F. (1980). Anorexia and cancer in animals and man. *Cancer Treat. Rev.*, **7**, 115–40.
Heymsfield, S.B. and McManus, C.B. (1985). Tissue components of weight loss in cancer patients. *Cancer*, **55** (Suppl.), 238–94.
Hollander, M.D., Ebert, E.C., Roberts, A.I. and Devereux, D.F. (1986). Effects of tumor type and burden on carcass lipid depletion in mice. *Surgery*, **100**, 292–6.
Kitada, S., Hays, E.F. and Mead, J.F. (1980). A lipid mobilizing factor in serum of tumor-bearing mice. *Lipids*, **15**, 168–74.
Kitada, S., Hays, E.F. and Mead, J.F. (1981). Characterization of a lipid mobilizing factor from tumors. *Prog. Lipid Res.*, **28**, 823–6.
Kitada, S., Hays, E.F., Mead, J.F. and Zabin, I. (1982). Lipolysis induction in adipocytes by a protein from tumor cells. *J. Cell Biochem.*, **20**, 409–16.
Mahony, S.M., Beck, S.A. and Tisdale, M.J. (1988). Comparison of weight loss induced by recombinant tumour necrosis factor with that produced by a cachexia-inducing tumour. *Br. J. Cancer*, **57**, 385–9.
Masuno, H., Yamasaki, N. and Okuda, H. (1981). Purification and characterization of lipolytic factor (toxohormone-L) from cell-free fluid of ascites Sarcoma 180. *Cancer Res.*, **42**, 284–8.
Masuno, H., Yoshimura, H., Ogawa, N. and Okuda, H. (1984). Isolation of lipolytic

factor (toxohormone-L) from ascites fluid of patients with hepatoma and its effect on feeding behaviour. *Eur. J. Cancer Clin. Oncol.*, **20**, 1177–85.

Oliff, A., Defo-Jones, D., Boyer, M., Martinez, D., Kiefer, D., Vuocolo, G., Wolfe, A. and Socher, S.H. (1987). Tumors secreting human TNF/cachectin induce cachexia in mice. *Cell*, **50**, 555–63.

Rouzer, C.A. and Cerami, A. (1980). Hypertriglyceridemia associated with *Trypanosoma brucei brucei* infection in rabbits: roll of defective triglyceride removal. *Mol. Biochem. Parasitol.*, **2**, 31–8.

Saarinen, U.M., Koskelo, E.-K., Teppo, A.-M. and Siimes, M.A. (1990). Tumor necrosis factor in children with malignancies. *Cancer Res.*, **50**, 592–5.

Selby, P., Hobbs, S., Viner, C., Jackson, E., Jones, A., Newell, D., Calvert, A.H., McElwain, T., Fearon K. and Shiga, T. (1987). Tumor necrosis factor in man: clinical and biological observations. *Br. J. Cancer.*, **56**, 803–8.

Sherman, M.L., Spriggs, D.R., Arthur, K.A., Imumura, K., Frei III, E. and Kufe, D.N. (1988). Recombinant human tumor necrosis factor administered as a five day continuous infusion in cancer patients: phase I toxicity and effects on lipid metabolism. *J. Clin. Oncol.*, **6**, 344–8.

Sherry, B.A., Gelin, J., Fong, Y., Marano, M., Wei, H., Cerami, A., Lowry, S.F., Lundholm, K.G. and Modawer, L.L. (1989). Anticachetin tumor necrosis factor-α antibodies attenuate development of cachexia in tumor models. *FASEB J.*, **3**, 1956–62.

Socher, S.H., Martinez, D., Craig, J.B., Kuhn, J.G. and Oliff, A. (1988). Tumor necrosis factor not detectable in patients with clinical cancer cachexia. *J. Natl Cancer Inst.*, **80**, 595–8.

Spector, A.A. and Burns, C.P. (1987). Biological and therapeutic potential of membrane lipid modification in tumors. *Cancer Res.*, **47**, 4529–37.

Strain, A.J. (1979). Cancer cachexia in man: A review. *Invest. Cell Pathol.*, **2**, 181–93.

Strain, A.J., Easty, G.C. and Neville, A.M. (1980). An experimental model of cachexia induced by a xenografted human tumor. *Cancer Res.*, **64**, 217–21.

Tisdale, M.J. and Beck, S.A. (1991). Inhibition of tumour-induced lipolysis *in vitro* and cachexia and tumour growth *in vivo* by eicosapentaenoic acid. *Biochem. Pharmacol.* **41**, 103–7.

Tisdale, M.J. and Dhesi, J.K. (1990). Inhibition of weight loss by Co-3 fatty acids in an experimental cachexia model. *Cancer Res.*, **50**, 5022–6.

Tisdale, M.J., Brennan, R.A. and Fearon, K.C. (1987). Reduction of weight loss and tumour size in a cachexia model by a high fat diet. *Br. J. Cancer*, **56**, 39–43.

Warnold, I., Lundholm, K. and Scherstein, T. (1978). Energy balance and body composition in cancer patients. *Cancer Res.*, **38**, 1801–7.

Warren, R.S., Starnes, H.F., Gabrilove, J.L., Oettgen, H.F. and Brennan, M.F. (1987). The acute metabolic effects of tumor necrosis factor administration in humans. *Arch. Surg.*, **122**, 1396–400.

Chapter Nine

Fever and Sepsis

Matthew J. Kluger
Department of Physiology, University of Michigan Medical School, Ann Arbor, MI 48109, USA

The picture of the lethargic, depressed, anorexic, and febrile individual is not specific to any particular animal species, but is seen in humans and a variety of animals, and the behavioral signs are seen with a variety of systemic diseases as well as with some more localized infections

(Hart, 1988)

THE ACUTE PHASE RESPONSE

The 'acute phase response' (APR) is defined as the non-specific stereotyped host defense sequelae that occur in response to disease and tissue injury. These physiologic, immunologic and behavioral responses include neutrophilia, increases in the plasma concentrations of haptoglobin, C-reactive protein, ceruloplasmin, and other 'acute-phase proteins,' the loss of food appetite, a decrease in activity, an increase in slow-wave sleep, and the development of fever (for reviews see Gauldie *et al.*, 1985; Stahl, 1987; Powanda *et al.*, 1988). Most components of the APR are thought to be adaptive responses, which slow the growth rate of the infectious agents or tumors (see, for example, Kluger, 1979; Roberts, 1979; Weinberg, 1984; Hart, 1988).

The systemic manifestations of disease mentioned above are thought to be the result of the release of endogenously produced proteins which orginally went by many names including 'endogenous pyrogen', 'leukocyte endogenous mediator', and 'hepatocyte stimulating factor'. Within the past decade, many data have supported the hypothesis that these proteins belong

Obesity and Cachexia. Edited by N.J. Rothwell and M.J. Stock

to a broad class of factors called cytokines (see, for example, Kluger *et al.*, 1985; Powanda *et al.*, 1988; Oppenheim *et al.*, 1990; Dinarello *et al.*, 1990). Considerable data have been generated implicating the cytokines interleukin-1α and IL-1β (IL-1α, IL-1β), tumor necrosis factor α (TNF) and interleukin-6 (IL-6) in orchestrating the APR.

One of the difficulties in studying the precise roles of cytokines in the APR has been the inability to block selectively the biological actions of specific cytokines. With the advent of recombinant cytokines and the development of antibodies that are capable of blocking their biological action, it has now become possible to assess their roles in the APR. For example, at the recently held Second International Workshop on Cytokines (Hilton Head, SC, December 10–14, 1989) many investigators were describing antibodies to cytokines and to their receptors, which are now available for *in vivo* investigation (see, for example, Dinarello *et al.*, 1990).

The purpose of this paper is to review briefly the evidence that specific cytokines are responsible for the induction of fever and anorexia. In addition, the causal relationship between fever and anorexia will be examined.

THE ROLE OF CYTOKINES IN THE FEVER AND ANOREXIA OF SEPSIS

Although most reviews on cytokines identify IL-1α, IL-1β, TNF and other cytokines as responsible for the fever and loss of food appetite that accompany infection or cancer, this conclusion is based largely on studies demonstrating that injections of these cytokines produce these acute phase responses (Tables 9.1 and 9.2).

What the data presented in Tables 9.1 and 9.2 indicate is that injections of cytokines produce fever and anorexia. However, it is not known whether these injections produce plasma or central nervous system concentrations of cytokines that are similar to those that occur during diseases known to produce fever and anorexia. Many of the results presented in these tables, including those from my laboratory, may reflect pharmacological effects. Simply demonstrating that injection of a cytokine results in a specific APR does not prove that that cytokine is involved in that APR during disease. Before any factor can be established to have a physiologic role (i.e. truly be an *endogenous* mediator of fever or anorexia) several criteria must be fulfilled (Kluger, 1990). Some of these Koch's-postulates-like criteria for defining a mediator as an endogenous pyrogen, or 'anorexigen' are listed below.

Table 9.1 Induction of fever following injection of recombinant cytokines (from Kluger, 1990). Included are only cytokines for which the source (i.e. species) has been identified in the publications

Cytokine	Species	Route of administration	Reference
mrIL-1α	Rat	i.p.	McCarthy *et al.* (1985)
	Rat	i.v.	Tocco-Bradley *et al.* (1986)
	Rat	i.p.	Otterness *et al.* (1988)
mrIL-1β	Rat	i.p.	LeMay *et al.* (1990b)
hrIL-1α	Rat	i.c.v.	Busbridge *et al.* (1989a)
	Rabbit	i.c.v. and i.v.	Hashimoto *et al.* (1988)
	Rabbit	i.c.v. and i.v.	Morimoto *et al.* (1989)
hrIL-1β	Rat	i.c.v.	Plata-Salaman *et al.* (1988)
	Rat	i.c.v.	Busbridge *et al.* (1989a)
	Rat	i.c.v. and i.v.	Busbridge *et al.* (1989b)
	Rabbit	i.v.	Dinarello *et al.* (1986a)
	Rabbit	i.c.v.	Opp *et al.* (1989)
	Mouse	i.p.	Dinarello *et al.* (1986)
rrIL-1α and β	Rabbit	i.v.	Cannon *et al.* (1989)
hrIL-2	Human	i.v.	Michie *et al.* (1988a)
hrIL-6	Rabbit	i.c.v.	Opp *et al.* (1989)
	Rat	i.c.v.	LeMay *et al.* (1990a)
hrTNFα	Rat	s.c.	Kettelhut and Goldberg (1988)
	Rat	i.c.v.	Plata-Salaman *et al.* (1988)
	Rat	i.c.v.	Rothwell (1988)
	Rabbit	i.v.	Dinarello *et al.* (1986b)
	Rabbit	i.c.v. and i.v.	Morimoto *et al.* (1989)
	Rabbit	i.v.	Nakamura *et al.* (1988)
	Human	i.v.	Creaven *et al.* (1987)
	Human	i.v.	Kimura *et al.* (1987)
	Human	i.v.	Starnes *et al.* (1988)
	Human	i.v.	Michie *et al.* (1988c)
	Human	i.v.	Steinmetz *et al.* (1988)
hrINFα	Rabbit, mouse	i.v.	Dinarello *et al.* (1984)
	Rabbit	i.c.v. and i.v.	Morimoto *et al.* (1987)
rhGM CSF	Human	i.v.	Peters *et al.*, (1988)

mr, murine recombinant; hr, human recombinant; rr, rabbit recombinant; poa, preoptic/anterior hypothalamus; i.c.v., intracerebroventricular; i.v., intravenous; i.p., intraperitoneal; s.c., subcutaneous.

Table 9.2 Induction of anorexia following injection of recombinant cytokines. Included are only cytokines for which the source (i.e. species) has been identified in the publication

Cytokine	Species	Route of administration	Reference
mrIL-α	Rat	i.p.	McCarthy *et al.* (1985)
	Rat	i.p.	Otterness *et al.* (1988)
	Rat	s.c.	Mrosovsky *et al.* (1990)
	Mouse	i.p.	Moldawer *et al.* (1988)
hrIL-1α	Mouse	i.p.	Moldawer *et al.* (1988)
	Rat	i.p.	Fong *et al.* (1989)
	Rat	i.p.	Moldawer *et al.* (1989)
hrIL-1β	Rat	i.c.v.	Plata-Salaman *et al.* (1988)
	Rat	i.p.	Hellerstein *et al.* (1989)
	Rat	i.p.	Uehara *et al.* (1989a, b)
	Mouse	i.p.	Moldawer *et al.* (1988)
hrIL-2	Rats	i.c.v.	Plata-Salaman (1988)
hrIL-6	Rats	i.c.v.	Plata-Salaman (1988)
	Rats	i.c.v.	Busbridge *et al.* (1989b)
hrTNFα	Mouse	i.p.	Moldawer *et al.* (1988)
	Mouse	i.v.	Mahony and Tisdale (1988)
	Mouse	i.p.	Socher *et al.* (1988)
	Rat	s.c.	Kettelhut and Goldberg (1988)
	Rat	i.p.	Stovroff *et al.* (1988)
	Rat	i.p.	Tracey *et al.* (1988)
	Rat	i.p.	Fraker *et al.* (1988)
	Rat	i.p.	Fong *et al.* (1989)
	Rat	i.c.v.	Plata-Salaman *et al.* (1988)
	Rat	i.p.	Moldawer *et al.* (1989)
	Human	i.v.	Creaven *et al.* (1987)
	Human	i.v.	Kimura *et al.* (1987)

mr, murine recombinant; hr, human recombinant; rr, rabbit recombinant; poa, preoptic/anterior hypothalamus; i.c.v., intracerebroventricular; i.v. intravenous; i.p. intraperitoneal; s.c. subcutaneous.

Criterion 1. Application of the putative endogenous mediator (e.g. pyrogen) at the hypothetical site of action results in the response observed during infection (e.g. a rise in body temperature)

For example, if the anterior hypothalamus is the theoretical site of action of an endogenous pyrogen, then injection of the putative endogenous pyrogen at this locus should induce fever.

Criterion 2. When an exogenous factor (e.g. bacteria) induces an acute phase response (e.g. fever), (a) the released putative endogenous mediator has some quantitative relationship to the response (e.g. rise in temperature), and (b) the putative mediator is released in amounts similar to those that cause an equivalent response (e.g. rise in temperature) when the putative endogenous mediator is given at the hypothetical site of action

Fevers and anorexia vary in magnitude and duration. Suppose a putative mediator is hypothesized as being secreted from fixed macrophages in the liver into the circulation. Then, when an animal is injected with a pathogen, the mediator should be released from the liver into the circulation roughly in proportion to the resultant acute phase response. Furthermore, when a putative mediator is infused into an experimental animal to achieve the plasma concentration observed during infection, the rise in body temperature or decrease in activity or loss of food appetite should be virtually indistinguishable from that following injection of the exogenous mediator.

It is not essential that criterion 2 be fulfilled because it assumes that the administration of an exogenous mediator does not influence (e.g. up-regulate) the receptor for a putative endogenous mediator. For example, were lipopolysaccharide (LPS) either directly or indirectly to increase the numbers of receptors for some endogenous pyrogen, then the plasma concentration of that endogenous pyrogen might vary in an 'all-or-none' fashion, rather than in proportion to the fever. Furthermore, this criterion assumes that during the course of the fever other factors (e.g. endogenous antipyretics or cryogens) are not influencing the fever. Nevertheless, it is probable that whatever turns out to be the 'circulating' mediator of fever or other APR, will fulfill this criterion, at least to some extent.

Criterion 3. Substances that prevent the production of the putative endogenous mediator should block the response (e.g. rise in body temperature)

If drug X blocks the production of a putative endogenous mediator of fever, decrease in activity, or loss of food appetite, this should result in the complete blockade of these acute phase responses. If drug X blocks the production of a putative endogenous mediator, but does not block these acute phase responses, or only attenuates them, then clearly other mediators are also involved.

Criterion 4. Substances that prevent the action of the putative endogenous mediator should block the response (e.g. rise in body temperature)

As for criterion 3, if drug X blocks the action of a putative endogenous mediator of fever or loss of food appetite or other APR, this should result in the complete blockade of the response. If drug X blocks the action of a putative EP, but does not block these acute phase responses, or only attenuates them, then clearly other mediators are also involved.

In large part, only criterion 1 listed above has been fulfilled for any cytokine implicated in fever or anorexia. Within the past few years there have been some attempts to assess the roles of cytokines in fever and anorexia using criteria 2, 3 and 4. These are described below.

CYTOKINES AND FEVER

Data relevant to criterion 2

Michie *et al.*, (1988b,c) found that intravenous injection of LPS to human patients resulted in fever and an increase in plasma concentration of TNF. They then infused these patients with TNF at a dosage intended to simulate the plasma concentrations measured following injection of LPS. Their results support the hypothesis that TNF is a pyrogen in human subjects. Although perhaps not directly impacting on their impressive results, two aspects of their study are difficult to interpret. One is that despite continuous infusion of TNF, the plasma concentration of TNF (as determined by ELISA) fell to non-detectable levels after 12 h. Perhaps the presence of high levels of TNF induces its rapid clearance. Since there are reports of discrepancies between immunoassays and bioassays for measurement of TNF (Duncombe and Brenner, 1988; Fomsgaard *et al.*, 1988; Petersen and Moller, 1988), it is also possible that the measurements of TNF in the study by Michie *et al.*, (1988c) did not accurately reflect the biologically active TNF. A second aspect of the study by Michie *et al.* that is unresolved is that the patients developed fevers despite being pre-treated with indomethacin, a nonsteroidal anti-inflammatory drug that is generally considered to have potent antipyretic properties.

Data relevant to criterion 3

Zabel *et al.* (1989a, b) showed that the drug pentoxifylline totally abolished the rise in plasma TNF concentration in human volunteers given LPS intravenously without influencing the course of their fever, their 'flu-like symptoms' and leukocytosis. We have found similar results in our experiments in LPS-injected rats that were pretreated with varying doses of pentoxifylline (LeMay *et al.*, 1990c). At moderate doses, this drug did not influence fever, but did significantly reduce the plasma concentration of TNF in LPS-injected rats. Taken together, these data support the hypothesis that TNF is not essential for the development of LPS-induced fever.

Data relevant to criterion 4

Work from my laboratory has shown that administration of antiserum to recombinant murine IL-1α had no effect on LPS-fever in the rat (Long *et al.*, 1989). This latter finding might not be surprising, since IL-1α is thought to be primarily cell-associated; IL-1β may be the secreted form of IL-1. In recent studies, we have found that pretreatment of rats with antiserum to IL-1β did lead to a 57% attenuation in the magnitude of LPS-induced fever (Long *et al.*, 1990a). Rothwell *et al.* (1989) reported similar findings. The precise role of IL-6 in LPS-induced fever is currently being evaluated (LeMay *et al.*, 1990a, b). In these studies, we have measured the plasma and cerebrospinal fluid concentrations of IL-6 following injections of LPS, and have found significant correlations between IL-6 and the magnitude of fever. As mentioned above, the fevers produced by injection of LPS are significantly attenuated by injection of antiserum to IL-1β. This antiserum also significantly reduced the rise in plasma IL-6 (LeMay *et al.*, 1990b). Experiments are currently in progress to assess whether administration of antibodies to IL-6 will attenuate LPS-fever.

Nagai *et al.* (1988) and Kawasaki *et al.* (1989) found that administration of *monoclonal* antibody to TNF resulted in a significant attenuation of LPS-fever in the rabbit. In these studies, rabbits that were injected with LPS and antibody to TNF developed similar fevers during the first 2 h, but there was a significant reduction in the fever during hours 2–4. Mathison *et al.* (1988) found that pretreatment of rabbits with *polyclonal* antibody to TNF did not result in any attenuation in LPS-induced fever. Based on the data presented in figure 13 of their paper, it appears that the antibody-treated rabbits actually developed larger fevers. Similar results were obtained when we pretreated rats with antiserum to TNF in amounts that completely blocked any rise in plasma TNF, as determined by the sensitive WEHI assay (Long *et al.*, 1990a,b). When rats injected with antiserum either intraperitoneally or intravenously were then injected with LPS, the resultant fevers were significantly enhanced, rather than suppressed. Our data, along with those of Mathison *et al.* (1988) are consistent with the hypothesis that TNF may be an endogenous antipyretic that limits the magnitude of fever. The mechanism of this TNF antipyresis is unknown. For example, it might be via the release of such endogenous antipyretics as arginine vasopressin (e.g. Kasting, 1989) or α-MSH (e.g. Shih *et al.*, 1986), or it may be a direct action on the hypothalamus. By neutralizing the biological activity of TNF, this negative feedback signal was removed, resulting in an exaggerated fever. An alternative possibility that could explain our TNF antipyretic action is that the antiserum, by complexing with endogenous TNF, produces immune complexes that actually enhance TNF's fever-promoting actions. This, however, is unlikely, since the same antiserum has been used by Colletti *et al.* (1989) to protect rats from TNF-induced liver and lung damage caused by hepatic ischaemia followed by reperfusion.

How can we explain the differences between those experiments using monoclonal antibodies (Nagai *et al.*, 1988; Kawasaki *et al.*, 1989) and those using polyclonal antibodies (Mathison *et al.*, 1988; Long *et al.*, 1990a,b)? In the studies involving injection of monoclonal antibody to TNF in rabbits no data were presented demonstrating that the antibody neutralized TNF bioactivity *in vivo*. One possibility is that injection with this antibody might have led to increased levels of TNF, producing greater antipyresis—thus, resulting in smaller fevers.

Overall, the data support the hypothesis that the cytokine IL-1β is involved in LPS-induced fever. Its actions might be mediated in part via IL-6 and other cytokines. TNF, rather than being an endogenous pyrogen, at physiological concentrations might be an endogenous antipyretic or cryogen. Clearly, more studies using different animal models and different fever-inducing agents (e.g. viruses, tumors) need to be done before any unified theory of cytokines in fever induction and defervescence can be postulated.

CYTOKINES AND ANOREXIA

Data relevant to criterion 2

Data from my laboratory were the first to demonstrate that IL-1 could lead to a reduction in food intake (McCarthy *et al.*, 1985, 1986). We have continued these studies to assess the effects of long-term infusions of IL-1α on food intake and body weight (Mrosovsky *et al.*, 1990). However, in this later study, we found that tolerance developed to the continuous infusion of IL-1. Does this mean that anorexia during chronic infection is not mediated by IL-1? Perhaps pulsatile releases of this cytokine might prevent in the development of tolerance.

It is not surprising that TNF, whose alternate name is 'cachectin', has often been claimed to be a principal cause of the cachexia observed during cancer and infection (eg. see Socher *et al.*, 1988; Oliff *et al.*, 1987; Stovroff *et al.*, 1988). As with IL-1, one problem with attributing all cachexias to TNF is that in many of the above models, tolerance to TNF injections or infusions developed. In addition, it is known that plasma TNF concentrations are not always elevated in cachectic animals. Furthermore, Vaisman *et al.* (1989) have reported that starvation, itself, increases the sensitivity of peripheral blood mononuclear cells to PHA- or LPS-induced production of

TNF. These data might indicate that the elevated TNF concentrations occasionally seen in cachectic patients might be the result of cachexia, rather than the cause of this syndrome.

Data relevant to criterion 4

One study has been reported in which anti-TNF antibodies have been used in an attempt to reverse the anorexia/cachexia of cancer (Sherry *et al.*, 1989). Antibodies to TNF were administered to mice that had previously been implanted with a methylcholanthrene-induced sarcoma. Although food intake decreased in tumor-bearing mice given either the anti-TNF antibodies or sham-injections, the decrease in food intake and body weight was not as great in the mice given the anti-TNF antibodies.

Gershenwald *et al.* (1989) presented some exciting data at the Cytokine Workshop indicating that an anti-IL-1 receptor monoclonal antibody (35SF5, Hoffman-La Roche) attenuated the anorexia induced by injection of mice subcutaneously with turpentine, which produces local inflammation.

I interpret the above data to support the hypothesis that the cytokines IL-1 and TNF are involved in some anorexias, but clearly other mediators must also be contributing to the marked reduction in food intake. In addition, as discussed above for fever, it will be necessary to assess other animal models and inducers of anorexia before postulating any comprehensive theory of cytokines and anorexia.

DOES ANOREXIA RESULT FROM FEVER?

For many years it was assumed that the loss of food appetite with infection was the result of an elevation in body temperature—the 'thermostatic theory of food intake' (Brobeck, 1981). However, in studies using the antipyretic drug sodium salicylate, McCarthy *et al.* (1984) showed that the anorexia produced by injection of LPS in rats was not a direct result of the development of fever. In addition, a subsequent study showed that intracerebroventricular injection of LPS or crude human endogenous pyrogen led to fevers in rats, but not to loss of food appetite (McCarthy *et al.* 1986). Van Miert *et al.* (1986) also found that flurbiprofen, while inhibiting fever, did not restore food intake in goats injected with LPS. Hellerstein *et al.* (1989) found that administration of ibuprofen (10 mg/kg) to rats injected with an anorexic dose of recombinant human IL-1β completely restored food intake, and the suggestion was made that IL-1-induced anorexia is mediated by prostaglandins. Body temperature was not monitored in this study, so it is not known whether the dose of ibuprofen used produced antipyresis. A similar finding has been reported by Uehara *et al.* (1989b).

In their study, rats were injected with a dose of recombinant human IL-1β that produced a marked reduction in food intake. Injection of indomethacin (2.5 mg/kg) led to the complete reversal of the IL-1-induced anorexia. Body temperature was not measured. Based on the above data, three conclusions can be drawn:

1. Anorexia induced by LPS is not the result of fever, and probably not mediated by prostaglandins.
2. Anorexia induced by IL-1β may be mediated either directly (e.g. on brain feeding 'centers') or indirectly (e.g. by causing fever) by prostaglandins.
3. Anorexia induced by LPS is not the result of IL-1β.

In order to test these hypotheses it will be necessary to evaluate in the same animal model (e.g. rats of same age, strain, diet, etc.) the effects of ibuprofen or idomethacin on LPS-induced and IL-1β-induced fever and anorexia. As mentioned earlier in this review, experiments using neutralizing antibodies to rat IL-1β in LPS-injected rats would lead to a direct test of the hypothesis that IL-1β is responsible for the anorexia caused by injection of LPS.

SUMMARY

Although there are considerable data supporting the hypothesis that cytokines are responsible for the fever and anorexia with infection and cancer, most of these data have been obtained simply by injection of cytokines. Some information is now beginning to be published regarding the roles of cytokines in LPS-induced fever that is based on more physiologically relevant criteria, but virtually nothing is known about the causes of fever during infections with live bacteria, viruses, or in response to cancer. With the exception of the study by Sherry *et al.* (1989), little is known about the causes of decreased activity or anorexia based on physiologically relevant criteria during infections with live bacteria, viruses, or in response to cancer. Lastly, although the loss of food appetite during infection may be caused by the release of specific cytokines, there is little evidence to support the hypothesis that this anorexia is the result of the development of fever.

REFERENCES

Brobeck, J.R. (1981). In: *Body Weight Regulatory System: Normal and Disturbed Mechanisms* (eds L.A. Cioffi, W.P.T. James and T.B. Van Itallie), Raven Press, New York, pp. 1–9.

Busbridge, N.J., Dascombe, M.J., Tilders, F.J.H., van Oers, J.W.A.M., Linton, E.A. and Rothwell, N.J. (1989a). Central activation of thermogenesis and fever by interleukin-1β and interleukin-1α involves different mechanisms, *Biochem. Biophys. Res. Commun.*, **162**, 591–6.

Busbridge, N.J., Dascome, J.J., Hopkins, S. and Rothwell, N.J. (1989b). Acute central effects of interleukin-6 on body temperature, thermogenesis and food intake in the rat. *Proc. Nutr. Soc.*, **48**, 48A.

Busbridge, M.J., Rothwell, N.J., Sagay, B.O. and Stock, M.J. (1989c). Pyrogenic and thermogenic effects of interleukin 1β in the rat. *Am. J. Physiol.*, **256**, E7–11.

Cannon, J.G., Clark, B.D., Wingfield, P., Schmeissner, U., Losberger, C., Dinarello, C.A. and Shaw, A.R. (1989). Rabbit IL-1. Cloning, expression, biologic properties, and transcription during endotoxemia. *J. Immunol.*, **142**, 2299–306.

Colletti, L.M., Remick, D.G., Burtch, G.D., Kunkel, S.L., Strieter, R.M. and Campbell, D.A. Jr. (1989). The role of TNF in pathophysiological alterations following hepatic ischemia/reperfusion. *Cytokine*, **1**, 150.

Creaven, P.J., Plager, J.E., Huben, R.P., Takita, H., Mittelman, A. and Proefrock, A. (1987). Phase 1 clinical trial of recombinant human tumor necrosis factor. *Cancer Chemother. Pharmacol.*, **20**, 137–44.

Dinarello, C.A., Bernheim, H.A., Duff, G.W., Le, H.V., Nagabhushan, T.L., Hamilton, N.C. and Coceani, F. (1984). Mechanisms of fever induced by recombinant human interferon. *J. Clin. Invest.*, **74**, 906–13.

Dinarello, C.A., Cannon, J.G., Mier, J.W., Bernheim, H.A., LoPreste, G., Lynn, D.L., Love, R.N., Webb, A.C., Auron, P.E., Reuben, R.C., Rich, A., Wolff, S.M. and Putney, S.D. (1986a). Multiple biological activities of human recombinant interleukin 1. *J. Clin. Invest.*, **77**, 1734–9.

Dinarello, C.A., Cannon, J.G., Wolff, S.M., Bernheim, H.A., Beutler, B., Cerami, A., Figari, I.S., Palladino, M.A. Jr and O'Connor, J.V. (1986b). Tumor necrosis factor (cachectin) is an endogenous pyrogen and induces production of interleukin 1. *J. Exp. Med.*, **163**, 1433–50.

Dinarello, C.A., Kluger, M.J., Oppenheim, J.J. and Powanda, M. (1990). The pathophysiological actions of cytokines. In: *Proceedings of Second International Workshop on Cytokines, December 11–14, 1989. Series: Progress in Leukocyte Biology*, Alan R. Liss, New York.

Duncombe, A.S. and Brenner, M.K. (1988). Is circulating tumor necrosis factor bioactive? *N. Engl. J. Med.*, **319**, 1227.

Fomsgaard, A., Worsaae, H. and Bendtzen, K. (1988). Detection of tumor necrosis factor from lipopolysaccharide-stimulated human mononuclear cells by enzyme-linked immunosorbent assay and cytotoxicity bioassay. *Scand. J. Immunol.*, **27**, 143–7.

Fong, Y., Moldawer, L.L., Marono, M., Wei, H., Barber, A., Manogue, K., Tracey, K.J., Juo, G., Fischman, D.A., Cerami, A. and Lowry, S.F. (1989). Cachectin/TNF or IL-1α induces cachexia with redistribution of body proteins. *Am. J. Physiol.*, **256**, R659–65.

Fraker, D.L., Stovroff, M.C., Merino, M.J. and Norton, J.A. (1988). Tolerance to tumor necrosis factor in rats and the relationship to endotoxin tolerance and toxicity. *J. Exp. Med.*, **168**, 95–105.

Gauldie, J., Lamontagne, L. and Stadnyk, A. (1985). Acute phase response in infectious disease. *Surv. Synth. Path. Res.*, **4**, 126–51.

Gershenwald, J.E., Fong, Y., Chizzonite, R., Lowry, S.F. and Moldawer, L.L. (1989). Anti-IL-1 receptor monoclonal antibodies attenuate metabolic changes in inflammation. *Cytokine*, **1**, 150.

Hart, B.L. (1988). The biological basis of the behavior of sick animals. *Neurosci. Biobehav. Rev.*, **12**, 123–37.

Hashimoto, M., Bando, T., Iriki, M. and Hashimoto, K. (1988) Effect of indomethacin on febrile response to recombinant human interleukin 1-α in rabbits. *Am. J. Physiol.*, **255**, R527–33.

Hellerstein, M.K., Meydani, S.N., Meydani, M., Wu, K. and Dinarello, C.A. (1989). Interleukin-1-induced anorexia in the rat. Influence of prostaglandins. *J. Clin. Invest.*, **84**, 228–35.

Kasting, N.W. (1989). Criteria for establishing a physiological role for brain peptides. A case in point: the role of vasopressin in thermoregulation during fever and antipyresis. *Brain Res. Rev.*, **14**, 143–53.

Kawasaki, H., Moriyama, M., Ohtani, Y., Naitoh, M., Tanaka, A. and Nariuchi, H. (1989). Analysis of endotoxin fever in rabbits by using a monoclonal antibody to tumor necrosis factor (cachectin). *Infect. Immun.*, **57**, 3131–5.

Kettelhut, I.C. and Goldberg, A.L. (1988). Tumor necrosis factor can induce fever in rats without activating protein breakdown in muscle or lipolysis in adipose tissue. *J. Clin. Invest.*, **81**, 1384–9.

Kimura, K., Taguchi, T., Uruskizaki, I., Ohno, R., Abe, O. *et al.*, (1987). Phase I study of recombinant human tumor necrosis factor. *Cancer Chemother. Pharmacol.*, **20**, 223–9.

Kluger, M.J. (1979). *Fever: Its Biology, Evolution, and Function* Princeton University Press.

Kluger, M.J. (1991). Fever: role of endogenous pyrogens and cryogens. *Physiol. Rev.*, **71**, 93–127.

Kluger, M.J., Oppenheim, J.J. and Powanda, M.C. (eds) (1985). The physiologic, metabolic and immunologic actions of interleukin-1. *Proceedings of International Symposium held in Ann Arbor. Michigan, June 4–6, 1985. Series: Progress in Leukocyte Biology* Vol. 2. Alan R. Liss, Inc, New York.

LeMay, L. Vander, A.J. and Kluger, M.J. (1990a). The role of IL-6 in fever in the rat. *Am. J. Physiol.*, **258**, R798–803.

LeMay, L., Otterness, I., Vander, A.J. and Kluger, M.J. (1990b). *In vivo* evidence that the rise in plasma IL-6 following injection of a fever-inducing dose of LPS is mediated by IL-1β. *Cytokine*, **2**, 199–204.

LeMay, L., Vander, A.J. and Kluger, M.J. (1990c). The effects of pentoxifylline on lipopolysaccharide (LPS) fever and plasma interleukin-6 (IL-6) and tumor necrosis factor (TNF) in the rat. *Cytokine*, **2**, 300–6.

Long, N.C., Kluger, M.J. and Vander, A.J. (1989). Antiserum against mouse IL-1 does not block stress hyperthermia or LPS fever in the rat. In: *Thermoregulation: Research and Clinical Applications. 7th International Symposium of Pharmacology of Thermoregulation, Odense, Denmark 1988.* (eds P. Lomax and E. Schonbaum) Karger, Basel, pp. 78–84.

Long, N.C., Otterness, I., Kunkel, S.L., Vander, A.J. and Kluger, M.J. (1990a). The roles of interleukin-1β and tumor necrosis factor in lipopolysaccharide-fever in the rat. *Am. J. Physiol.*, **259**, R724–8.

Long, N.C., Kunkel, S.L., Vander, A.J. and Kluger, M.J. (1990b). Antiserum against TNF enhances LPS fever in the rat. *Am. J. Physiol.*, **258**, R332–7.

Mahony, S.M. and Tisdale, M.J. (1988). Induction of weight loss and metabolic alterations by human recombinant tumour necrosis factor. *Br. J. Cancer*, **58**, 345–9.

Mathison, J.C., Wolfson, E. and Ulevitch, R.J. (1988). Participation of tumor necrosis factor in the mediation of gram negative bacterial lipopolysaccharide-induced injury in rabbits. *J. Clin. Invest.*, **81**, 1925–37.

McCarthy, D.O., Kluger, M.J. and Vander, A.J. (1984). The role of fever in

appetite suppression after endotoxin administration. *Am. J. Clin. Nutr.* **40**, 310–16.
McCarthy, D.O., Kluger, M.J. and Vander, A.J. (1985). Interleukin-1 suppresses food intake. *Am. J. Clin. Nutr.*, **42**, 1179–82.
McCarthy, D.O., Kluger, M.J. and Vander, A.J. (1986). Effect of centrally administered interleukin-1 and endotoxin on food intake of fasted rats. *Physiol. Behav.*, **36**, 745–9.
Michie, H.R., Eberlein, T.J., Spriggs, D.R., Manogue, K.R., Cerami, A. and Wilmore, D.W. (1988a). Interleukin-2 initiates metabolic responses associated with critical illness in humans. *Ann. Surg.*, **208**, 493–503.
Michie, H.R., Manogue, K.R., Spriggs, D.R., Revhaug, A., O'Dwyer, S., Dinarello, C.A., Cerami, A., Wolff, S.M. and Wilmore, D.W. (1988b). Detection of circulating tumor necrosis factor after endotoxin administration. *N. Engl. J. Med.*, **318**, 1481–6.
Michie, H.R., Spriggs, D.R., Manogue, K.R., Sherman, M.L., Revhaug, A., O'Dwyer, S.T., Arthur, K., Dinarello, C.A., Cerami, A., Wolff, S.M., Kufe, D.W. and Wilmore, D.W. (1988c). Tumor necrosis factor and endotoxin induce similar metabolic responses in human beings. *Surgery*, **104**, 280–6.
Moldawer, L.L., Andersson, C., Gelin, J. and Lundholm, K.G. (1988). Regulation of food intake and hepatic protein synthesis by recombinant-derived cytokines. *Am. J. Physiol.*, **254**, G450–6.
Moldawer, L.L., Marono, M.A., Wei, H., Fong, Y., Silen, M.L., Kuo, G., Manogue, K.R., Vlassara, H., Cohen, H., Cerami, A. and Lowry, S.F. (1989). Cachectin/tumor necrosis factor α alters red blood cell kinetics and induces anemia *in vivo*. *FASEB J.*, **3**, 1637–43.
Morimoto, A., Murakami, N., Takada, M., Teshirogi, S. and Watanabe, T. (1987). Fever and acute phase response induced in rabbits by human recombinant interferon-gamma. *J. Physiol.*, **391**, 209–18.
Morimoto, A., Sakata, Y., Watanabe, T. and Murakami, N. (1989). Characteristics of fever and acute-phase response induced in rabbits by IL-1 and TNF. *Am. J. Physiol.*, **256**, R35–41.
Mrosovsky, N., Molony, L.A., Conn, C.A. and Kluger, M.J. (1990). The anorexic effects of interleukin-1 in the rat. *Am. J. Physiol.*, **257**, R1315–21.
Nagai, M., Saigusa, T., Shimada, Y., Inagawa, H., Oshima, H. and Iriki, M. (1988). Antibody to tumor necrosis factor (TNF) reduces endotoxin fever. *Experientia*, **44**, 606–7.
Nakamura, H., Seto, Y., Motoyoshi, S., Kadokawa, T. and Sunahara, N. (1988). Recombinant human tumor necrosis factor causes long-lasting and prostaglandin-mediated fever, with little tolerance in rabbits. *J. Pharm. Exp. Ther.*, **245**, 336–41.
Oliff, A., Defeo-Jones, D., Boyer, M., Martinez, D., Kiefer, D., Vuocolo, G., Wolfe, A. and Socher, S.H. (1987). Tumor secreting human TNF/cachectin induced cachexia in mice. *Cell*, **50**, 555–63.
Opp, M., Obal, R. Jr., Cady, A.B., Johannsen, L. and Kreuger, J.M. (1989). Interleukin-6 ispyrogenic but not somnogenic. *Physiol. Behav.*, **45**, 1069–72.
Oppenheim, J.J., Powanda, M., Kluger, M.J. and Dinarello, C.A. (1990). Molecular and cellular biology of cytokines. *Proceedings of Second International Workshop on Cytokines*, December 11–14, 1989. *Series: Progress in Leukocyte Biology*, Alan R. Liss, New York.
Otterness, I., Seymour, P.A., Golden, H.W., Reynolds, J.A. and Daumy, Gaston, O. (1988). The effects of continuous administration of murine interleukin-1α in the rat. *Physiol. Behav.*, **43**, 797–804.

Peters, W.P., Shogan, J., Shpall, E.J., Jones, R.B. and Kim, S.C. (1988). Recombinant human granulocyte-macrophage colony-stimulating factor produces fever. *Lancet*, **i**, 950.

Petersen, C.M. and Moller, B.K. (1988). Immunological reactivity and bioactivity of tumour necrosis factor. *Lancet*, **i**, 934–5.

Plata-Salaman, C.R. (1988). Food intake suppression by immunomodulators. *Neurosci. Res. Commun.*, **3**, 159–65.

Plata-Salaman, C.R., Oomura, Y. and Kai, Y. (1988). Tumor necrosis factor and interleukin-1β: suppression of food intake by direct action in the central nervous system. *Brain Res.*, **448**, 106–14.

Powanda, M.C., Oppenheim, J.J., Kluger, M.J. and Dinarello, C.A. (eds) (1988). *International Workshop on Monokines and Nonlymphocytic Cytokines. Proceedings of International Symposium held in Hilton Head, S.C., December, 1987. Series: Progress in Leukocyte Biology*, Vol. 8. Alan R. Liss, New York.

Roberts, N.J., Jr. (1979). Temperature and host defense. *Microbiol. Rev.* , **43**, 241–59.

Rothwell, N.J. (1988). Central effects of TNFα on thermogenesis and fever in the rat. *Biosci. Rep.*, **8**, 345–51.

Rothwell, N.J., Busbridge, N.J., Humphray, H. and Hissey. (1989). Central actions of interleukin-1β on fever and thermogenesis. *Cytokine*, **1**, 153.

Sherry, B.A., Gelin, J., Fong, Y., Marono, M., Wei, H., Cerami, A., Lowry, S.F., Lundholm, K.G. and Moldawer, L.L. (1989). Anticachetin/tumor necrosis factor-α antibodies attenuate the development of cachexia in tumor models. *FASEB J.*, **3**, 1956–62.

Shih, S.T., Khorram, O., Lipton, J.M. and McCann, S.M. (1986). Central administration of α-MSH antiserum augments fever in the rabbit. *Am. J. Physiol.*, **250**, R803–6.

Socher, S.H., Friedman, A. and Martinez, D. (1988). Recombinant human tumor necrosis factor induces reductions in food intake and body weight in mice. *J. Exp. Med.*, **167**, 1957–62.

Stahl, W.M. (1987). Acute phase protein response to tissue injury. *Crit. Care Med.*, **15**, 545–50.

Starnes, H.F., Warren, R.S., Jeevanandam, M., Gabrilove, J.L., Larchian, W., Oettgen, H.F. and Brennan, M.F. (1988). Tumor necrosis factor and the acute metabolic response to injury in man. *J. Clin. Invest.*, **82**, 1321–5.

Stovroff, M.C., Fraker, D.L., Swedenborg, J.A. and Norton, J.A. (1988) Cachectin/-tumor necrosis factor: a possible mediator of cancer anorexia in the rat. *Cancer Res.*, **48**, 4567–72.

Steinmetz, T., Schaadt, M., Gahl, R., Schenk, V., Diehl, V. and Pfreundschuh, M. (1988). Phase 1 study of 24-hour continuous intravenous infusion of recombinant human tumor necrosis factor. *J. Biol. Resp. Modi.*, **7**, 417–23.

Tocco-Bradley, R.T., Moldawer, L.L., Jones, C.T., Gerson, B., Blackburn, G.L. and Bistrian, B.R. (1986). The biological activity *in vivo* of recombinant murine interleukin 1 in the rat. *Proc. Soc. Exp. Biol. Med.*, **182**, 263–71.

Tracey, K.J., Wei, H., Manogue, K.R., Fong, Y., Hesse, D.G., *et al.*, (1988). Cachetin/TNF induces cachexia, anorexia and inflammation. *J. Exp. Med.*, **167**, 1211–17.

Uehara, A., Sekiya, C., Takasugi, Y., Namiki, M. and Arimura, A. (1989a). Anorexia induced by interleukin 1: involvement of corticotropin-releasing factor. *Am. J. Physiol.*, **257**, R613–7.

Uehara, A., Ishikawa, Y., Okumura, T., Okamura, K., Sekiya, C., Takasugi, Y.

and Namiki, M. (1989b). Indomethacin blocks the anorexic action of interleukin-1. *Eur. J. Pharmacol.*, 170, 257–60.

Vaisman, N., Schattner, A. and Hahn, T. (1989). Tumor necrosis factor production during starvation. *Am. J. Med.*, **87**, 115.

Van Miert, A.S.J.P.A.M., van Duin, C.T.M. and Anika, S.M. (1986). Anorexia during febrile conditions in dwarf goats. The effect of diazepam, flurbiprofen and naloxone. *Vet. Quart.*, **8**, 266–73.

Weinberg, E.D. (1984). Iron withholding: a defense against infection and neoplasia. *Physiol. Rev.*, **64**, 65–102.

Zabel, P., Kreiker, C., Schonharting, M.M. and Schade, U.F. (1989a). Endotoxin-induced tumor necrosis factor in man and its reversal by pentoxifylline. *Cytokine*, **1**, 151.

Zabel, P., Wolter, D.T., Schonharting, M.M. and Schade, U.F. (1989b). Oxpentifylline in endotoxaemia. *Lancet*, **ii**, 1472–7.

Chapter Ten

Role of the Sympathetic Nervous System in Obesity

Arne Astrup, Benjamin Buemann and
Niels Juel Christensen†
Research Department of Human Nutrition, The Royal Veterinary and Agricultural University, 25 Rolighedsvej, DK-1958 Frederiksberg, Copenhagen, Denmark, and †Department of Internal Medicine and Endocrinology, Herlev Hospital, University of Copenhagen, Denmark

INTRODUCTION

Obesity is often a serious condition because it is commonly accompanied by complications such as type 2 diabetes, hypertension and hyperlipidaemia causing ischaemic heart disease and stroke. In the case of upper body obesity the incidence of somatic complications is increased, even at small degrees of overweight (Larsson *et al.*, 1984). Many efforts have been made to prevent and to treat obesity. The results are, however, modest and unsatisfactory. Even after combined treatment with diet, anorectics and behaviour therapy a large part of the weight loss is regained within a few years. This fact may be due in part to the development of a subnormal resting energy expenditure (Geissler *et al.*, 1987), and a defective diet-induced thermogenesis in individuals prone to become obese (Astrup *et al.*, 1990a). To establish the required knowledge for prevention and treatment of obesity it is necessary to gain increased insight into the physiological mechanisms of the disorder. Naturally, a crucial part is the factors controlling energy balance.

Obesity and Cachexia. Edited by N.J. Rothwell and M.J. Stock

GENETIC OBESITY IN ANIMAL MODELS

A large number of different types of inherited and experimentally induced obesity syndromes have been described in rodents (Trayhurn, 1984). The genetic obesities are generally accompanied by increased food intake, but the animals are hyperphagic only from about 4 weeks of age, by which time the obese state is already established. They still become obese when pair-fed to lean controls (Cleary and Vasselli, 1980). This indicates that a low energy expenditure is of primary importance for the development of their obesity. The high energetic efficiency can be ascribed both to diminished diet-induced thermogenesis and thermoregulatory thermogenesis, caused by an inability of the defective sympathetic nervous system (SNS) to increase activity when challenged with stimuli such as food and cold (Rothwell and Stock, 1979, 1984).

In rodents the increased thermogenesis may take place in brown fat. This is mainly caused by an activation of the SNS via the release of noradrenaline acting on β_3-receptors on brown adipocytes. Studies have demonstrated that brown fat thermogenesis may also be involved in different pathophysiological conditions such as obesity, sepsis and cachexia in rodents. The sympathoadrenal system may also control body composition, e.g. the ratio of lean body tissue to fat mass, by influencing skeletal muscle and heart protein synthesis as well as energy expenditure and lipolysis (Astrup, 1986).

In accordance with this concept, chronic administration of various agents stimulating SNS is effective in increasing energy expenditure and in reducing body fat content in genetically obese rodents (Dulloo and Miller, 1984). In particular, ephedrine in combination with the methylxanthines, caffeine and theophylline, has been shown to exhibit potent thermogenic and anti-obesity properties. Indeed, with this treatment obesity may be entirely reversed (Dulloo and Miller, 1987).

GENETICS IN HUMAN OBESITY

A number of recent studies have indicated that human obesity is also genetically determined. It does not mean that obesity is determined at conception, but rather that a vulnerability to certain environmental factors is inherited. Obesity becomes manifest when subjects with the predisposed liability are exposed to the environmental factors (Stunkard, 1989). Using the Danish Adoption Register the body mass index (BMI) of 540 adult adoptees was compared with the BMI of both the biological and adoptive parents. The weight class of the adoptees was strongly related to the BMI of their biological parents, while the weight class of the adoptees was not related to the BMI of their adoptive parents (Stunkard *et al.*, 1986). An extension of this study examining the relationship between the weights of the adoptees and their siblings supports the concept of a strong genetic influence. The exact mode of inheritance remains to be determined.

But what component of the energy balance is inherited? Is it an increased appetite leading to an increased energy intake or a subnormal energy expenditure, or perhaps a combination?

ENERGY BALANCE IN NORMAL SUBJECTS

The daily energy expenditure (24-h EE) and its components are shown in Figure 10.1. It is well established that the effect of age, sex and body weight on basal energy expenditure (BEE) and sleeping energy expenditure (SEE) in normal weight subjects is mainly due to variations in lean body mass (LBM). A pronounced variation in BEE and SEE measured in a respiration chamber was found between normal weight subjects (CV = 17–20%), females having significantly lower figures than males (Astrup *et al.*, 1991a). By contrast, the effect of age and sex disappears when the energy expenditures are standardized on the basis of LBM, which is an estimate of the metabolically active tissue. When adjusted for differences in LBM the between-subject variation in 24-h EE (on a fixed physical activity), BEE and SEE were reduced to 4.1–6.0%. Furthermore, differences in LBM account for 91–93% of the variation in BEE, SEE and 24-h EE (Astrup *et al.*, 1991a). On a 24-h basis BEE may account for 60–75% of the total 24-h EE.

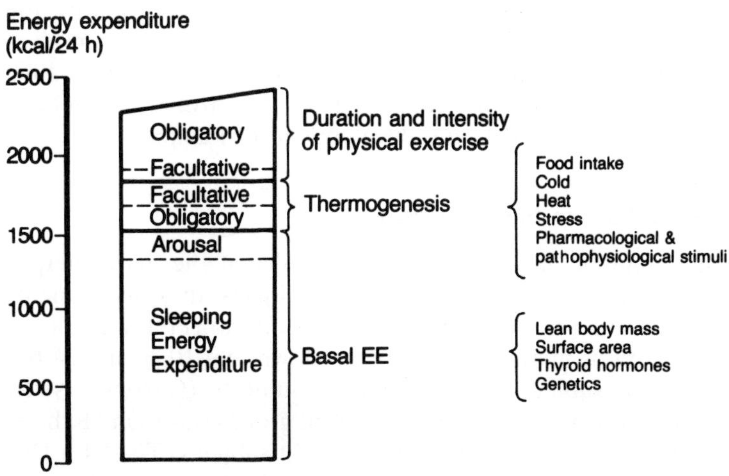

Figure 10.1 The 24-h energy expenditure and its components in a normal sedentary subject

Together with the energy expended for physical activity, thermogenesis accounts for the remaining part of the 24-h EE (Figure 10.1). The most important part is thermogenesis caused by food intake, but also exposure to warm and cold environments, stress, pharmacological (caffeine, nicotine, etc.) and pathophysiological stimuli (fever, infection, trauma, etc.) contribute. These stimuli activate the sympathoadrenal system, and this elicits a number of cardiovascular and metabolic effects, including increased thermogenesis.

SNS and energy expenditure

These normal, everyday physiological activations of the sympathoadrenal system have an impact on energy expenditure, and hence on energy balance and body weight. Stimuli, such as upright posture, food intake, cold exposure, heat and even mild mental stress, activate the sympathoadrenal system and result in a thermogenic response. A part of the thermogenic response may be unrelated to sympathoadrenal activity and is named 'obligatory thermogenesis'. Examples are the specific dynamic action of food, muscle shivering during cold exposure, and motor activity during exercise. By contrast, the sympathoadrenal component is named 'facultative thermogenesis'. Facultative thermogenesis varies between subjects, genetic factors may play an important role. Differences in the capacity for facultative thermogenesis seem to have sufficient quantitative importance to be involved in the propensity to obesity in humans (Astrup, 1990a). That the impairment of the sympathetic nervous system may result in obesity, is supported by experiments with chronic pharmacological blockade by β-adrenergic antagonists. β-Blockade is known to lower basal metabolic rate (Vernet *et al.*, 1987) and to diminish the increase in energy expenditure following carbohydrate-rich meals (Figure 10.2). Treatment with β-blockers also results in weight gain due to increased fat deposition (Astrup *et al.*, 1990b). In analogy with this concept it is probable that a defective sympathetic nervous system activity is involved in the development of human obesity.

Energy expenditure in human obesity

Although a number of studies have reported that the daily energy intake of obese subjects is lower than that of lean subjects, these assessments made by dietary records are not valid in obese subjects (Prentice *et al.*, 1986). From a diversity of methods it has been shown that BEE and 24-h EE are higher in obese subjects than in normal subjects (Garrow, 1990). BEE measured by indirect calorimetry in different groups was found to be 18–36% higher in obese groups than in normal weight subjects. The 24-h EE of the obese groups was also greater than that of the lean controls by 19–51%

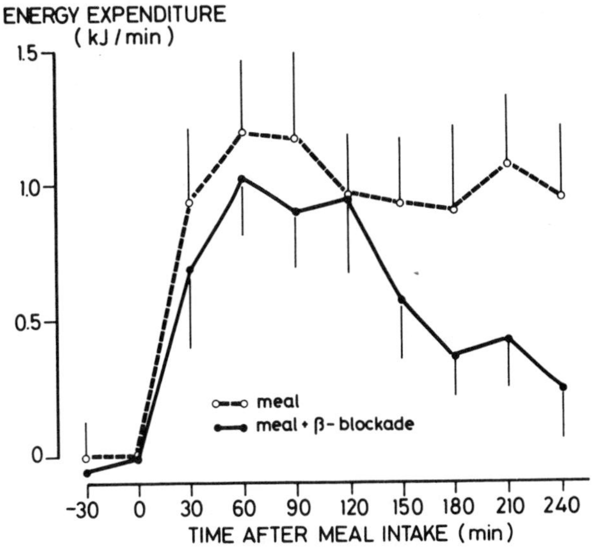

Figure 10.2 The effect of β-blockade on meal-induced thermogenesis. Normal subjects were given a mixed carbohydrate-rich meal with and without acute β-blockade with propranolol (Astrup *et al.* 1989b). Reproduced with permission of the American Physiological Society

(Garrow, 1990). This indicates that the obese patients had to maintain a higher energy intake than the lean controls to avoid weight loss, pointing also to a defective regulation of appetite. The fact that obese subjects have a higher energy expenditure than lean controls is not unexpected: about 25% of the excess body weight is LBM, which has its own energy requirements. This LBM is synthesized along with weight gain in humans and lost during slimming. However, this is a consequence of the obese state, and not a primary finding.

We have studied 24-h EE and its components in 41 obese women and in 16 lean controls in a respiration chamber. A positive correlation was found between free thyroxine index and the unadjusted SEE (Figure 10.3), but the effect of thyroxine on SEE was lost in the step-wise regression analysis. By contrast, in this analysis free T3-index was positively correlated to SEE. Step-wise regression analysis indicated that a total of 81% of the variation of SEE could be accounted for by differences in LBM, fat mass and free T3 index, while no effects were exerted by catecholamines. No difference in adjusted SEE could be found between obese and normal weight subjects (Figure 10.4). Indeed, SEE among obese patients varied by nearly 500 kcal/d (2 MJ/d), but the variation among lean control subjects was similar (Figure 10.4).

A group of 41 obese women have entered a weight reducing programme,

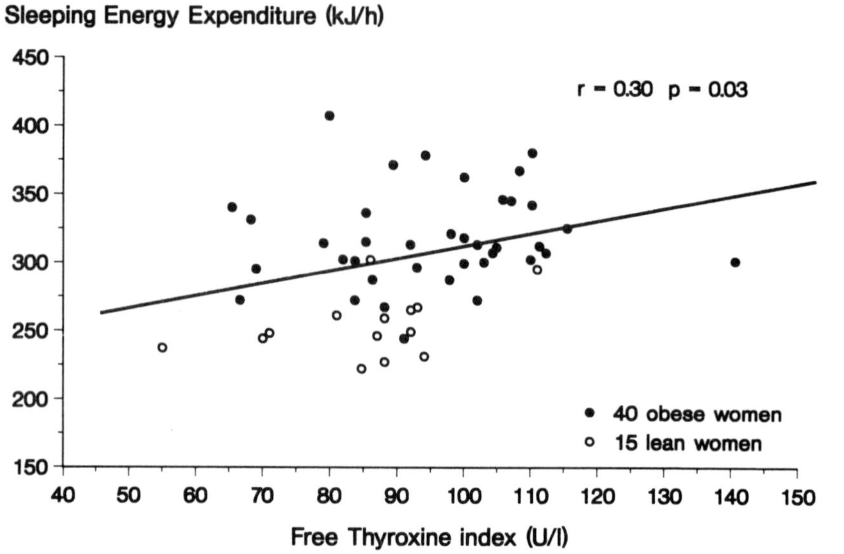

Figure 10.3 Correlation between free T4 index and sleeping energy expenditure in 40 obese women and 15 controls, $r = 0.30$, $p = 0.03$

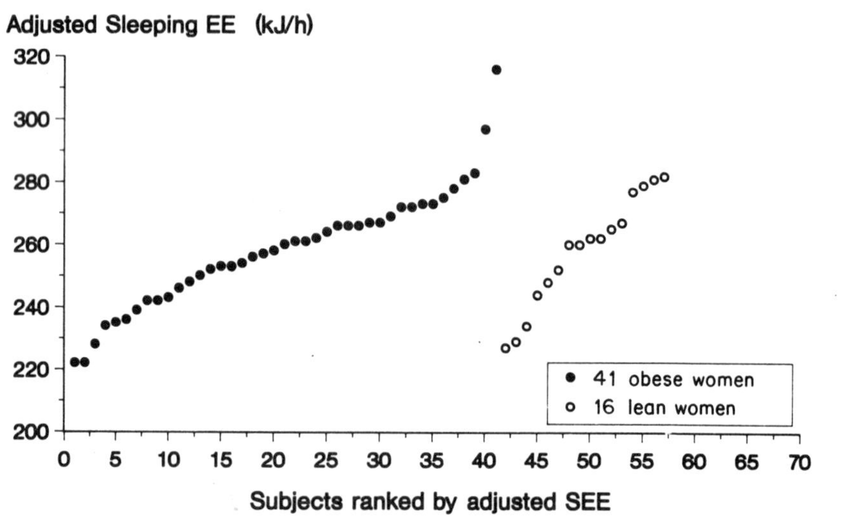

Figure 10.4 Sleeping energy expenditure adjusted for differences in lean body mass and fat mass separately for obese and lean individuals. Mean ± SE: obese vs. lean, 259 ± 3 vs. 258 ± 4 kJ/h, $p = 0.94$

and measurements of body composition, EE and thermogenic hormones will be repeated when ideal body weight is reached. Measurements on the first post-obese subjects suggest that their SNS activity and EE are highly susceptible to the dietary composition. A high-fat diet suppresses SNS activity and hence EE, while a high-carbohydrate diet enhances SNS activity and EE above the levels of a matched control group. Therefore, a normal adjusted EE in the obese state does not necessarily mean that EE is normal in the pre-obese and post-obese state.

Energy intake and EE in pre-obese and post-obese

The increased LBM and accompanying increased EE does not exclude that a subnormal EE may play an important role in the development of obesity and contribute to relapse after weight reduction. Indeed, several studies support this hypothesis. In a prospective study, Roberts and co-workers measured EE in infants at the age of 3 months (Roberts *et al.*, 1988). Despite similar body weight and composition, a group of babies had a 21% lower EE than that of the others. At 2 years of age the group with low EE had gained significantly more weight than the others. The lower EE could account entirely for positive energy balance. Bogardus *et al.* (1986) found in adult Pima Indians that BEE adjusted for differences in LBM among families varied by 500 kcal/d, but only by 60 kcal/d within families. Differences in LBM could account for only 83%, while family membership accounted for an additional 11%. In a follow-up study they found a markedly increased risk of becoming obese in subjects with a low adjusted BEE (Ravussin *et al.*, 1988). In fact, an inverse correlation between EE and subsequent weight gain was found ($r = -0.39$, $p<0.001$). It is tempting to speculate whether the 11% variation of BEE accounted for by family membership was due to differences in thyroid activity or varying SNS tone. However, these studies show that a low BEE predisposes to obesity.

Studies of normal weight, post-obese subjects have shown that at three different levels of activity they had a 15% lower 24-h EE than matched controls (Geissler *et al.*, 1987). A subsequent analysis further showed that BEE was 10% lower and thermogenesis 50% lower in the post-obese subjects (Shah *et al.*, 1988). It is, however, not known whether this is caused by a lower sympathetic tone. However, the finding that β-blockade may reduce BEE in normal subjects, but not in obese subjects, supports this explanation (Vernet *et al.*, 1987). β-Blockade in lower-body obese patients reduces BEE, while the effect is blunted in upper-body obesity (Christin *et al.*, 1989).

Low basal SNS activity in obesity

There are conflicting reports on the fasting levels of noradrenaline in obese patients compared with lean controls. Some claim that it is elevated or normal, while others have found it to be reduced. These studies, however, have been conducted in small groups of subjects. Recently, a more extensive study has been carried out, where several indexes of sympathetic activity, including plasma noradrenaline concentration, were measured in obese subjects (Peterson *et al.*, 1988). SNS activity was inversely correlated with body fat mass. However, it is still unknown in what way obesity and sympathetic tone are interrelated. Insulin has been shown to stimulate the sympathetic nervous system and insulin resistance is a feature of obesity and NIDDM, so an interaction between the degree of insulin resistance and the sympathetic activity seems possible. We have conducted a number of studies to elucidate SNS activity in obesity, and we have consistently found reduced levels of noradrenaline in obese subjects compared with normal weight controls (Astrup *et al.*, 1989a). Two major pitfalls should be avoided in comparative studies: (1) SNS activity and plasma noradrenaline concentration increases with age by 5–6 pg/ml per year (0.03 nmol/l), which is important when matching with a control group (Astrup *et al.*, 1991b). (2) The seasonal variation in noradrenaline concentration with high levels in the winter and low levels in the summer and autumn is important in obese and post-obese subjects (Jung *et al.*, 1982). Body fat topography may also have an impact on SNS via differences in insulin resistance (Christin *et al.*, 1989).

A defective SNS responsiveness obesity

Human obesity is accompanied by a reduced thermogenic response to glucose and carbohydrate-rich meals (Astrup *et al.*, 1990a), cold (Astrup *et al.*, 1989a), and hypoglycaemia (Swaminathan *et al.*, 1986), all of which seem to be caused by a deficiency of the sympathoadrenal system. Also, post-obese subjects exhibit an impaired thermogenic response to pharmacological SNS activation (Dolloo *et al.*, 1986).

The thermic effect of carbohydrate is partly facultative, mediated by the sympathoadrenal system via catecholamines acting on β-adrenoceptors (Astrup *et al.*, 1989b). A decreased glucose-induced activation of the sympathetic nervous system has been reported in obese individuals (Astrup *et al.*, 1987). It is likely that an impaired sympathetic activity may be responsible for the lower carbohydrate-induced thermogenesis in obese and post-obese subjects and that the reduced energy expenditure may predispose to weight regain.

We have found a low thermogenic effect and a blunted increase in SNS activity in response to glucose in obese patients. After a 30 kg weight loss glucose and lipid metabolism were almost normalized, but the thermogenic response to glucose was still lower in the obese patients than in the control group (Astrup *et al.*, 1990a). Also the glucose-induced increase in noradrenaline concentration remained diminished in the reduced obese patients. Jung *et al.* also reported a blunted noradrenaline response to hypoglycaemia in formerly obese patients (Jung *et al.*, 1983). These abnormalities point to a central defect, and it is conceivable that they are causative in the development of obesity and in promoting relapse after weight reduction.

SNS activity in obese diabetics

As insulin activates SNS, the degree of insulin resistance of obesity and overt diabetes may influence SNS activity. We studied fasting and post-glucose noradrenaline concentrations in 50 obese patients with and without diabetes (Astrup *et al.*, 1991b). The results of this study indicate that obese subjects without signs of autonomic neuropathy, have lower fasting noradrenaline concentrations, both in non-diabetics and in diabetics, than in a normal weight control group. Step-wise regression analysis showed that the inverse relation between noradrenaline level and the degree of obesity were partly due to co-variation of the degree of obesity with fasting hyperglycaemia. In an other study, we found that SNS activity inversely correlated to plasma glucose concentration in obese non-diabetics (Astrup *et al.*, unpublished results). The increase in noradrenaline 2 h after an oral glucose load was preserved in the non-diabetic obese group, whereas it was abolished in the diabetic group (Figure 10.5). In response to oral glucose we found an inverse correlation between the increment in plasma noradrenaline concentration and the increment in plasma glucose concentration.

This study suggests that impairment of glucose metabolism is an independent correlate of sympathetic nervous system activity in obesity, and heterogeneity in glucose metabolism may contribute to explaining the diverging results on plasma noradrenaline levels reported in obese subjects.

It is unknown whether this deficiency is a cause or a consequence of human obesity, but the lower sympathetic activity may result in lower energy expenditure and thus contribute to a positive energy balance. The finding that an effect of obesity on base-line noradrenaline remained after adjustment for hyperglycaemia may suggest that both primary and secondary mechanisms are operating.

CONCLUSION

There is accumulating evidence showing that obesity has a strong genetic influence. The inherited abnormality involves both energy expenditure and energy intake. A subnormal basal energy expenditure and thermogenic response to food, and other stimuli activating the sympathetic nervous

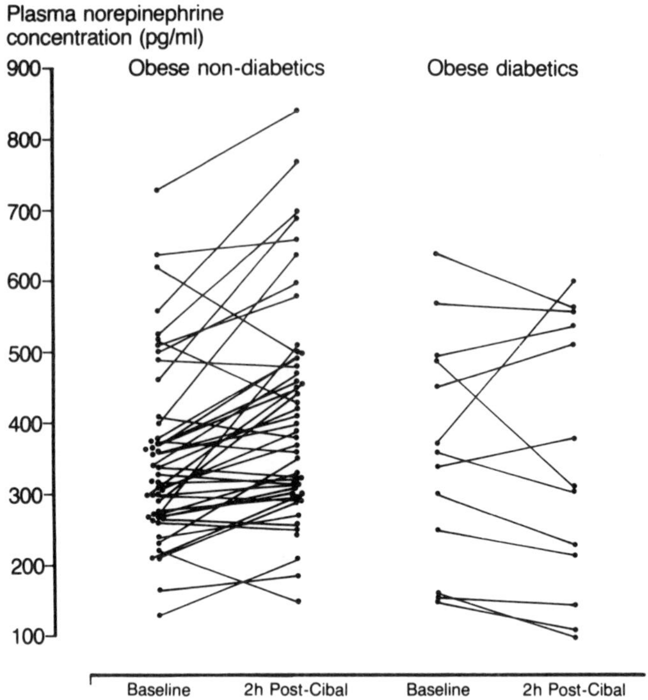

Figure 10.5 Changes in plasma norepinephrine concentration after an oral glucose load in obese patients with and without type 2 diabetes. In the obese non-diabetics the normal increment is preserved, while the response is blunted in the obese diabetics (Astrup *et al.* 1991b).

system, are present in the post-obese and perhaps also in the pre-obese state. When the obese state is established the energy intake and energy expenditure are higher than in normal weight subjects. The heterogeneity of obese subjects has produced conflicting results on SNS activity in obesity. However, generally SNS activity is reduced in obese patients, but the increase in SNS activity to carbohydrate is blunted only in the presence of a marked insulin resistance or diabetes. More studies on regulation of appetite and energy expenditure are needed in pre-obese and post-obese individuals.

ACKNOWLEDGEMENTS

The authors thank Grete Thorbek, Björn Isaksson and Flemming Quaade and the staff of the Metabolic Unit; Tina Cuthbertson, Lene Kristiansen, Inge Timmermann and John Lind. The work was supported by grants from the Danish Medical Research Council (12-8510 and 12-9084), the Danish Veterinary and Agricultural Research Council (13-4268) and the Ib Henriksen Foundation.

REFERENCES

Astrup, A. (1986). Thermogenesis in human brown adipose tissue and skeletal muscle induced by sympathomimetic stimulation. *Acta Endocrinol.*, **112** (suppl. 278), 1–32.

Astrup, A., Andersen, T. and Henriksen, O. (1987). Impaired glucose-induced thermogenesis in skeletal muscle in obesity. The role of the sympathoadrenal system. *Int. J. Obesity*, **11**, 51–67.

Astrup, A., Simonsen, L., Bülow, J. and Christensen, N.J. (1989a). The contribution of skeletal muscle to carbohydrate-induced thermogenesis in man: the role of the sympathoadrenal system. In: *Hormones, Thermogenesis and Obesity* (eds H. Lardy and F. Stratman), Elsevier, New York, pp. 187–96.

Astrup, A., Simonsen, L., Bülow, J., Madsen, J. and Christensen, N.J. (1989b). Epinephrine mediates facultative carbohydrate-induced thermogenesis in human skeletal muscle. *Am. J. Physiol.*, **257**, E340–5.

Astrup, A., Andersen, T., Christensen, N.J., Henriksen, J. and Quaade, F. (1990a). Impaired glucose-induced thermogenesis and arterial norepinephrine response persist after weight reduction in human obesity. *Am. J. Clin. Nutr.*, **51**, 331–7.

Astrup, A., Christensen, N.J., Simonsen, L. and Bülow, J. (1990b). Effects of nutrient intake on sympathoadrenal activity and thermogenic mechanisms. *J. Neurosci. Methods*, **34**, 187–92.

Astrup, A., Thorbek, G., Lind, J. and Isaksson, B. (1990c). Prediction of 24 hours energy expenditure and its components from physical characteristics and body composition in normal weight subjects. *Am. J. Clin. Nutr.*, **52**, 777–83.

Astrup, A., Christensen, N.J. and Breum, L. (1991b). Reduced activity of the sympathetic nervous system in obese patients with and without NIDDM. *Clin. Sci.*, **80**, 53–8.

Bogardus, C., Lillioja, S., Ravussin, E., Abbott, W., Zawadzki, J.K., Young, A., Knowler, W.C., Jacobowitz, R. and Moll. P.P. (1986). Familial dependence of resting metabolic rate. *N. Engl. J. Med.*, **315**, 96–100.

Cleary, P.M. and Vasselli, J.R. (1980). Development of obesity in Zucker obese (fa/fa) rat in absence of hyperfagia. *Am. J. Physiol.*, **238**, E284–92.

Christin, L., Ravussin, E., Bogardus, C. and Howard, B. (1989). The effect of propranolol on free fatty acid mobilization and resting metabolic rate. *Metabolism*, **38**, 439–44.

Dulloo, A.G. and Miller, D.S. (1984). Thermogenic drugs for the treatment of obesity: sympathetic stimulants in animal models. *Br. J. Nutr.*, **52**, 179–96.

Dulloo, A.G. and Miller, D.S. (1986). The thermogenic properties of ephedrine/methylxanthine mixtures: human studies. *Int. J. Obesity*, **10**, 467–81.

Dulloo, A.G. and Miller, D.S. (1987). Reversal of obesity in the genetically obese fa/fa Zucker rat with an ephedrine/methylxanthines thermogenic mixture. *J. Nutr.*, **117**, 383–9.

Garrow, J.S. (1990). Hormonal and nutritional factors contributary to obesity. In: *Hormones and Nutrition in Obesity and Cachexia* (eds M.J. Muller *et al.*) Springer-Verlag, Berlin, pp. 69–76.

Geissler, C.A., Miller, D.S. and Shah, M. (1987). The daily metabolic rate of the post-obese and the lean. *Am. J. Clin. Nutr.*, **45**, 914–20.
Jung, R.T., Shetty, P.S., James, W.P.T., Barrand, M.A. and Callingham, B.A. (1982). Plasma catecholamines and autonomic responsiveness in obesity. *Int. J. Obesity*, **6**, 131–41.
Jung, R.T., Campbell, R.G., James, W.P.T. and Callingham, B.A. (1983). Altered hypothalmic and sympathetic responses to hypoglycaemia in familial obesity. *Lancet*, **ii**, 1043–6.
Larsson, B., Svardsudd, K., Welin, L., Wilhelmsen, L., Björntorp, P. and Tibblin, G. (1984). Abdominal adipose tissue distribution, obesity, and risk of cardiovascular disease and death: 13 year follow up of participants in the study of men born in 1913. *Br. Med. J.*, **288**, 1401–40.
Peterson, H., Rothschild, M. and Weinberg, C.R. (1988). Body fat and the activity of the autonomic nervous system. *N. Engl. J. Med.*, **318**, 1077–83.
Prentice, A.M., Black, E.A., Coward, W.A., Davies, H.L. and Goldberg, G.R. (1986). High levels of energy expenditure in obese women. *Br. Med. J.*, **292**, 983–7.
Ravussin, E., Lillioja, S., Knowler, W.C. *et al.* (1988). Reduced rate of energy expenditure as a risk factor for body-weight gain. *N. Engl. J. Med.*, **318**, 467–72.
Roberts, S.B., Savage, S., Coward, W.A., Chew, B. and Lucas, A. (1988). Energy expenditure and intake in infants born to lean and overweight mothers. *N. Engl. J. Med.*, **318**, 461–6.
Rothwell, N.J. and Stock, M.J. (1979). A role for brown adipose tissue in diet-induced thermogenesis. *Nature*, **281**, 31–5.
Rothwell, N.J. and Stock, M.J. (1984). Brown adipose tissue. In: *Recent Advances in Physiology* (ed. P.F. Baker), Churchill Livingstone, Edinburgh, pp. 349–84.
Shah, M., Miller, D.S. and Geissler, C.A. (1988). Lower metabolic rates of post-obese versus lean women: thermogenesis, basal metabolic rate and genetics. *Eur. J. Clin. Nutr.*, **42**, 741–52.
Stunkard, A.J., Sørensen, T.I.A., Hanis, C. *et al.* (1986). An adoption study of human obesity. *N. Engl. J. Med.*, **314**, 193–8.
Stunkard, A.J. (1989). Some perspectives on human obesity: its causes. *Bull. NY Acad. Med.*, **64**, 902–23.
Swaminathan, R., Andersen, E., Dean, H. and Wales, J.K. (1986). Metabolic response to insulin induced hypoglycaemia in lean and obese subjects. *Horm. Metab. Res.*, **18**, 45–8.
Trayhurn, P. (1984). The development of obesity in animals: The role of genetic susceptibility. In: *Clinics in Endocrinology and Metabolism* (ed. W.P.T. James), W.B. Saunders, London, pp. 451–73.
Vernet, O., Nacht, C.A. and Christin, L. (1987). β-Adrenergic blockade and intravenous nutrient-induced thermogenesis in lean and obese women. *Am. J. Physiol.*, **253**, E65–71.

Chapter Eleven

Gender Hormonal Response in the Human Adipocyte

Marielle Rebuffé-Scrive
Department of Psychology, Yale University, New Haven, CT 06520-7447, USA

INTRODUCTION

Artists have since long recognized that fat is differently distributed between men and women: the typical female fat depots being in the lower part of the body (around the hips and thighs), while men tend to have fat in the central part of the body (abdominal regions). These differences in fat distribution which might seem, at first, to be only of aesthetical importance have in fact major unwanted health consequences. In 1956, Professor Jean Vague noted the importance of fat distribution as a determinant of various diseases. He pointed out that android (abdominal) obesity was a factor determining predisposition for diabetes, atherosclerosis, gout and uric calculous diseases. However, it was not until recently that this whole area of physiology has received renewed interest, when a number of prospective studies have shown that, in contrast to fat localized in the lower part of the body (gluteofemoral regions), abdominal distribution of fat was a strong independent risk factor for cardiovascular diseases, premature death, stroke, diabetes mellitus and female carcinomas (Larsson *et al.*, 1984; Lapidus *et al.*, 1984; Ohlsson *et al.*, 1985; Björntorp, 1987).

In these studies, the ratio of waist to hip circumferences (WHR) was used as an index of fat distribution. A low WHR, indicating a fat distribution in the gluteofemoral regions was found particularly in women, while a high WHR (abdominal distribution of fat) was more typical for men. It is

Obesity and Cachexia. Edited by N.J. Rothwell and M.J. Stock

important to note, however, that the prospective study of women in Göteborg (Lapidus *et al.*, 1984) has shown that women who have a male type of fat distribution (in the abdominal regions) have as many risks as men for the metabolic diseases mentioned earlier. The WHR is a simple, inexpensive and very powerful predictor for various diseases; however, it may not necessarily reflect differences in the distribution of fat among various intra-abdominal regions. Recently, using more sophisticated methods such as computed tomography (CT) and magnetic resonance imaging (MRI), it was possible to assess intra-abdominal fat. It was shown (Kvist *et al.*, 1988) that men have two to three times more intra-abdominal fat than women and that with ageing, men and women tend to accumulate more fat in these regions. Moreover, the ratio of intra- to subcutaneous abdominal fat was even more strongly correlated to risk factors for cardiovascular diseases and diabetes than the WHR itself (Fujioka *et al.*, 1987; Deprés *et al.*, 1985; Peiris *et al.*, 1989).

Whether the enlargement of abdominal adipose tissues is directly causal to the increased risk factors for cardiovascular diseases and diabetes or is only a parallel phenomenon to the diseases, is not known. It therefore becomes very important to study the factors which regulate the distribution of fat in different regions of the body. Several recent studies have shown that genetic (Bouchard, 1987) but also hormonal factors, particularly sex steroid and glucocorticoid hormones (Rebuffé-Scrive *et al.*, 1985a, 1986, 1987, 1989a, 1989b, 1990a) seem to play a major role in the regulation of adipose tissue distribution and function.

In the following, the current knowledge of the morphological and functional responses of subcutaneous and intra-abdominal adipocytes to various concentrations of sex steroid and glucocorticoid hormones as well as their consequences for fat distribution, will be reviewed.

RESPONSES TO PHYSIOLOGICAL CONCENTRATIONS OF STEROID HORMONES

Women

Results obtained from descriptive studies from non-obese young women with functioning ovaries have shown that these women have a low WHR, indicating a preferential distribution of fat in the gluteofemoral regions. In comparison with fat cells obtained from the abdominal regions, femoral adipocytes were enlarged and lipoprotein lipase (LPL) activity, the key enzyme for fat accumulation, was higher in this region (Rebuffé-Scrive *et al.*, 1985a). In contrast, fat mobilization measured as glycerol release from isolated adipocytes in the presence or absence of noradrenaline at various concentrations was lower in the femoral region (Rebuffé-Scrive *et al.*, 1986). The combination of a high LPL and a low lipolysis in the femoral region seems to favour a distribution of fat in the gluteofemoral regions as observed *in vivo* in these women (low WHR).

High femoral LPL activity and low lipolysis were found during both parts of the menstrual cycle as well as during early pregnancy, while no regional differences were observed in either parameter after one month of lactation (Rebuffé-Scrive *et al.*, 1985a). This suggested that during lactation other factors regulate adipose tissue metabolism. A potential role played by prolactin, which increases during lactation, was investigated in patients with hyperprolactinaemia. It was found that prolactin, at least alone, did not seem to have any specific effect on regional adipose tissue metabolism in humans (Rebuffé-Scrive *et al.*, 1991). It has been suggested, in rat studies, that changes in insulin levels might be the trigger for a low LPL activity in adipose tissue and a high LPL activity in the mammary gland (see Chapter 5 by D.H. Williamson in this book).

In post-menopausal women who have no or very low levels of female sex hormones, regional differences were found neither in LPL activity nor in lipolysis. However, when post-menopausal women received gestagen and progestogen replacement therapy, a reappearance of a high LPL activity was found only in the femoral region, suggesting a regulation of femoral LPL activity by female sex hormones (Rebuffé-Scrive *et al.*, 1987).

Thus, it seems that in women, female sex hormones regulate adipose tissue metabolism in the gluteofemoral regions, where fat is preferentially accumulated and difficult to mobilize except during lactation.

Men

Studies of adipocyte morphology in men of different ages have shown that men tend to have large adipocytes in the subcutaneous abdominal (Rebuffé-Scrive *et al.*, 1987) and intra-abdominal regions (Rebuffé-Scrive *et al.*, 1989a), particularly when they get older. They also have larger abdominal fat depots as shown by CT scans (Kvist *et al.*, 1988). No regional differences were observed in LPL activity between depots. However, testosterone undecanoate given daily to middle aged men for 6–8 weeks dramatically inhibits LPL activity in the abdominal region (Rebuffé-Scrive *et al.*, 1989b). This suggests that when men get older and have lower levels of testosterone, abdominal LPL activity might be less inhibited and therefore fat might be more easily accumulated in the abdominal regions.

In men, lipolysis was high in subcutaneous abdominal adipose tissue in comparison with femoral fat (Rebuffé-Scrive *et al.*, 1987) and was even higher in intra-abdominal fat tissues (Rebuffé-Scrive *et al.*, 1989a). When testosterone supplementation (as described above) was given to middle aged men, noradrenaline-stimulated lipolysis increased in the abdominal region. Important regional differences were also observed between portal (omental and mesenteric) and non-portal (retroperitoneal and subcutaneous abdominal) adipose tissues when lipolysis was stimulated by both noradrenaline and isoproterenol, a pure β agonist, suggesting a regulation of lipolysis mainly through the β-adrenergic receptors. In addition, studies performed under adipose tissue culture conditions have shown that testosterone seemed to express an increased lipolytic potential and depress LPL activity (Xu *et al.*, 1990). These results contrast with those obtained in women showing that portal lipolysis was lower than non-portal. The high portal lipolysis found in men might have a specific physiological significance and might be an important trigger for the pathogenesis of abdominal obesity, as suggested by Björntorp (Björntorp, 1987). A high release of free fatty acids (FFA) from these tissues into the portal vein might lead to disturbances in hepatic metabolism, such as increasing VLDL, stimulating gluconeogenesis and inhibiting insulin clearance and therefore might lead to increased risk factors for a number of prevalent diseases.

In summary, male sex hormones seem to regulate adipose tissue metabolism preferentially in the abdominal, particularly intra-abdominal, regions by inhibiting LPL activity and stimulating lipolysis.

RESPONSES TO ABNORMAL STEROID HORMONE LEVEL

A number of clinical conditions where the balance between sex steroid and glucocorticoid hormones is disturbed (with low levels of sex steroid and/or high levels of glucocorticoid hormones) seem to be associated with an abdominal type of fat distribution. In the following such examples will be described.

Low sex steroid hormone levels

Normal physiological ageing is accompanied by a decreased secretion of sex steroid hormones. In men, as mentioned earlier, this might mean a decreased inhibition of LPL and of noradrenaline-stimulated lipolysis in the abdominal regions. In fact, it was observed that abdominal lipolysis was lower in older men (55 years <38 years, Rebuffé-Scrive unpublished data). It is therefore likely that the lower level of testosterone found in older men will favour a fat deposition in the abdominal regions. CT scan measurements showed that this was, in fact, the case. Furthermore, a recent report has shown that the mass of intra-abdominal fat was strongly negatively associated with plasma testosterone levels (Seidell *et al.*, 1990).

As described earlier, in older women, menopause seems to be associated with a disappearance of the female characteristics of adipose tissue (low femoral LPL activity and abdominal lipolysis), suggesting that older women will accumulate fat without regional specificity (proportionally more in the abdominal regions in comparison with young women). Indeed it was shown in the prospective study of women in Göteborg (Lapidus *et al.*, 1984) where the distribution of the WHR was measured in three age groups of women (38, 50, 60 years), that older women (60 years) have a higher WHR indicating an increased fat accumulation in the abdominal regions. In addition, Enzi *et al.* (1986) have shown that post-menopausal women have more intra-abdominal fat.

Android (abdominal) obesity, in women, is not only associated with a lower level of female sex hormones but also with an increased level of male sex hormones and decreased sex hormone binding globulin (SHBG) (Kissebah *et al.*, 1982). These women have male adipose tissue characteristics. In contrast to gynoid obese women who had similar adipose tissue characteristics as young fertile women, no regional differences in LPL activity or adipose tissue morphology were observed in android obese women (Rebuffé-Scrive, unpublished data). Furthermore, these women had large intra-abdominal adipocytes and had a high portal lipolysis as shown in men (Rebuffé-Scrive *et al.*, 1990a). They also have increased risk factors for prevalent diseases (Lapidus *et al.*, 1984).

Polycystic ovarian syndrome (PCO) is another condition associated with sex steroid hormones disturbances (low female sex hormones, high testosterone, high SHBG) and typical male characteristics of adipose tissue function and distribution (Rebuffé-Scrive *et al.*, 1989c).

All together these data suggest that decreased secretion of sex steroid hormones, such as during ageing, both in men and women and increased secretion of male sex hormones in women (android obesity and PCO syndrome) lead to preferential fat accumulation in the abdominal (intra-abdominal) regions.

High glucocorticoid and low sex steroid hormones

Several studies have shown that glucocorticoid hormones regulate fat metabolism and distribution. In *in vitro* culture conditions, cortisol in the presence of insulin induces a four-fold increase in LPL activity in abdominal human adipose tissue (Cigolini and Smith, 1979). *In vivo*, patients with Cushing's syndrome (hypersecretion of cortisol) have a 2–3 times higher LPL activity than controls, specifically in abdominal adipose tissues (Rebuffé-Scrive *et al.*, 1988). They also have a central fat deposition. Moreover, recently Mayo-Smith *et al.* (1989) have shown that Cushing's syndrome patients have five times more intra-abdominal fat than controls.

Glucocorticoid hormones seem to affect adipose tissue directly via specific cytosolic receptors. The presence of these receptors (Rebuffé-Scrive *et al.*, 1985b) as well as their messenger RNA (mRNA) (Rebuffé-Scrive *et al.*, 1990b) has been demonstrated in human adipose tissue, and interestingly regional differences were observed with a highest density in intra-abdominal > subcutaneous abdominal > femoral fat tissues, respectively. It has been suggested that these regional differences might explain the specific fat accumulation in the abdominal regions found in Cushing's syndrome patients (Rebuffé-Scrive *et al.*, 1988). The low level of female sex hormones found in these patients might also contribute to this specific fat deposition (see above).

Finally, low sex hormones and high glucocorticoid hormones are typical neuroendocrine responses to a number of environmental factors such as, for example, alcohol and stress.

The association between these conditions, fat distribution and increased risk factors for metabolic disorders has only been recently investigated. Alcoholic men and women have high cortisol levels and decreased levels of sex hormones. Furthermore, alcoholic women have increased levels of free testosterone and low SHBG (Petterson *et al.*, personal communication). Alcoholic men and women have a central distribution of fat (Larsson *et al.*, 1989) and CT scan measurements have shown that they have a dramatic increase of intra-abdominal fat tissues (Sjöström, personal communication).

A positive correlation was found between feeling of stress as well as with other psychological maladjustments and WHR (Lapidus *et al.*, 1989). Stress, particularly uncontrollable or submissive stress, as shown by Henry *et al.* (Henry and Stephens, 1977), in mice, stimulates the secretion of glucocorticoid hormones and decreases the production of sex hormones. It is therefore possible that these neuroendocrine disturbances following stress might lead to specific changes in fat deposition towards the abdominal regions. This hypothesis was tested recently in a rat model, where it was possible to demonstrate that after exposure to a chronic uncontrollable stress, rats increased specifically their intra-abdominal (portal) fat tissue (Rebuffé-Scrive *et al.*, 1990d). Furthermore, it was shown that this increase was positively correlated with plasma corticosterone concentrations. Further studies are currently being performed in order to develop this model which seems to be very appropriate to study further the regulation of abdominal obesity and its associated metabolic disorders.

The combination of high glucocorticoid and low sex steroid hormones secretion seem therefore to favour fat accumulation in the abdominal, particularly intra-abdominal, regions.

CONCLUSION

The results of the studies described here clearly indicated that steroid hormones affect human adipocyte metabolism. These effects vary with the type and concentration of hormone. Female sex hormones seem to favour fat accumulation in the lower part of the body (gluteofemoral region) via increased femoral LPL activity and decreased lipolysis. In young men, male sex hormones seem to inhibit fat accumulation in the abdominal region via inhibition of LPL activity and stimulation of lipolysis. Glucocorticoid hormones seem to favour abdominal (intra-abdominal) fat deposition by stimulation of LPL activity.

These metabolic adaptations seem to be followed by parallel changes in fat distribution. In consequence, it seems that in conditions where hormonal disturbances are present, such as decreased sex hormone and increased glucocorticoid hormone secretion, fat is preferentially accumulated in the abdominal and particularly in the intra-abdominal regions. This in turn might result in increased risk factors for several prevalent metabolic disorders (Larsson *et al.*, 1984; Lapidus *et al.*, 1984; Fujioka *et al.*, 1987; Peiris *et al.*, 1989).

ACKNOWLEDGEMENTS

All my collaborators at the University of Göteborg (Sweden) in the Departments of Medicine, Surgery and Gynecology, at the University of Stockholm (Sweden) Department of Clinical Nutrition, and at the University of Yale (USA), Department of Psychology are gratefully acknowledged for their contribution to the work presented here.

REFERENCES

Björntorp, P. (1987). The association between obesity, tissue distribution and disease. *Acta Med. Scand.*, **723**, 121–34.

Bouchard, C. (1987). Genetic factors in the regulation of adipose tissue distribution. *Acta Med. Scand.*, **723**, 135–41.

Cigolini, M. and Smith, U. (1979). Human adipose tissue in culture. VIII. Studies on the insulin antagonistic effect of glucocorticoid. *Metabolism*, **28**, 502–10.

Després, J.P., Allard, C., Tremblay, A., Talbot, J. and Bouchard, C. (1985). Evidence for a regional component of body fatness in the association with serum lipids in men and women. *Metabolism*, **34**, 967–73.

Enzi, G., Gasparo, M., Biondetti, P.R., Fiore, D., Semisa, M. and Zurlo. (1986). Subcutaneous and visceral fat distribution according to sex, age and overweight, evaluated by computed tomography. *Am. J. Clin. Nutr.*, **44**, 739–46.

Fujioka, S., Matsuzawa, Y., Tokunaga, K. and Tarui, S. (1987). Contribution of intraabdominal fat accumulation to the impairment of glucose and lipid metabolism in human obesity. *Metabolism*, **36**, 54–9.

Henry, J.P. and Stephens, P.M. (1977). *Stress, Health, and the Social Environment: A Sociobiological Approach to Medicine*, Spring, New York.

Kissebah, A.H., Vydelingum, N., Murray, R., Evans, D.J., Hartz, A.J., Kalkhoff, R.K. and Adams, P.W. (1982). Relation of body fat distribution to metabolic complications of obesity. *J. Clin. Endocrinol. Metab.*, **542**, 54–60.

Kvist, J., Chowdhur, B., Grangard, U., Tylèn, U. and Sjöström, L. (1988). Predictive equations of total and visceral adipose tissue volumes derived from measurements with computed tomography in adult men and women. *Am. J. Clin. Nutr.*, **48**, 1351–61.

Lapidus, L., Bengtsson, C., Larsson, B. *et al.* (1984). Distribution of adipose tissue and risk of cardiovascular disease and death: A 12 year follow-up of participants in the population study of women in Gothenburg, Sweden. *Med. J.*, **289**, 1257–61.

Lapidus, L., Bengtsson, C., Hallstrom, T. and Björntorp, P. (1989). Obesity adipose tissue distribution and health in women: Results from a population study in Gothenburg, Sweden. *Appetite*, **13**, 25–35.

Larsson, B., Svardsudd, K., Welin, L., Wilhelmsen, L., Björntorp, P. and Tibblin, G. (1984). Abdominal adipose tissue distribution, obesity and risk of cardiovascular disease and death: A 13 year follow-up of participants in the study of men born in 1913. *Diabetes*, **34**, 1055–8.

Larsson, B., Seidell, J., Svardsudd, K., Welin, L., Tibblin, G., Wilhelmsen, L. and Björntorp, P. (1989). Obesity, adipose tissue distribution and health in men: The study of men born in 1913. *Appetite*, **13**, 37–44.

Mayo-Smith, W., Hayes, C., Biller, B., Klibanski, A., Rosenthal, H. and Rosenthal, D. (1989). Body fat distribution measured with CT: Correlations in healthy subjects, patients with anorexia nervosa, and patients with Cushing's syndrome. *Radiology*, **170**, 515–8.

Ohlsson, L.O., Larsson, B., Svardsudd, K. *et al.* (1985). The influence of body fat distribution on the incidence of diabetes mellitus: 13.5 years of follow-up of the participants in the study of men born in 1913. *Diabetes*, **34**, 1055–8.

Peiris, A.N., Sothmann, M.S., Hoffmann, R.G., Hennes, M.I., Wilson, C.R., Gustafson, A.B. and Kissebah, A.H. (1989). Adiposity, fat distribution, and cardiovascular risk. *Ann. Intern. Med.*, **110**, 867–72.

Rebuffé-Scrive, M., Enk, L., Crona, N., Lönnroth, P., Abrahamsson, L., Smith, U. and Björntorp, P. (1985a). Fat cell metabolism in different regions in women: Effect of menstrual cycle, pregnancy and lactation. *J. Clin. Invest.*, **75**, 1973.

Rebuffé-Scrive, M., Lundholm, K. and Björntorp, P. (1985b). Glucocorticoid hormone binding to human adipose tissue. *Eur. J. Clin. Invest.*, **15**, 267–71.

Rebuffé-Scrive, M., Eldh, J., Hafström, L. and Björntorp, P. (1986). Metabolism of mammary, abdominal and femoral adipocytes in women before and after menopause. *Metabolism*, **35**, 792–7.

Rebuffé-Scrive, M., Lönnroth, P., Wesslau, C., Björntorp, P. and Smith, U. (1987). Regional adipose tissue metabolism in men and in postmenopausal women. *Int. J. Obesity*, **11**, 347–55.

Rebuffé-Scrive, M., Krotkiewski, M., Elfverson, J. and Björntorp, P. (1988). Muscle and adipose tissue morphology and metabolism in Cushing's syndrome. *J. Clin. Endocrinol Metab.*, **67**, 1122.

Rebuffé-Scrive, M., Andersson, B., Olbe, L. and Björntorp, P. (1989a). Metabolism of adipose tissue in intraabdominal depots of nonobese men and women. *Metabolism*, **38**, 453–8.

Rebuffé-Scrive, M., Mårin, P. and Björntorp, P. (1989b). Male adipose tissue. *Int. J. Obesity*, **13** (suppl. 1), 187. *NMCD*, in press.

Rebuffé-Scrive, M., Cullberg, G., Lindstedt, G. and Björntorp, P. (1989c). Anthropometric variables and metabolism in polycystic ovarian disease. *Horm. Metab. Res.*, **21**, 391–7.

Rebuffé-Scrive, M., Andersson, B., Olbe, L. and Björntorp, P. (1990a). Metabolism of adipose tissue in intraabdominal depots in severely obese men and women. *Metabolism*, **39**, 1021–5.

Rebuffé-Scrive, M., Brönnegard, M., Nilsson, A., Eldh, J., Gustafsson, J.A. and Björntorp, P. (1990b). Steroid hormone receptors in human adipose tissue. *J. Clin. Endocrinol. Metab.*, **71**, 1215–9.

Rebuffé-Scrive, M., Walsh, U. and Rodin, J. (1990c). Effect of chronic stress on regional adiposity. Submitted for publication.

Rebuffé-Scrive, M., Nyström, E., Lundberg, P.A., Lindstedt, G. and Björntorp, P. (1991). Adipose tissue metabolism in women with hyperprolactinemia.

Seidell, J., Björntorp, P., Sjöström, L., Kvist, H. and Sannerstedt, R. (1990). Visceral fat accumulation positively associated with insulin, glucose and C-peptide levels but negatively with testosterone levels. *Metabolism*, **39**, 897–901.

Vague, J. (1956). The degree of masculine differentiation of obesities: A factor determining predisposition to diabetes, artherosclerosis, gout, and uric calculous disease. *Am. J. Clin. Nutr.*, **4**, 20–34.

Xu, X., de Pergola, G. and Björntorp, P. (1990). The effects of androgens on the regulation of lipolysis in adipose precursor cells. *Endocrinology*, **2**, 1229–34.

Chapter Twelve

Metabolic Rate and its Control after Accidental Injury

Roderick A. Little
North Western Injury Research Centre, University of Manchester, Stopford Building, Oxford Road, Manchester M13 9PT, UK

INTRODUCTION

Changes in body temperature after injury have been documented for many years. For example a fall in temperature was noted acutely after severe injury by Ambroise Paré (1582) whilst at later times after injury a rise, reminiscent of traumatic fever following surgery (Malcolm, 1893), has been recorded. This biphasic pattern of temperature change formed the basis for the division of the response to injury into the early 'ebb' and the delayed 'flow' phases by Sir David Cuthbertson (1942).

In this chapter these temperature changes and their pathogenesis will be discussed in some detail. Inevitably much of the available data are from animal studies, especially during the period immediately following injury, but whenever possible the emphasis will be on clinical studies.

STUDIES IN ANIMALS

The first systematic experimental studies were done by Rosenthal and his colleagues in the 1940s. They noted that following a 5.5–6.0% body weight haemorrhage over 1 h in unanaesthetized mice treated with saline (15% body weight i.p.) the animals were unable to maintain body temperature

Obesity and Cachexia. Edited by N.J. Rothwell and M.J. Stock

when transferred from an ambient temperature of 26–29°C to 18°C. The fall in temperature could be prevented or reversed by the administration of 100% oxygen; a similar beneficial effect was seen with intravenous injections of whole blood although saline was without effect (Tabor and Rosenthal, 1947) (Figure 12.1).

The conclusion was that the effect of haemorrhage on body temperature was secondary to changes in oxygen delivery, confirming earlier studies in unanaesthetized cats which showed a reduction in whole body oxygen consumption following severe haemorrhage which could be reversed by transfusion (Aub and Cunningham, 1920; Aub, 1920). However, a very different picture was seen following tissue injury when the fall in body temperature (and oxygen consumption) on transfer to the lower ambient temperature could not be corrected by increasing tissue oxygen delivery (Tabor and Rosenthal, 1947). These studies suggested that tissue injury in some way impaired heat production on cold exposure which could not be related to a change in oxygen delivery.

A more complete description of these changes in body temperature and heat production during the ebb phase was provided by Stoner and his colleagues (e.g. reviews: Stoner, 1986; Little, 1988). They used a model of bilateral hindlimb ischaemia in the rat and showed an inhibition of thermoregulatory heat production resulting from an activation of noradrenergic neurones in the hindbrain, from which axons ascend in the ventral noradrenergic bundle to liberate noradrenaline in the region of the dorsomedial nucleus of the hypothalamus. The changes are initiated by nociceptive stimuli triggered by the application of the tourniquets, however,

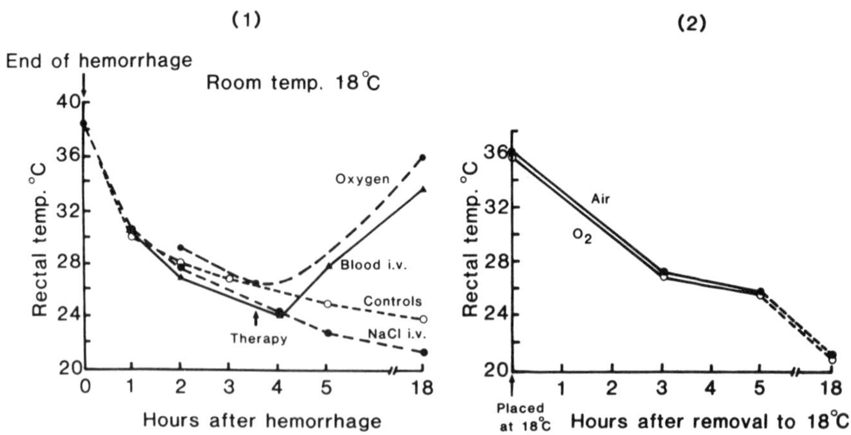

Figure 12.1 Changes in rectal temperature of mice exposed to an ambient temperature of 18°C following haemorrhage (1) or a 2 h period of bilateral hind limb ischaemia (2) the tourniquets were removed shortly before transfer of 18°C, and the effects of treatment with either 100% inspired oxygen, blood (i.v.) or saline (i.v.) (Figures redrawn from Tabor and Rosenthal, 1947)

they are exacerbated by fluid loss from the circulation so that at ambient temperatures below the normal thermoneutral range oxygen consumption and core temperature fall. It must be emphasized that the changes are not secondary to a failure of oxygen transport. Thermoregulatory heat production can be stimulated in such injured animals if the ambient temperature is lowered sufficiently and the sensitivity of the relationship between oxygen consumption and ambient temperature is the same as in uninjured animals (Stoner, 1969).

The inhibition of thermoregulatory heat production by limb ischaemia in the rat involves both skeletal muscle and brown adipose tissue. Lower temperatures have to be applied to the skin or the hypothalamus to induce shivering in skeletal muscle although once shivering has started the proportionality constant (sensitivity) for the response is normal (Stoner, 1971). The ambient temperature at which heat transfer from interscapular brown adipose tissue commenced is reduced in injured animals although once again the slope of the relationship between heat production and ambient temperature is unaltered (Stoner, 1974) (Figure 12.2).

The thermogenic activity of brown adipose tissue is also decreased shortly after scalding injury in the rat (Rothwell *et al.*, 1991). However, after both types of injury brown adipose tissue thermogenesis could be activated by the injection of exogenous noradrenaline, strongly supporting the suggestion that the reduction in heat production in the 'ebb' phase of the response to injury in experimental animals is due to a change in central control rather than to an impairment of peripheral effector mechanisms.

THE EBB PHASE IN MAN

The evidence for a similar pattern of 'ebb' phase response in man is not nearly so convincing. Core and whole body (calculated from the core and mean skin values) temperatures are reduced acutely after accidental injury and the reduction is directly related to the severity of injury (Little and Stoner, 1981). It was also noted that such patients did not shiver when their body temperatures fell below the normal threshold for the onset of shivering, a situation reminiscent of that seen in the rat (*vide supra*). It is difficult to be sure that the changes in man are central in origin because in the most severely injured plasma lactate concentrations are elevated and impairment of oxygen transport cannot be excluded. There is, however, some evidence for a change in thermoregulatory control at this time in man, the normal relationship between core temperature and the selection of an ambient temperature which maximizes thermal comfort is lost and all subjects select a high temperature irrespective of body temperature (Little *et al.*, 1986).

There is no convincing evidence for a reduction in metabolic rate in man

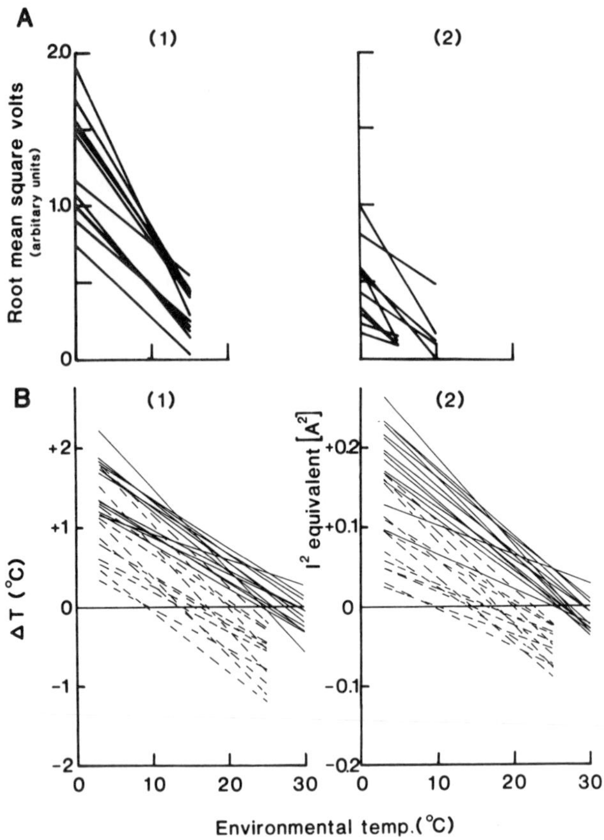

Figure 12.2 (A) Relationships between intensity of shivering (electromyographic activity measured as root mean square volts) and environmental temperature in uninjured control rats (1) and injured rats (2) 1–1.5 h after a 4 h period of bilateral hind limb ischaemia (redrawn from Stoner, 1971). (B) Relationships between (1) the corrected temperature difference between aortic blood and interscapular brown adipose tissue (ΔT°C) and (2) heat production (I^2 equivalent) and the environmental temperature in uninjured control 3°C acclimated rats (continuous lines) and injured 3°C acclimated rats (interrupted lines) 1–2.5 h after a 4 h period of bilateral hind limb ischaemia (redrawn from Stoner, 1974)

shortly after accidental trauma, although values as low as 55% of predicted have been measured after severe injuries (Little *et al.*, 1981). Oxygen consumptions calculated from the data provided by Cournand and his colleagues (Cournand *et al.*, 1943) suggested that heat production is maintained in critically injured patients, even before resuscitation, at higher than normal values (Edwards *et al.*, 1988). This latter group have also calculated oxygen consumption in their own patients using a modification of the Fick equation and shown that it was above normal in 10 and reduced in only 2 of 16 patients vigorously resuscitated shortly after severe multiple injuries (Edwards *et al.*, 1988) (Table 12.1). A trend that was supported by direct measurements of oxygen consumption in a group of critically ill patients intubated and ventilated in the emergency room (Skootsky and Abraham, 1988). There is also no evidence for a reduction in metabolic rate immediately following surgery (data reviewed by Barton *et al.*, 1990). In fact there is often a rise in metabolic rate at this time which cannot be simply explained as a response to intraoperative cooling (Carli and Aber, 1987).

A recent study has shown that after burn injury children rapidly develop a pyrexia and not a fall in body temperature as would be expected both from experimental work (Little, 1974) and the clinical studies in adults described above. The increase in body temperature and heat content is maximal within the first 12 h following injury and is due to an upward resetting of metabolic control rather than to the use of occlusive dressings and a high ambient temperature (Childs, 1988; Childs *et al.*, 1989). In such

Table 12.1 Oxygen consumption (V_{O_2}) calculated early (mean 1.6 h, range 0.8–0.9 h) after severe injury in man. All patients received fluid replacement and arterial oxyhaemoglobin saturation was maintained ≥95%. Ambient temperature 24°C. (Data taken from Edwards *et al.*, 1988)

Case number	ISS	V_{O_2}
Values less than predicted		
4	32	105
12	34	94
Values equal to predicted		
1	25	145
2	34*	119
3	34	155
7	24	110
Values higher than predicted		
5	32	176
6	32	176
8	66*	308
9	38	175
10	48	176
11	48	190
13	32*	238
14	34	198
15	41*	186
16	16	190

*Patient died.
ISS, injury severity score.

children body temperature was positively related to plasma levels of interleukin-6 (IL-6) (Figure 12.3). Interleukin-1 (IL-1) was not detectable in plasma at this time but was found, together with IL-6, in high concentrations in blister fluid (Childs *et al.*, 1990). Thus it is possible that IL-6 produced by macrophages and endothelial cells activated by burning injury may have a role in the mediation of the acute hypermetabolic response to such injuries in children, perhaps involving the central release of prostaglandins which in turn influence hypothalamic thermoregulatory control. A suggestion which is supported by the reduction in core temperature elicited in burned children by acetaminophen, a cyclo-oxygenase inhibitor (Childs and Little, 1988).

In summary there does seem to be good experimental evidence for a reduction in body temperature and heat production during the ebb phase response to tissue injury which is not due to a failure of oxygen transport but to a change in central metabolic control. However, the clinical evidence for such a response is not nearly so convincing and what data there are suggest that metabolic rate is increased, not decreased at this time.

THE FLOW PHASE IN MAN

Increases in metabolic rate and core temperature are characteristic features of the 'flow' phase. There seems to be a consensus that the increase in

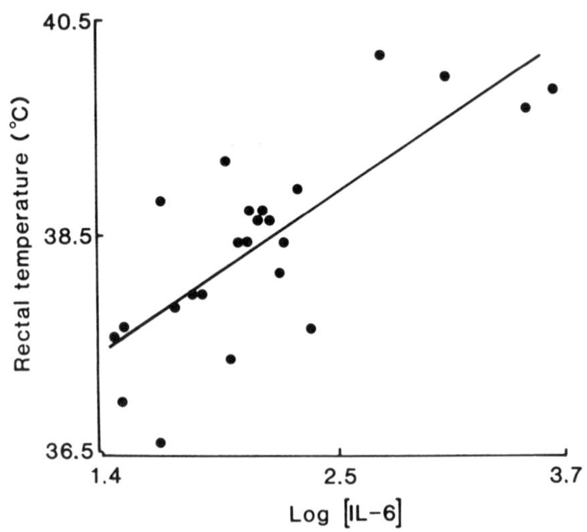

Figure 12.3 Relationship between log plasma interleukin-6 concentration and rectal temperature in six children (age 9–84 months, median 21 months) during the 24 h period immediately following burn injury (10–44%, median 17% body surface area). ($r = 0.780$; $p<0.001$) (redrawn from Childs *et al.*, 1990)

metabolic rate is directly related to the severity of injury (cf. Wilmore, 1977; Matsuda *et al.*, 1987). The largest increases are seen after major burns (>40% total body surface area) with a sustained doubling of the predicted metabolic rate being elicited. Any such increase in metabolic rate has to be met by an increase in tissue oxygen delivery secondary to increases in minute ventilation, cardiac output and tissue blood flow. An increase in the extraction of the delivered oxygen by the tissues is also important, although it has been calculated that if the oxygen saturation of mixed venous blood falls below 30% then re-oxygenation in the lungs may not be complete (Dantzker, 1990) with the development of a progressive arterial hypoxaemia. A problem that would be exacerbated if pulmonary gas exchange was impaired, a problem common in the critically ill.

When metabolic rates are measured in patients who would be expected to show a marked flow phase response (e.g. 1 week after a major head injury or with sepsis complicating intra-abdominal surgery) they are often close to predicted or 'normal' (Table 12.2). This apparent discrepancy is related to problems with predicting energy expenditure in patients who have been bed-ridden for a period and/or have lost weight (especially lean body mass) because of their catabolic state and inadequate nutritional support. Thus a hypermetabolic state is superimposed on conditions which tend to lower energy expenditure. Indeed, in some septic patients who have had a prolonged period of reduced nutritional intake due to, for example, inflammatory bowel disease, metabolic rates as low as 700 Kcal/day have been measured (Heath and Little, 1987). If it is accepted that a number of injured or septic patients will exhibit a hypermetabolic state then it is important to understand what its pathogenesis is. Such knowledge will

Table 12.2 Metabolic rates measured by indirect calorimetry in a number of hypermetabolic/hypercatabolic states in man

	Metabolic rate ($kcal/m^2$ h) Median (range)
Sepsis	
(Sepsis score 8–29)	34.2 (22.3–57.4)
Injury (days 5–10)	
Non-head (ISS 9–47)	43.1 (36.6–48.9)
Head (ISS 25–35)	39.4 (27.0–47.3)
Acute renal failure	43.6 (33.4–51.8)
Control	37.7 (32.8–45.6)

ISS, injury severity score.

enable treatment to be soundly based should modification of the response be considered necessary. A number of factors have been proposed (Table 12.3) and it is quite possible that they may all contribute although their relative roles may depend on the underlying clinical condition.

An upward central resetting of metabolic control is strongly suggested by the studies of Wilmore and his colleagues (Wilmore, 1977). They have shown an increase in energy expenditure in burned patients even at ambient temperatures within the thermoneutral range. This was not secondary to an increased afferent cold input stimulating metabolism, in fact the burned patient at this time is warm both peripherally and centrally. It seems most likely that the burned wound triggers a cytokine/prostanoid 'cascade' which elicits an increase in sympathetic nervous system activity. Plasma catecholamine concentrations are raised during the 'flow' phase following burns and combined α- and β-adrenergic receptor blockade reduces the hypermetabolic response to such injuries (Wilmore *et al.*, 1974). It should be noted, however, that although plasma catecholamine levels are raised during the hypermetabolic phase of the response to burning injury and also to a lesser extent following severe head injuries, they are close to normal at the same time after major non-head injuries (Frayn, 1986).

It has been suggested that one mechanism by which catecholamines may stimulate energy expenditure is by an increase in substrate cycling. This is a process in which although there is no change in the amount of either substrate or product ATP is used and energy expenditure has to be increased for its resynthesis. Two such cycles, glycolysis–gluconeogenesis and triglyceride–fatty acid, have been shown to be increased during the period 9–48 days following severe 60–95% burns. The increase in the

Table 12.3 Factors which may be involved in the pathogenesis of the hypermetabolism associated with the 'flow' phase response to trauma/sepsis

(a)	Upward central resetting of metabolic control—cytokine/prostanoid 'cascade'
(b)	Increased sympathetic efferent activity
(c)	Substrate cycling
(d)	Protein oxidation/synthesis
(e)	Wound
	O_2 consumption
	Lactate/glucose conversion in liver
	Production of cytokines
	Increase cardiac work to maintain hyperdynamic circulation
(f)	Q_{10} effect of raised body temperature
(g)	Evaporative water loss (especially after burns)
(h)	Brown adipose tissue

triglyceride–fatty acid cycle is due to β-adrenergic stimulation (Wolfe *et al.*, 1987). The contribution that such 'cycling' makes to total energy expenditure is difficult to assess, but an estimate of at least 15% has been made (Wolfe *et al.*, 1987).

An increase in glucose cycling may not, however, be a general feature of the flow phase following injury. In severely non-thermally injured patients there was no increase in glucose recycling 48–72 h after injury when metabolic rate would be expected to be elevated, although perhaps not maximally (Jeevanandan *et al.*, 1990).

An increase in sympathetic activity will also stimulate brown adipose tissue thermogenesis. This tissue is an important site of heat production in the small mammal and its activity is increased concomitantly with an increase in whole body oxygen consumption following a scald injury in the rat (Rothwell *et al.*, 1991). Its role in energy balance in adult man is controversial. There is little evidence for its importance under normal conditions although it has been reported to become more so in a number of situations where circulating catecholamine levels are increased for some time, for example, during chronic cold exposure or after the development of a phaeochromocytoma (Lean and James, 1986) but the situation after severe injury or sepsis is unknown.

As alluded to previously, the severely injured are frequently not fed for a number of days immediately following their accident. This may, perhaps surprisingly, contribute to the development of the flow phase increase in heat production as a period of starvation enhances the thermogenic response to adrenaline (Mansell *et al.*, 1990).

A feature of the flow phase is increased catabolism of lean body mass and there has been considerable debate about the contribution this makes to the increase in heat production (Little, 1985). Early predictions that the hypermetabolism could be fully explained by the thermic effect of protein catabolism (cf. Cuthbertson, 1980) have not been substantiated. It now seems that the protein calorie contribution to heat production is no more than 10–20% even in multiply injured patients or in those following burns of up to 60% of body surface area (Duke *et al.*, 1970). The situation may be different after head injuries where values of up to 34% have been calculated (Clifton *et al.*, 1984; Dickerson *et al.*, 1990), although these data have to be interpreted with caution as the patients were all treated with dexamethasone which augments the catabolic response to head injury (Greenblatt *et al.*, 1989).

The evaporation of water from the surface of a burned wound or an area of granulation tissue will increase energy expenditure. For example, the energy cost of the latent heat of evaporation of a transcutaneous water loss of 3 litres/day, which is not unusual after severe burns (Davies, 1982), will increase energy expenditure by over 1500 kcal/day. The impact of this evaporative water loss on energy balance can be reduced by the application of impermeable dressings or by raising ambient temperature to reduce dry heat losses. However, although such manoeuvres may ameliorate the hypermetabolism associated with burning injury, they do not abolish it and indeed Wilmore has argued that the evaporative water loss is not the cause of the hypermetabolism, but instead acts as a convenient route for the dissipation of the excess calories generated by an upward resetting of metabolic control (*vide supra*; Wilmore, 1977).

The wound itself may be the most important influence on heat production in the 'flow' phase. It has been suggested that the wound (e.g. burn, fracture site, abscess, etc) should be considered as an extra organ 'grafted' onto the body by injury/infection (Wilmore, 1986). The wound is a heterogeneous tissue consisting of polymorphonuclear leucocytes, monocytes, fibroblasts, endothelial and epithelial cells all of which are metabolically very active. The wound has a large blood supply which is not under neural control and indeed much of the increase in cardiac output during the 'flow' phase may be directed to the wound (Wilmore, 1986). One feature of the wound very relevant to a discussion of metabolic rate is the ability of monocytes and endothelial cells to produce cytokines (Fong *et al.*, 1990). The burn wound, involving a limb, has been studied extensively and has been shown to be a consumer of large amounts of glucose and to a lesser extent of oxygen (Wilmore *et al.*, 1977). The glucose is converted to lactate, in the presence of oxygen, i.e. aerobic glycolysis, and the lactate is carried to the liver where it is reconverted to glucose. This is an energy consuming process which is reflected in an increase in hepatic oxygen consumption (Wilmore *et al.*, 1980). Thus the wound becomes a user of glucose at a time when other tissues such as skeletal muscle reduce their uptake of glucose secondary to insulin resistance (Henderson *et al.*, 1991). Thus it can be seen that the wound may increase whole body heat production in two ways. First, there is the increase in the peripheral glucose/lactate cycle and second, the wound acts as a focus for triggering the cytokine/prostanoid 'cascade' which, as discussed above, may lead to a central upward resetting of metabolic control.

REFERENCES

Aub, J.C. (1920). Studies in experimental traumatic shock. I. The basal metabolism. *Am. J. Physiol.*, **54**, 388–407.

Aub, J.C. and Cunningham, T.D. (1920). Studies in experimental shock. II. The oxygen content of the blood. *Am. J. Physiol.*, **54**, 408–15.

Barton, R.N., Frayn, K.N. and Little, R.A. (1991). Trauma, burns and surgery. In: *The Metabolic and Molecular Basis of Acquired Disease* (eds R.D. Cohen, K.G.M.M. Alberti, B. Lewis and A.M. Denman), Baillière Tindall, London.

Carli, F. and Aber, V.R. (1987). Thermogenesis after major elective surgical procedures. *Br. J. Surg.*, **74**, 1041–5.

Childs, C. (1988). Fever in burned children. *Burns*, **14**, 1–6.

Childs, C. and Little, R.A. (1988). Acetaminophen (paracetamol) in the management of burned children with fever. *Burns*, **14**, 343–8.

Childs, C., Ratcliffe, R.J., Holt, I., Little, R.A. and Hopkins, S.J. (1990). The relationship between interleukin-1, interleukin-6 and pyrexia in burned children. In: *Physiological and Pathological Effects of Cytokines — Progress in Leukocyte, Biology 10b* (eds C.A. Dinarello, M. Kluger, M. Powanda and J. Oppenheim), Alan R. Liss, New York, pp. 295–300.

Childs, C., Stoner, H.B., Little, R.A. and Davenport, P.J. (1989). A comparison of some thermoregulatory responses in healthy children and in children with burn injury. *Clin. Sci.*, **77**, 425–9.

Clifton, G.L., Robertson, C.S., Grossman, R.G., Hodge, S., Foltz, R. and Garza, C. (1984). The metabolic response to severe head injury. *J. Neurosurg.*, **60**, 687–96.

Cournand, A., Riley, R.L., Bradley, S.E., Breed, E.S., Noble, R.P., Lauson, H.D., Gregersen, M.I. and Richards, D.W. (1943). Studies of the circulation in clinical shock. *Surgery*, **13**, 964–95.

Cuthbertson, D.P. (1942). Post-shock metabolic response. *Lancet*, **i**, 433–7.

Cuthbertson, D.P. (1980). Alterations in metabolism following injury: Part 1. *Injury*, **11**, 175–89.

Dantzker, D.R. (1990). The influence of mixed venous PO_2 on arterial oxygenation. In: *Update in Intensive Care and Emergency Medicine — Update 1990* (ed. J.L. Vincent), Springer-Verlag, Berlin, pp. 131–7.

Davies, J.W.L. (1982). *Physiological Responses to Burning Injury*, Academic Press, London.

Dickerson, R.N., Guenter, P.A., Gennarelli, T.A., Dempsey, D.T. and Mullen, J.L. (1990). Increased contribution of protein oxidation to energy expenditure in head-injured patients. *J. Am. Coll. Nutr.*, **9**, 86–8.

Duke, J.H., Jorgensen, S.V., Broell, J.R., Long, C.L. and Kinney, J.M. (1970). Contribution of protein to caloric expenditure following injury. *Surgery*, **68**, 168–74.

Edwards, J.D., Redmond, A.D., Nightingale, P. and Wilkins, R.G. (1988). Oxygen consumption following trauma: A reappraisal in severely injured patients requiring mechanical ventilation. *Br. J. Surg.*, **75**, 690–2.

Fong, Y., Moldawer, L.L., Shires, G.T. and Lowry, S.F. (1990). The biologic characteristics of cytokines and their implications in surgical injury. *Surg. Gynecol. Obstet.*, **170**, 363–78.

Frayn, K.N. (1986). Hormonal control of metabolism in trauma and sepsis. *Clin. Endocrinol.*, **24**, 577–99.

Greenblatt, S.H., Long, C.L., Blakemore, W.S., Dennis, R.S., Rayport, M. and Geiger, J.W. (1989). Catabolic effect of dexamethasone in patients with major head injuries. *JPEN*, **13**, 372–6.

Heath, D.F. and Little, R.A. (1987). Whole body energy metabolism in the hyperdynamic phase of sepsis. In: *Update in Intensive Care and Emergency Medicine—Septic Shock* (eds J.L. Vincent and L.G. Thijs), Springer-Verlag, Berlin, pp. 197–213.

Henderson, A.A., Frayn, K.N., Galasko, C.S.B. and Little, R.A. (1991). Dose response relationships for the effect of insulin on glucose and fat metabolism in injured patients and controls. *Clin. Sci.*, **80**, 25–32.

Jeevanandam, M., Young, D.H. and Schiller, W.R. (1990). Glucose turnove, oxidation and indices of recycling in severely traumatised patients. *J. Trauma,* **30**, 582–9.

Lean, M.E.J. and James, W.P.T. (1986). Brown adipose tissue man. In: *Brown Adipose Tissue* (ed. P. Trayhurn and D.G. Nichols), pp. 339–65.

Little, R.A. (1974). The impairment of thermoregulation by trauma during the first days of life of the rabbit. *Biol. Neonate*, **24**, 363–74.
Little, R.A. (1985). Heat production after injury. *Br. Med. Bull.*, **41**, 226–31.
Little, R.A. (1988). Metabolic rate and thermoregulation after injury. In: *Recent Advances in Critical Care Medicine—3* (ed. I. McA. Ledingham), Churchill Livingstone, Edinburgh, pp. 159–73.
Little, R.A. and Stoner, H.B. (1981). Body temperature after accidental injury. *Br. J. Surg.*, **68**, 221–4.
Little, R.A., Stoner, H.B. and Frayn, K.N. (1981). Substrate oxidation shortly after accidental injury in man. *Clin. Sci.*, **61**, 789–91.
Little, R.A., Stoner, H.B., Randall, P. and Carlson, G. (1986). An effect of injury on thermoregulation in man. *Q. J. Exp. Physiol.*, **71**, 295–306.
Malcolm, J.D. (1983). *The Physiology of Death from Traumatic Fever—A Study in Abdominal Surgery*, Churchill, London.
Mansell, P.I., Fellows, I.W. and MacDonald, I.A. (1990). Enhanced thermogenic response to epinephrine after 48-h starvation in humans. *Am. J. Physiol.*, **258** (Regulatory, Integrative and Comparative Physiology, 27), R87–93.
Matsuda, T., Clarke, N., Hariyani, G.D., Bryant, R.S., Hanumadass, M.L. and Kagan, R.J. (1987). The effect of burn wound size on resting energy expenditure. *J. Trauma*, **27**, 115–18.
Paré, A. (1582). *Les Oeuvres d'Ambroise Paré*, Dupuye, Paris.
Rothwell, N.J., Little, R.A. and Rose, J.G. (1991). Brown adipose tissue activity and oxygen consumption after scald injury in the rat. *Circ. Shock*, **33**, 33–6.
Skootsky, S.A. and Abraham, E. (1988). Continuous oxygen consumption measurement during initial emergency department resuscitation of critically ill patients. *Crit. Care Med.*, **16**, 706–9.
Stoner, H.B. (1969). Studies on the mechanism of shock: The impairment of thermoregulation by trauma. *Br. J. Exp. Pathol.*, **50**, 125–38.
Stoner, H.B. (1971). Effect of injury on shivering thermogenesis in the rat. *J. Physiol.*, **214**, 599–615.
Stoner, H.B. (1974). Inhibition of thermoregulatory non-shivering thermogenesis by trauma in cold acclimated rats. *J. Physiol.*, **238**, 657–70.
Stoner, H.B. (1986). Metabolism after trauma and in sepsis. *Circ. Shock*, **19**, 75–87.
Tabor, H. and Rosenthal, S.M. (1947). Body temperature and oxygen consumption in traumatic shock and hemorrhage in mice. *Am. J. Physiol.*, **149**, 449–64.
Wilmore, D.W. (1977). *The Metabolic Measurement of the Critically Ill*, Plenum Press, New York.
Wilmore, D.W. (1986). The wound as an organ. In: *The Scientific Basis for the Care of the Critically Ill* (eds R.A. Little and K.N. Frayn), Manchester University Press, Manchester, pp. 45–59.
Wilmore, D.W., Aulick, L.H., Mason, A.D. and Pruitt, B.A. (1977). Influence of the burn wound on local and systemic responses to injury. *Ann. Surg.*, **186**, 444–58.
Wilmore, D.W., Goodwin, C.W., Aulick, L.H., Powanda, M.C., Mason, A.D. and Pruitt, B.A. (1980). Effect of injury and infection on visceral metabolism and circulation. *Ann. Surg.*, **192**, 491–502.
Wilmore, D.W., Long, J.M., Mason, A.D., Skreen, R.W. and Pruitt, B.A. (1974). Catecholamines: Mediator of the hypermetabolic response to thermal injury. *Ann. Surg.*, **180**, 653–68.
Wolfe, R.R., Herndon, D.N., Jahoor, F., Miyoshi, H. and Wolfe, M. (1987). Effect of severe burn injury on substrate cycling by glucose and fatty acids. *N. Engl. J. Med.*, **317**, 403–8.

Chapter Thirteen

Cancer Cachexia and Cytokines

Johan Gelin, Anders Hyltander and Kent Lundholm
Department of Surgery, Sahlgrenska Hospital, University of Göteborg, S41345 Göteborg, Sweden

INTRODUCTION

Cancer cachexia is characterized by weight loss and anorexia. The cause of cachexia remains an enigma, but recently cytokines have been implicated as possible endogenous mediators. This proposition is based on the observed similarity in host response to invasive stimuli, irrespective of causative agent, where weight loss is a frequent manifestation. Furthermore, the metabolic alterations observed when cytokines are administered under experimental conditions are similar to the metabolic reactions observed during tissue injury. The two most important factors giving rise to cachexia are anorexia and increased energy expenditure and this presentation will discuss these two aspects.

ENERGY EXPENDITURE

There has been a controversy in the literature as to whether cancer patients exhibit elevated energy expenditure or not when compared with equally malnourished patients without malignancy. Recent studies in our laboratory have repeatedly confirmed that many cancer patients lose weight in part due to elevated energy expenditure both in the resting and non-resting state (Lindmark *et al.*, 1984; Hyltander *et al.*, 1991a) (Figure 13.1). Thus, elevated energy expenditure in cancer patients is a significant metabolic alteration,

Obesity and Cachexia. Edited by N.J. Rothwell and M.J. Stock

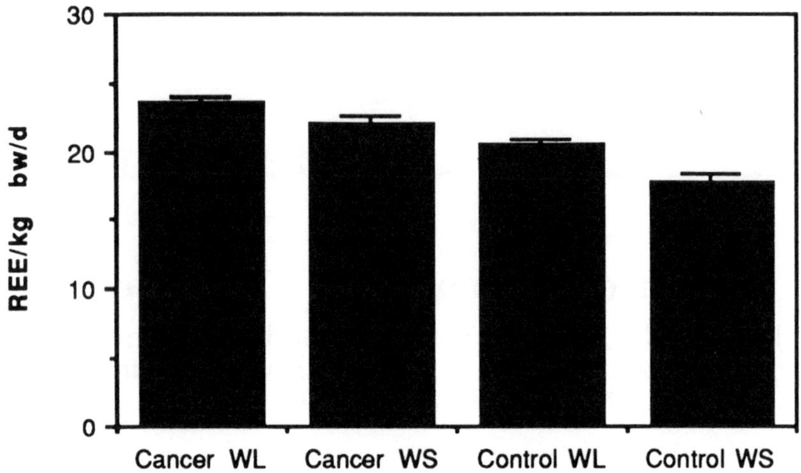

Figure 13.1 Resting energy expenditure in 215 cancer and non-cancer patients. WL, weight losing; WS, weight stable. The differences are significant in both groups of cancer patients compared with non-cancer patients ($p<0.01$). (Redrawn from Hyltander *et al.*, 1991a)

sometimes already present in the early course of the disease. Actually, it may occur even before it is possible to identify any measurable changes in body composition. It has also been confirmed that increased energy expenditure in cancer patients is explained in part by changes in fuel utilization. Measurements of substrate oxidation have shown a significantly increased fat oxidation which is consistent with reports of increased fat turnover and lipolysis in cancer patients (Figure 13.2). Such findings point to the possibility that increased fat oxidation is one of the initial metabolic alterations in the host response to a growing tumour. Although the increase in energy expenditure is generally small, 10–15% in several studies, it could explain the loss of several kilograms of body weight over a period of months. Few investigations have attempted to identify the cause of this increased metabolism. Our own experiments have pointed to the possibility of an increased adrenergic tone in weight-losing cancer patients (Hyltander *et al.*, 1991a). The elevated sympathetic activity may also be facilitated by an up-regulation in the sensitivity to adrenergic agonists in malnourished patients (Drott *et al.*, 1989). In this respect, the heart has been studied in particular, and our own experiments suggest that it is one compartment which consumes disproportionally large amounts of energy for the degree of malnutrition (Drott *et al.*, 1987). Clinical signs of such alterations are increased heart rate and significantly elevated urinary excretion of stress hormones. An abnormal cardiac sensitivity to adrenergic compounds may increase the risk of cardiac arrythmia in malnourished cancer patients, particularly in situations of stress such as operation and general anaesthesia.

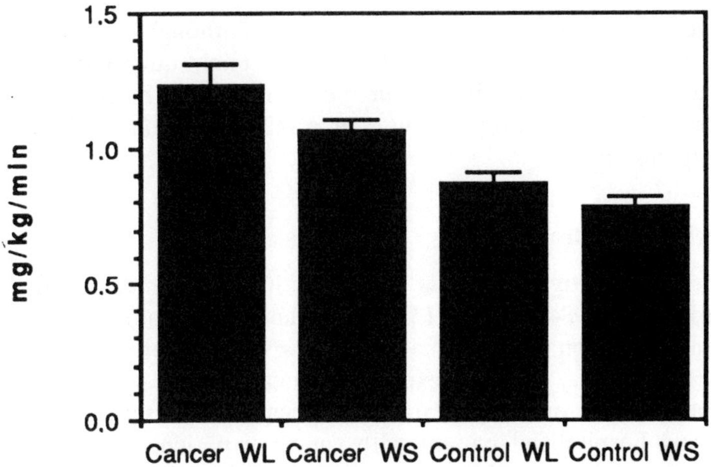

Figure 13.2 Fat oxidation in 215 cancer and non-cancer patients who were weight-losing (WL) or weight-stable (WS). Fat oxidation was significantly higher in both groups of cancer patients compared with non-cancer patients ($p<0.01$) (Hyltander *et al.*, 1991a)

Lipid metabolism

Cancer patients have altered lipid metabolism, partly explained by increased fat oxidation irrespective of weight loss. This has been confirmed in several studies showing increased plasma concentrations of free fatty acids and glycerol without evidence of increased fractional re-esterification of free fatty acids (Bennegård *et al.*, 1982). Measurements of glycerol dynamics have also suggested an increased plasma glycerol turnover in weight-losing cancer patients partly explaining an increase in whole body lipolysis (Edén *et al.*, 1985). Such alterations in lipid metabolism are explained by increased insulin resistance due to both anorexia and elevated sympathetic adrenergic activities promoting lipolysis. However, such metabolic alterations are generally not entirely specific for cancer.

Carbohydrate metabolism

Normally, caloric deprivation is associated with a decreased turnover of glucose, which is a physiological response to conserve fuel reserves. However, several studies have confirmed increased glucose turnover in a variety of cancer patients who lose weight (Edén *et al.*, 1984). The major explanation behind increased glucose turnover is elevated recycling of glucose carbons, since glucose oxidation is simultaneously decreased. Increased recycling of glucose carbons is essentially explained by elevated gluconeogenesis to which both alanine, lactate and glycerol contribute (Holroyde *et al.*, 1984). Lactate is quantitatively the most important precursor to glucose in human beings, while glycerol may be important in murines. Although, it was originally believed that an increase in such 'futile cycling' might quantitatively explain the increased energy expenditure seen in cancer patients, more recent studies have shown that it can only account for a minor fraction of the increased energy expenditure.

Amino acid metabolism

The turnover of body proteins is frequently increased in cachectic patients (Lindmark *et al.*, 1984) (Figure 13.3), and the energy required to synthesize and degrade additional protein can explain approximately one-third of this increase (Edén *et al.*, 1984). In experimental cancer, the site of this increase in protein synthesis and degradation is principally in visceral and immune tissues. Any hospitalized patient with cancer experiences an acute phase protein response, comparable with that seen during acute inflammation. It has also been assumed that weight-losing cancer patients have increased skeletal muscle breakdown. However, direct measurements of amino acid

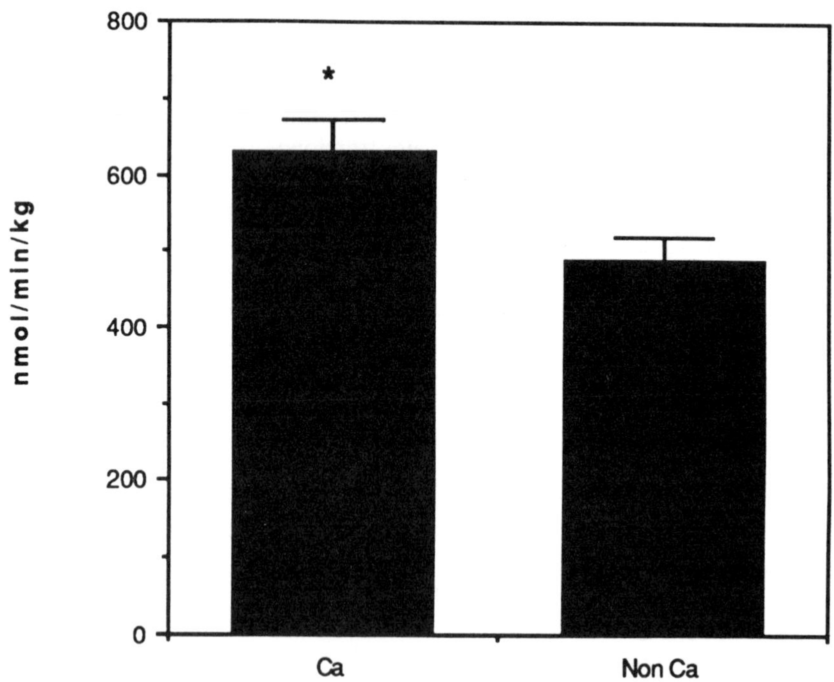

Figure 13.3 Whole body tyrosine flux after an overnight fast in cancer patients and controls with comparable body composition. Flux rates were significantly higher in cancer patients ($p<0.01$) (Hyltander *et al.*, 1991b)

balances across leg tissues have not confirmed this. Therefore, it is more likely that decreased protein synthesis, both in the fasted and fed state, explains the muscle loss in the majority of cancer patients (Svaninger *et al.*, 1983).

Hormonal alterations

Characteristic changes in the levels of most hormones have been observed during tumour growth. These include insulin, growth hormone and glucocorticoids, as well as the characteristic decline in both thyroxine and triiodothyronine concentrations. It is quite clear that these changes represent in part the effector system, which explains the metabolic alterations (Svaninger, 1987). However, in many patients significant aberrant tumour production of hormone-like substances and peptides may further contribute to the metabolic alterations.

Endogenous cytokine production and cachexia

Several lines of evidence support the notion that endogenous production of cytokines may contribute to the catabolic effects observed in cancer. Prolonged exogenous administration of recombinant IL-1 and TNF can reproduce most of the metabolic changes that occur in the host as a consequence of cancer. Loss of appetite, loss of body weight, including loss of body fat and protein, and induction of hepatic acute phase protein synthesis have all been documented to occur following cytokine administration to healthy experimental animals (Moldawer, 1986; Tracey *et al.*, 1988; Oliff, 1988; Mahoney and Tisdale, 1988; Socher *et al.*, 1988). Although increased circulating TNF and IL-1 concentrations have not been a regular finding in human or experimental cancer, indirect evidence supports endogenous TNF as a proximal mediator in the development of cachexia. In studies by Oliff, a genetically engineered cell line continuously secreted h-TNF. This genetic construct when transplanted to mice caused severe host wasting and progressive cachexia with eventual death (Oliff *et al.*, 1987). In such animals TNF was recently found in the circulation.

A role for IL-1 and TNF as mediators in the development of cachexia is further supported by studies in which passive immunization against the IL-1 receptor and the TNF molecule in the tumour-bearing state ameliorated part of the development of cachexia. Appetite was improved (Figure 13.4) and body composition, both lean body mass and fat, was preserved when compared with tumour-bearing animals receiving non-immune IgG (Sherry *et al.*, 1989). This illustrates a role for cytokines as endogenous mediators in cancer cachexia.

The experimental modes of replicating the production of endogenous

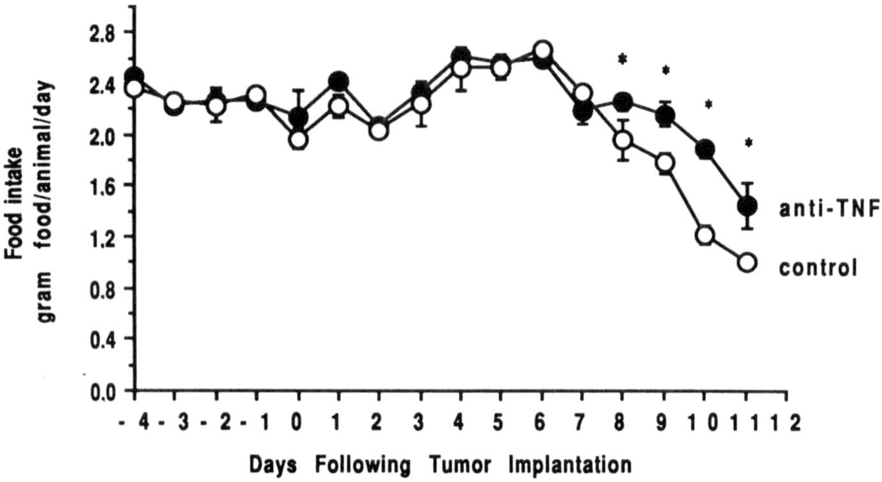

Figure 13.4 Food intake in tumour-bearing animals treated with antibodies against endogenously produced TNF (anti-TNF) compared with animals treated with non-specific IgG. Food intake was significantly improved in animals treated with specific antibodies. Similar results were obtained by treating the animals with specific antibodies against the IL-1 receptor. Redrawn from Sherry *et al.* (1989)

mediators have varied considerably. Norton and colleagues argue that the intermittent administration of cytokines to otherwise healthy animals does not reflect the tumour-bearing situation. The effects of repetitive TNF administration gives rise to successive tachyphylaxis. Comparing intermittent bolus i.v. injections with continuous i.v. infusions of the same sublethal dose of TNF, it was observed that continuous i.v. infusion produced anorexia, wasting and lethality that was attenuated with the bolus mode of administration, thus supporting a role for a low continuous release of TNF mediating cachexia (Darling *et al.*, 1990).

Using hybridization techniques and immunocytochemical methods we have measured mRNA for IL-1 and TNF concentrations *in vivo* in tumour and lymphoid tissues in cachectic tumour-bearing mice (Figure 13.5). Furthermore, utilizing immunofluorescent antibodies, the presence of TNF protein could be demonstrated in tumour cells *in vitro*, providing evidence that the mechanisms for continuous cytokine production exist at least in our cachectic tumour-bearing mice (Lönnroth *et al.*, 1990). Recent reports have added support to these observations, TNF and IL-6 production being found in human epithelial tumour cell lines (Spriggs *et al.*, 1988; Kirnbauer *et al.*, 1989), and studies on cytokine gene expression have demonstrated the presence of mRNA for TNF in the majority of biopsy samples in human colorectal cancer (Naylor *et al.*, 1990).

Circulating levels of cytokines in cancer patients have only been found

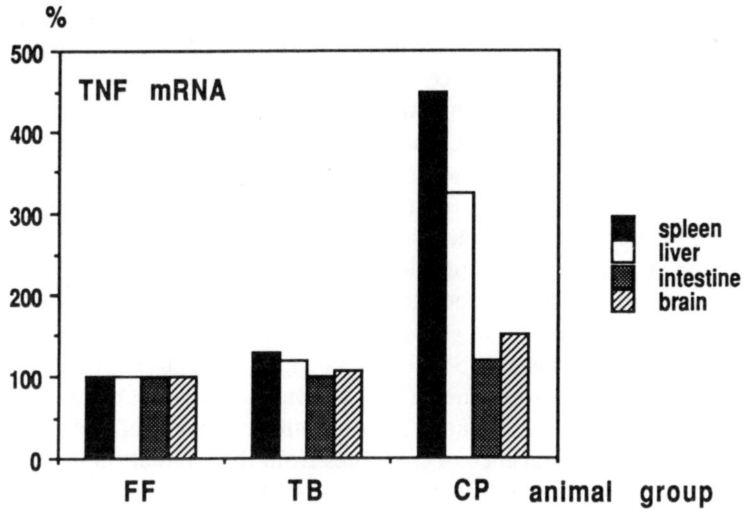

Figure 13.5 Relative concentrations of TNF mRNA in spleen, liver, intestine and brain from tumour-bearing (TB), antigen-stimulated (CP) and freely fed control (FF) mice (Lönnroth *et al.*, 1990)

inconsistently. Measurements of IL-1 and TNF in inflammation and sepsis have generally been more rewarding. There are numerous reports finding correlations between TNF levels and disease activity and for that matter mortality rates. Attempts to measure circulating IL-1 and TNF in our mouse model failed to detect elevated circulating levels with progression of cachexia, whereas both cytokines were found to be present at increased levels in benign inflammation, in particular when the RES had been primed before stimulation with LPS (Ternell *et al.*, 1987; Moldawer, 1986). Since TNF and IL-1 could not be detected in the plasma of our cachectic mice, we speculated that IL-6 could be the circulating messenger that induced the anorexia and the hepatic acute phase reaction. IL-6 has been shown to be an endogenous pyrogen, inducing the hepatic acute phase protein response. A close correlation could be shown for the synthesis of the acute phase reactant SAP and IL-6, lending support to the theory that IL-6 may well be the circulating cytokine induced by the local production of IL-1 and TNF, largely effectuating the acute liver protein synthesis (Gelin *et al.*, 1988; Jablons *et al.*, 1989). Many of the biological activities of IL-6 are shared by IL-1 and TNF and recent experiments administering IL-6 intracerebroventricularly have further pointed to a role for IL-6 as an anorectogenic agent (Busbridge *et al.*, 1989).

IL-6 has also been found to be circulating at increased concentrations following endotoxin challenge in humans (Fong *et al.*, 1989) and in sepsis

(Hack *et al.*, 1989), confirming its elaboration in inflammation and invasive lesions. Our own *in vitro* studies have established that our experimental tumour cells can directly produce IL-6. IL-6 mRNA was found in biopsies from primary liver cancer, supporting these findings to be a more general observation; in the study by Utsumi, IL-6 was found to be produced by cells other than tumour cells where T-cells were found to be a source of IL-6 (Utsumi *et al.*, 1990). IL-6 immune reactivity was further observed in a number of neoplastic elements of human origin and confirmed IL-6 participation in tumour–host interactions (Tabibzadeh *et al.*, 1989).

Previous reports have described inhibitors to IL-1 and TNF that interfered with measurements, these were originally observed in a febrile patient's urine as well as in plasma (Seckinger *et al.*, 1988). Recently, a natural interleukin-1 receptor antagonist (ILRA) was identified (Hannum *et al.*, 1990; Eisenberg *et al.*, 1990) and a TNF-a inhibitor characterized (Seckinger *et al.*, 1990). A rapid clearance of cytokines from the circulation may perhaps be accounted for by the presence of antagonists which may be regarded as a protective means against the deleterious side effects of biologically potent cytokines when they appear in the circulation. Natural inhibitors to IL-1 are released under similar circumstances and were recently documented to partially inhibit IL-1 stimulation of PGE_2 production (Arend *et al.*, 1990).

TUMOUR GROWTH AND ANOREXIA

The reaction of anorexia as part of the purposeful host response can of course be questioned since feeding is a prerequisite for the continuation of life, and hunger is one of the strongest forces in animal life. In infections, it has been proposed that lack of substrates, particularly zinc and iron, is a useful reaction for bacteriostasis since most bacteria require these trace elements for optimal growth (Hoffman-Goetz and Kluger, 1979). Studies in our laboratory by Westin (1991) have shown that the tumour cell cycle can be sensitive to substrate flow, and that complete starvation decreased cell cycling which upon re-feeding entered normal cell cycle dynamics. Starvation retarded tumour growth by arresting cells in the G_0G_1-phase and prolonged the potential doubling time of the tumour. Thus, anorexia can be regarded as part of the host's non-specific reaction to invasive tissue injury, the host being unable to discriminate between different insults. Anorexia could be an adequate non-specific response in an attempt to starve the tumour or infective agent.

In our attempts to influence tumour growth, administration of IL-1 and TNF to tumour-bearing animals produced inhibition of tumour growth. Animals treated with IL-1 and TNF on a continuous twice daily basis until spontaneous death, exhibited the smallest tumours, although this did not confer any advantage to these animals, either in improved survival or in improved body composition (Gelin *et al.*, 1991a). In fact, these mice had the poorest body composition, expressed as the lowest carcass fat content and lean body mass. Hence, retarded tumour growth did not confer any survival advantage and the effects of IL-1 in this situation could be explained entirely by the potentiation of the tumour-induced anorexia. Thus, existing reports of positive treatment effects of cytokines in the tumour-bearing state have not considered this anorectic property of cytokines in accounting for the observed treatment effects.

Passive immunization

The *in vivo* administration of polyclonal antibodies against the TNF molecule (anti-TNF) and monoclonal antibodies against the IL-1 receptor (anti-IL-1r) improved food intake in tumour-bearing animals (Figure 13.4). Despite this finding, there was retardation of tumour growth (Figure 13.6). This implies that the tumour inhibitory effect was not dependent on substrate flow to the tumour but was a more direct effect on the tumour, where IL-1 and

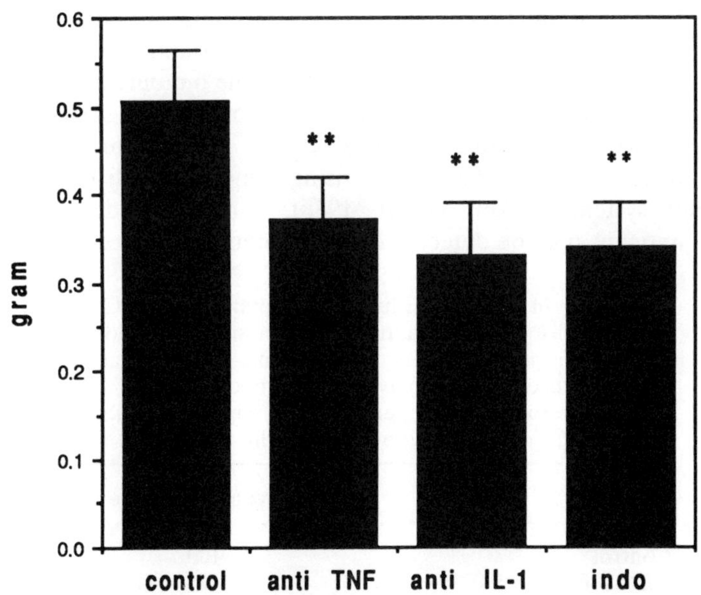

Figure 13.6 Tumour dry weight (tumour-growth) in tumour-bearing animals. The animal groups were treated with either saline or non-specific IgG as controls, polyvalent immunoglobulin raised against the TNF molecule (anti TNF); monoclonal antibodies against the IL-1 receptor (anti IL-1) or with indomethacin (indo). Treatments with specific antibodies and indomethacin then produced significant tumour inhibition

TNF may act as putative growth factors. Supporting this suggestion, *in vitro* experiments confirm our *in vivo* findings that anti-IL-1r and anti-TNF antibodies decreased [^{3}H]thymidine uptake by tumour cells in culture. In addition, supplementation of the culture medium with either recombinant IL-1 or TNF significantly increased [^{3}H]thymidine uptake (Table 13.1) (Gelin *et al.*, 1991b). Furthermore, the morphology of the tumour cells altered with cytokine administration in which treatment with anti-TNF antibodies behaved as a cytostatic agent to the cells, whereas the cells grew normally with non-immune IgG supplementation acting as a control (Gelin *et al.*, 1991b).

Role of the immune system in tumour growth

One possibility for the inhibitory effects on tumours of passive specific immunization could be due to a non-specific augmentation of the immune response, since the antibodies represent proteins from rabbit and rat. However, the use of pre-immune IgG as a control, producing no identifiable response, argues against such a mechanism. The concept of immune surveillance has generally been accepted as one of the central dogmas in tumour biology and the occurrence of tumours should, according to this concept, be a consequence of the tumour escaping immune recognition. Specific tumour antigens or altered transplantation antigen expression is a prerequisite for the immune system to recognize the tumour. This concept has recently been questioned for several reasons: tumours arising spontaneously in animals produce low immune reactions *in vitro*, athymic nude mice have a 'normal' incidence of spontaneously arising tumours which makes a T-cell surveillance system unlikely (Roitt, 1988). Virally coded, idiotypic and oncofetal antigens may be detected on experimentally induced tumour cells,

Table 13.1 *In vitro* [^{3}H]thymidine uptake in MCG 101 tumour cells grown in culture medium supplemented with cytokines as growth factors or monoclonal antibodies against the IL-1 receptor and polyclonal antibodies against TNF. Results are expressed as percentage of controls. Antibodies were produced as described in Gelin *et al.* (1991b)

Group	Percentage uptake
Control	100
anti-TNF	75
anti-IL-1	77
TNF	130
IL-1	128
IgG	99

but these tumours may represent special tumours that arise by different means from those of spontaneous tumours. These tumours are usually infected with oncogenic viruses that usually display new antigens on their surface. A further possibility for the tumour escaping immune recognition is antigen loss and down-regulation of MHC class I molecules that are required for T-cell recognition by cytolytic T-cells. This question was recently addressed in a murine tumour model and found not to be applicable (Ward *et al.*, 1990). In organ transplantation and in patients with medically suppressed immune functions, the incidence of solid tumours, mainly of epithelial and mesenchymal origin, does not increase, whereas the incidence of lymphoid tumours does (Davies *et al.*, 1990). One explanation for this phenomenon was recently proposed suggesting that, following immuno-suppression in the course of transplantation, there may be chronic T-cell activation in response to transplantation antigens with cytokine production. This may lead to paracrine stimulation inducing tumours among bystander cell populations such as adjacent B-cells (Davies *et al.*, 1990). In humans, the ubiquity of EBV may serve as a promoter for this transition. This reasoning gains further support from the observed effect of IL-6 molecules in human B-cell tumorigenesis. To test the functional importance of the specific immune system in our tumour-bearing hosts, cyclosporine A (CsA), a potent immuno-suppressive T-cell inhibitor, was used. CsA selectively blocks transcription of lymphokine mRNAs (Roitt, 1988) and in particular IL-2 production, although the precise mechanism of the immunosuppressive effects of CsA is incompletely understood. When employed on a continuous daily basis to our tumour-bearing mice, CsA did not affect tumour growth. Neither were survival nor body composition changes altered when compared with tumour-bearing controls at spontaneous death (Gelin *et al.*, 1991a). In additional experiments CsA in a dose of 30 mg/kg was given to *Listeria* monocytogenesis infected C57bl/6 mice in a dose of 5000 CFU known to be sublethal to this strain of mice (Havell, 1989). Mice treated with CsA experienced a 100% mortality rate within 3–4 days due to disseminated bacterial infection compared with all mice surviving with no apparent morbidity when the immune system was not medically manipulated. These results clearly indicated that the dose of CsA used in our tumour-bearing experiments was sufficient to compromise the immune system. Furthermore it illustrates the importance of the specific immune system for the eradication of bacterial infections. Recent publications on CsA activities have pointed to effects on diminished TNF potency (Remick *et al.*, 1989) which, in conjunction with decreased IL-2 production may in part explain the clinically beneficial effects observed with CsA in auto-immune disorders thought to be mediated by cytokines (Remick *et al.*, 1989).

Prostaglandins

Prostaglandins are now recognized as mediating many of the biological effects induced by cytokines (Mahoney and Tisdale, 1989) and several of the cytokines are known to stimulate prostaglandin E_2 production (Dayer *et al.*, 1985; Zucali *et al.*, 1986). Interleukin-6 production was recently shown to be regulated by PGE_2 (Callery *et al.*, 1990). In addition, cyclo-oxygenase inhibitors have been found to have protective effects against the deleterious complications that follow large doses of endotoxin or TNF administration (Revhaug *et al.*, 1988; Evans *et al.*, 1989) and the acute anorectogenic effect of administered IL-1 could be ameliorated by ibuprofen as a cyclo-oxygenase inhibitor (Hellerstein *et al.*, 1989).

The role of phospholipids in inflammation has recently attracted considerable attention. Dietary manipulation by supplementing a normal diet with omega-3 fatty acids was demonstrated to reduce the amounts of LPS inducible IL-1 and TNF production by mononuclear cells, which may thus contribute to decreased inflammation (Billiar *et al.*, 1988; Endres *et al.*, 1989). This effect is thought to be due to altered PGE_2 synthesis by reducing substrate availability for arachidonic acid. In order to evaluate a possible role for PGE_2 in the development of cachexia, the non-toxic cyclo-oxygenase inhibitor indomethacin was used in our tumour-bearing model. Indomethacin exerted several interesting effects in the tumour-bearing situation. The most striking effect was improved survival which increased from 14.7 days to 22.0 days (Figure 13.7). The effect on survival has also been reported by others

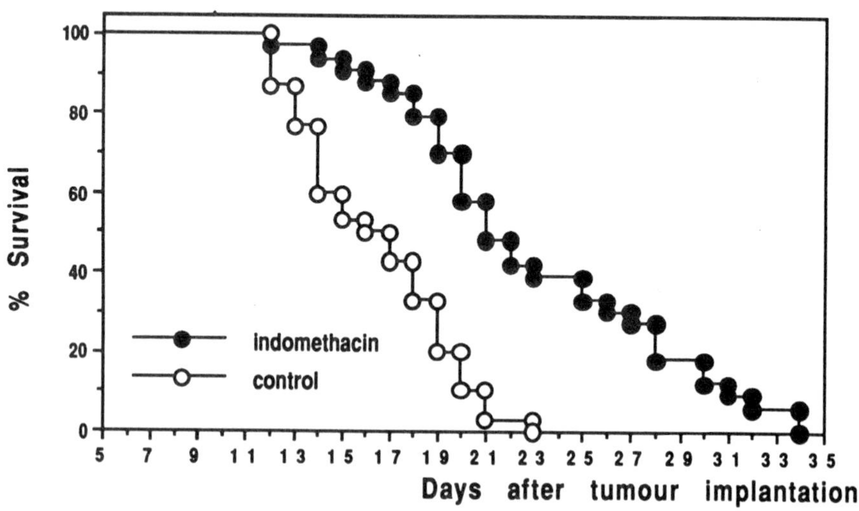

Figure 13.7 Survival in tumour-bearing animals treated with indomethacin compared with saline injected controls. Survival was significantly increased from 14.7 to 22.0 days following tumour implantation. (Reprinted from Gelin *et al.*, 1991a)

(Bennet *et al.*, 1982). In addition, the observation that indomethacin had a positive effect on survival even when instituted late in the disease process may offer a new dimension to treatment (Figure 13.8). In most reports on effects of various treatment modalities in cancer models, the agent used is commenced at the time of tumour inoculation or even beforehand. Such circumstances are hardly appropriate in the clinical setting. In our studies, indomethacin was given on a daily basis and treatment was started on day 1, the day after tumour implantation; on day 5 when the tumour was just becoming visible but the animals were not yet anorectic; on day 7 when these animals normally show the first signs of cachexia, and on day 9 when cachexia is overt in non-treated animals. Indomethacin still influenced survival and appetite when instituted late in the disease process. Statistically, there was no difference between when the treatment was commenced on day 1 or on days 5–9 (Figure 13.8) (Gelin *et al.*, 1991a).

The animals treated with indomethacin not only lived longer, their apparent health was also improved. This was judged from measuring food consumption and observing the general well-being of the animals. When treated with indomethacin, the food intake was maintained at normal consumption for a longer period and did not start to decline until late during the progression of cachexia (Figure 13.9). When indomethacin-treated animals died after 22 days, they did so with a better preserved body composition than when the controls died after 14–15 days. At spontaneous death, the body energy reserves were not totally depleted and hence body energy reserves could not be the only limiting factor for survival in these animals.

At spontaneous death, tumour dry weight was greater but did not differ statistically from controls although the treated animals carried their tumour

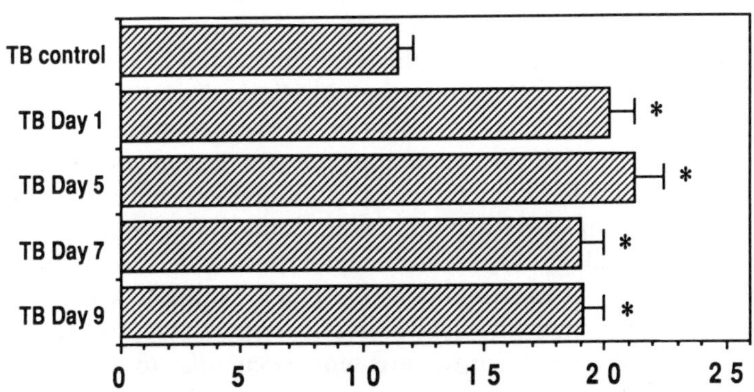

Figure 13.8 Mean survival in tumour-bearing animals treated with indomethacin when treatment was instituted on different days (days 1–9) following tumour implantation. (Reprinted from Gelin *et al.*, 1991a)

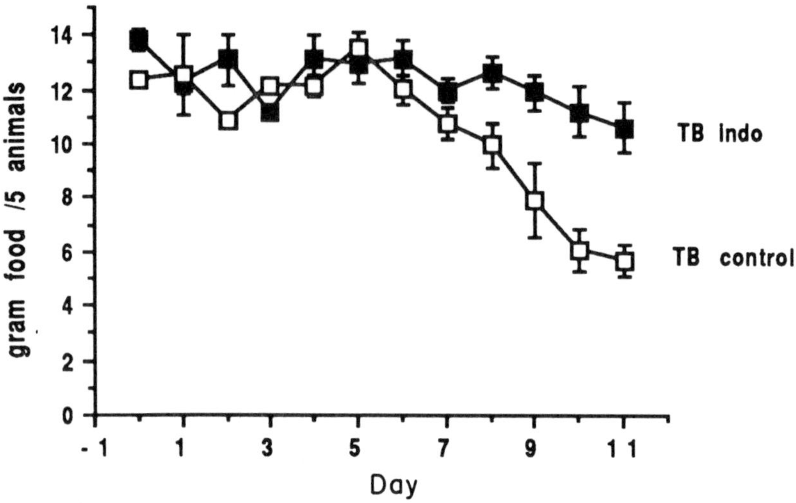

Figure 13.9 Time course of food intake in tumour-bearing mice treated with indomethacin 1 mg/kg/day (TB indo) or saline (TB control). From day 7 after tumour implantation, food intake was significantly better maintained in indomethacin treated animals. (Reprinted from Gelin *et al.*, 1991)

burden for a longer period. When treatment was evaluated over the same time period and compared with controls, i.e. receiving treatment for the same length of time; the tumour grew at a slower rate on indomethacin treatment (Gelin *et al.*, 1991a).

NUTRITION

It has been the hope of many clinicians that artificial nutrition could improve the situation for weight-losing cancer patients. However, controlled clinical studies have not given support for this expectation (Hyltander *et al.*, 1990). There may be several explanations for this limited effect of nutrition on weight-losing cancer patients. One may be the fear that nutrition itself will stimulate tumour growth. Although convincingly demonstrated in work on tumour-bearing animals, it has not been confirmed that nutrition accelerates tumour growth in cancer patients. However, this may in part be explained by limits in methodology. Our own recent experiments have demonstrated that even physiological nutrition controls DNA synthesis and tumour cell replication to some extent in experimental cancer. Therefore, it cannot be excluded that such mechanisms are not occurring in clinical cancer, particularly in rapidly growing tumours. Another explanation of the limited effect of artificial nutrition in cancer patients may be insufficient nutritional therapy intake itself. This suggestion has been addressed in our own recent studies. Cancer patients responded to the challenge of nutrition, whether given intravenously or enterally, with the same degree of substrate oxidation as malnourished non-cancer patients. Their thermogenic response was also normal. Therefore, we have no reason to suspect that the cellular capacity to utilize non-protein calories is impaired in cancer patients (Lindmark *et al.*, 1986). However, conventional nutrition fails to stimulate protein synthesis in either cancer or non-cancer patients, particularly in peripheral tissues such as skeletal muscles (Hyltander *et al.*, 1991b). If this is a more generalized phenomenon in conventional artificial nutrition it is not surprising that nutrition therapy in cancer patients has been less than rewarding. Basal tissue experiments have not confirmed a particular defect in the protein synthesis response to substrate and hormonal stimulation in skeletal muscles from tumour-bearing animals and cancer patients. Therefore, the explanation is that current conventional nutrition regimens are less than efficient in maintaining body weight.

REFERENCES

Arend, W.P., Welgus, H.G., Thompson, R.C. and Eisenberg, S.P. (1990). Biological properties of recombinant human monocyte-derived interleukin 1 receptor antagonist. *J. Clin. Invest.*, **85**, 1694–7.

Bennegård, K., Edén, E., Ekman, L., Schersten, T. and Lundholm, K. (1982). Metabolic balance across the leg in weight losing cancer patients compared to depleted patients without cancer. *Cancer Res.*, **42**, 4293–9.

Bennet, A., Berstack, D.A. and Carrol, M.A. (1982). Increased survival of cancer-bearing mice with inhibitors of prostaglandin synthesis, alone or with chemotherapy. *Br. J. Cancer*, **45**, 767–8.

Billiar, T.R., Bankey, P.E., Svingen, B.A., Curran, R.D., West, M.A., Holman, R.T., Simmons, R.L. and Cerra, F.B. (1988). Fatty acid intake and Kupffer cell function: Fish oil alters eicosanoid and monokine production to endotoxin stimulation. *Surgery*, **104**, 343–9.

Busbridge, J., Dascombe, M.J., Hoopkins, S. and Rothwell, N.J. (1989). Acute central effects of interleukin-6 on body temperature, thermogenesis and food intake in the rat. *Proc. Nutr.*, **38**, 48A.

Callery, M.P., Mangino, M.J., Kamei, T. and Flye, M.W. (1990). Interleukin-6 production by endotoxin-stimulated Kupffer cells is regulated by prostaglandin E_2. *J. Surg. Res.*, **48**, 523–7.

Darling, G., Fraker, D.L., Jensen, J.C., Gorschboth, C.M. and Norton, J.A. (1990). Cachectic effects of recombinant human tumor necrosis factor in rats. *Cancer Res.*, **50**, 4008–13.

Davies, A.J.S., Wallins, V.J. and Morrison, W.I. (1990). The trouble with T cells. *Lancet*, **335**, 1574–6.

Dayer, J.M., Beutler, B.M. and Cerami, A.C. (1985). Cachectin-tumor necrosis factor stimulates collagenase and prostaglandin E_2 production by human synovial cells and dermal fibroblasts. *J. Exp. Med.*, **162**, 2163–8.

Drott, C., Waldenström, A. and Lundholm, K. (1987). Cardiac sensitivity and responsiveness to beta-adrenergic stimulation in experimental cancer and undernutrition. *J. Mol. Cell Cardiol.*, **19**, 675–83.

Drott, C., Persson, H. and Lundholm, K. (1989). Cardiovascular and metabolic response to adrenaline infusion in undernourished patients with and without cancer. *Clin. Physiol.*, **9**, 427–39.

Edén, E., Ekman, L., Bennegård, K., Lindmark, L. and Lundholm, K. (1984a). Whole body tyrosine flux in relation to energy expenditure in weight-losing cancer patients. *Metabolism*, **33**, 1020–7.

Edén, E., Edström, S., Bennegård, K., Scherstén, T. and Lundholm, K. (1984b). Glucose flux in relation to energy expenditure in malnourished patients with and without cancer during periods of fasting and feeding. *Cancer Res.*, **44**, 1718–24.

Edén, E., Edstrom, S., Bennegård, K., Lindmark, L. and Lundholm, K. (1985). Glycerol dynamics in weight-losing cancer patients. *Surgery*, **97**, 176–84.

Eisenberg, S.P., Evans, R.J., Arend, W.P., Verderber, E., Brewer, M.T., Hannum, C.H. and Thompson, R.C. (1990). Primary structure and functional expression from complementary DNA of a human interleukin-1 receptor antagonist. *Nature*, **343**, 341–6.

Endres, S., Ghorbani, R., Kelley, V.E., Georgilis, K., Lonnemann, G., Van Der Meer, J.W.M., Cannon, J.G., Rogers, T.S., Klempner, M.S., Weber, P.C., Schaefer, E.J., Wolff, S.M. and Dinarello, C.A. (1989). The effect of dietary supplementation with n-3 polyunsaturated fatty acids on the synthesis of interleukin-1 and tumor necrosis factor by mononuclear cells. *N. Eng. J. Med.*, **320**, 265–71.

Evans, D.A., Jacobs, D.O., Revhaug, A. and Wilmore, D.W. (1989). The effects of tumor necrosis factor and their selective inhibition by Ibuprofen. *Ann. Surg.*, **209**, 312–21.

Fong, Y., Moldawer, L.L., Marano, M., Wei, H., Tatter, S.B., Clarick, R.H., Santhanam, U., Sherris, D., May, L.T., Sehgal, P.B. and Lowry, S.F. (1989). Endotoxemia elicits increased circulating beta2-IF/JL-6 in man. *J. Immunol.*, **142**, 2321–4.

Gelin, J., Moldawer, L.L., Lönnroth, C., deMan, P., Svanborg-Edén, C., Lowry, S.F. and Lundholm, K.G. (1988). Appearance of hybridoma growth factor/Interleukin-6 in the serum of mice bearing a methylcholanthrene-induced sarcoma. *Biochem. Biophys. Res. Commun.*, **157**, 575–9.

Gelin, J., Andersson, C. and Lundholm, K. (1991a). Effects of indomethacin, cytokines and cyclosporin A on tumor growth and the subsequent development of cancer cachexia. *Cancer Res.*, **51**, 880–5.

Gelin, J., Moldawer, L.L., Lönnroth, C., Sherry, B., Chizzonite, R. and Lundholm, K.G. (1991b). The role of endogenous tumour necrosis factor a and Interleukin-1 for experimental tumour growth and the development of cancer cachexia. *Cancer Res.*, **51**, 415–21.

Hack, C.E., De Groot, E.R., Felt-Bersma, R.F.J., Nuijens, J.H., Strack Van Schijndel, R.J.M., Eerenberg-Belmer, A.J.M., Thijs, L.G. and Aarden, L.A. (1989). Increased plasma levels of Interleukin-6 in sepsis. *Blood*, **74**, 1704–10.

Hannum, C.H., Wilcox, C.J., Arend, W.P., Joslin, F.G., Dripps, D.J., Heimdal, P.L., Armes, L.G., Sommer, A., Eisenberg, S.P. and Thompson, R.C. (1990). Interleukin-1 receptor antagonist activity of a human interleukin-1 inhibitor. *Nature*, **343**, 336–40.

Havell, E. (1989). Evidence that tumor necrosis factor has an important role in antibacterial resistance. *J. Immunol.*, **143**, 2894–9.

Hellerstein, M.K., Meydani, S.N., Meydani, J., Wu, K. and Dinarello, C.A. (1989). Interleukin-1 induced anorexia in the rat. Influence of prostaglandins. *J. Clin. Invest.*, **84**, 228–35.

Hoffman-Goetz, L. and Kluger, M. (1979). Protein deprivation: Its effects on fever and plasma iron during bacterial infections in rabbits. *J. Physiol. (Lond.)*, **295**, 419–30.

Holroyde, C.P., Skutches, C.L., Boden, G. and Reichard, G.A. (1984). Glucose metabolism in cachectic patients with colorectal cancer. *Cancer Res.*, **44**, 5910–13.

Hyltander, A., Drott, C., Unsgaard, B., Tolli, J., Korner, U., Arfvidsson, B. and Lundholm, K. (1990). The effect on body composition and exercise performance of home parenteral nutrition when given as adjunct to chemotherapy of testicular carcinoma. A randomized and prospective study. *J. Clin. Invest.*, in press.

Hyltander, A., Drott, C., Körner, U., Sandström, R. and Lundholm, K. (1991a). Elevated energy expenditure in cancer patients with solid tumor. Evaluation of co-variating factors. *Eur. J. Cancer*, in press.

Hyltander, A., Warnold, I., Edén, E. and Lundholm, K. (1991b). Stimulation of whole body protein synthesis following institution of intravenous nutrition in cancer and non-cancer patients who lose weight. *Eur. J. Cancer*, in press.

Jablons, D.M., Mulé, J.J., McIntosh, J.K., Sehgal, P.B., May, L.T., Huang, C.M., Rosenberg, S.A. and Lotzé, M.T. (1989). IL-6/IFN-beta-2 as a circulating hormone. Induction by cytokine administration in humans. *J. Immunol.*, **142**, 1542–7.

Kirnbauer, R., Köck, A., Schwarz, T., Urbanski, A., Krutmann, K., Borth, W., Damm, D., Shipley, G., Ansel, J.C. and Luger, T.A. (1989). IFN-beta 2, B cell differentiation factor 2, or hybridoma growth factor (IL-6) is expressed and released by human epidermal cells and epidermoid carcinoma cell lines. *J. Immunol.*, **142**, 1922–8.

Lindmark, L., Bennegård, K., Edén, E., Ekman, L., Scherstén, T., Svaninger, G. and Lundholm, K. (1984). Resting energy expenditure in malnourished patients with and without cancer. *Gastroenterology*, **87**, 402–8.

Lindmark, L., Edén, E., Ternell, M., Bennegård, K., Svaninger, G. and Lundholm, K. (1986). Thermic effect and substrate oxidation in response to intravenous nutrition in cancer patients who lose weight. *Ann. Surg.*, **204**, 628–36.

Lönnroth, C., Moldawer, L., Gelin, J., Kindblom, L.G., Sherry, B. and Lundholm, K. (1990). Tumor necrosis factor-alpha and interleukin-1 alpha production in cachetic, tumor-bearing mice. *Int. J. Cancer*, **46**, 889–96.

Mahony, S.M. and Tisdale, M.J. (1988). Induction of weight loss and metabolic alterations by human recombinant tumour necrosis factor. *Br. J. Cancer*, **58**, 345–9.

Mahony, S.M. and Tisdale, M.J. (1989). Role of prostaglandins in tumour necrosis factor induced weight loss. *Br. J. Cancer*, **60**, 51–5.

Moldawer, L.L. (1986). *Interleukin 1, Tumour Necrosis Factor-alpha (Cachectin) and the Pathogenesis of Cancer Cachexia*. Thesis. University of Göteborg, Sweden.

Naylor, M.S., Stamp, G.W.H. and Balkwill, F.R. (1990). Investigation of cytokine gene expression in human colorectal cancer. *Cancer Res.*, **50**, 4436–40.

Oliff, A. (1988). The role of tumor necrosis factor (cachectin) in cachexia. *Cell*, **54**, 141–2.

Oliff, A., Defeo-Jones, D., Boyer, M., Martinez, D., Kiefer, D., Vuocolo, G., Wolfe, A. and Socher, S.H. (1987). Tumors secreting human TNF-cachetin induce cachexia in mice. *Cell*, **50**, 555–63.

Remick, D.G., Njuyen, D.T., Eskandari, M.K., Strieter, R.M. and Kunkel, S.L. (1989). Cyclosporine A inhibits TNF production without decreasing TNF mRNA levels. *Biochem. Biophys. Res. Commun.*, **161**, 551–5.

Revhaug, A., Michie, H.R., Manson, J. Mc K., Watters, J.M., Dinarello, C.A., Wolff, S.M. and Wilmore, D.W. (1988). Inhibition of cyclo-oxygenase attenuates the metabolic response to endotoxin in humans. *Arch. Surg.*, **123**, 162–70.

Roitt, I.M. (1988). *Essential Immunology*, 6th edn., Blackwell Scientific Publications, London.

Seckinger, P., Isaaz, S. and Dayer, J.M. (1988). A human inhibitor of tumor necrosis factor alpha. *J. Exp. Med.*, **167**, 1511–6.

Seckinger, P., Zhang, J.H., Hauptmann, B. and Dayer, J.M. (1990). Characterization of a tumor necrosis factor a (TNF-a) inhibitor: Evidence of immunological cross-reactivity with the TNF receptor. *Proc. Natl Acad. Sci. (USA)*, **87**, 5188–92.

Sherry, B.A., Gelin, J., Fong, Y., Marano, M., Wei, H., Cerami, A., Lowry, S.F. and Lundholm, K.G. (1989). Anticachectin/tumor necrosis factor-alpha antibodies attenuate development of cachexia in tumor models. *FASEB J.*, **3**, 1956–62.

Socher, S.H., Friedman, A. and Martinez, D. (1988). Recombinant human tumor necrosis factor induces acute reductions in food intake and body weight in mice. *J. Exp. Med.*, **167**, 1957–62.

Spriggs, D.R., Imamura, K., Rodriguez, C., Sariban, E. and Kufe, D.W. (1988). Tumor necrosis factor expression in human epithelial tumor cell lines. *J. Clin. Invest.*, **81**, 455–60.

Svaninger, G. (1987). *The Role of Hormone Alterations in Development of Cancer Cachexia. An Experimental Study with Special Reference to Wasting of Skeletal Muscles.* Thesis: University of Göteborg, Sweden.

Svaninger, G., Bennegård, K., Ekman, L., Ternell, M. and Lundholm, K. (1983). Lack of evidence for elevated breakdown rate of skeletal muscles in weight-losing tumor-bearing mice. *J. Natl Cancer Inst.*, **71**, 341–6.

Tabibzadeh, S.S., Poubouridis, D., May, L.T. and Sehgal, P.B. (1989). Interleukin-6 immunoreactivity in human tumors. *Am. J. Pathol.*, **135**, 427–33.

Ternell, M., Moldawer, L.L., Lönnroth, C., Gelin, J. and Lundholm, K.G. (1987). Plasma protein synthesis in experimental cancer compared to paraneoplastic conditions, including monokine administration. *Cancer Res.*, **47**, 5825–30.

Tracey, K.J., Wei, H., Manogue, K.R., Fong, Y., Hesse, D.G., Nguyen, H.T., Kuo, G.C., Beutler, B., Cotran, R.S., Cerami, A. and Lowry, S.F. (1988). Cachectin-tumor necrosis factor induces cachexia, anemia, and inflammation. *J. Exp. Med.*, **167**, 1211–27.

Utsumi, K., Takai, Y., Tada, T., Ohzeki, S., Fujiwara, H. and Hamaoka, T. (1990). Enhanced production of IL-6 in tumor-bearing mice and determination of cells responsible for its augmented production. *J. Immunol.*, **145**, 397–403.

Ward, P.L., Koeppen, H.K., Hurteau, T., Rowley, D.A. and Schreiber, H. (1990). Major histocompatibility complex class I and unique antigen expression by murine tumors that escaped from CD8+ T-cell-dependent surveillance. *Cancer Res.*, **50**, 3851–8.

Westin, T., Gustavsson, B., Edström, S., Hellander, K., Reinholdtsen, L., Tibell, L. and Lundholm, K. (1991). Tumor cytokinetic effects of acute starvation versus polyamine depletion in tumor-bearing mice. *Cytometry*, in press.

Zucali, J.R., Dinarello, C.A., Oblon, D.J., Gross, M.A., Anderson, L. and Weiner, R.S. (1986). Interleukin 1 stimulates fibroblasts to produce granulocyte-macrophage colony-stimulating activity and prostaglandin E_2. *J. Clin. Invest.*, **77**, 1857–63.

Chapter Fourteen

Progress Report on Drugs Inducing Anorexia by Affecting Brain Serotonin

Silvio Garattini, Silvio Caccia, Tiziana Mennini and Rosario Samanin
Istituto di Ricerche Farmacologiche 'Mario Negri', Via Eritrea 62, 20157 Milan, Italy

INTRODUCTION

Anorexia can be induced not only by stimulant amphetamine drugs but also by non-stimulants such as fenfluramine (Garattini and Samanin, 1976). Fenfluramine differs in many respects from amphetamine (Le Douarec and Neveu, 1970) and under suitable experimental conditions it even shows anti-amphetamine properties (Berger *et al.*, 1973; Bendotti *et al.*, 1980). Fenfluramine and particularly its active isomer, d-fenfluramine (dF), are believed to reduce food intake by enhancing central serotoninergic transmission (Garattini *et al.*, 1988). Indeed at suitable doses dF lowers both serotonin (5HT) and 5-hydroxyindolacetic acid (5HIAA) in various brain regions (Garattini *et al.*, 1988) while raising extracellular 5HT as determined by microdialysis (Fantino *et al.*, 1988) or extracellular 5HIAA determined by pulse voltammetry (De Simoni *et al.*, 1988). The *in vivo* findings may be explained by the ability of dF to inhibit 5HT uptake (Garattini *et al.*, 1986) and at higher doses to enhance 5HT release from brain synaptosomes (Garattini *et al.*, 1986).

That enhancement of serotoninergic transmission may be responsible for the decrease of food intake is borne out by several reports that 5HT injected locally into the paraventricular nucleus of the hypothalamus reduces food intake and antagonizes the hyperphagia induced by noradrenaline (Weiss

Obesity and Cachexia. Edited by N.J. Rothwell and M.J. Stock

et al., 1986). The possibility of influencing food intake in normal and obese animals and the subsequent demonstration that dF may have an anorectic effect, probably therefore reducing body weight (Apfelbaum, 1988) in man, has stimulated numerous studies in this field.

Several drugs have been proposed as 5HT-anorectic agents; among them fluoxetine (Fl) and sertraline (ST) (see chemical structures in Figure 14.1) have been intensively investigated (Lucki *et al.*, 1988; Wong *et al.*, 1988). This report summarizes present knowledge on these two drugs, in comparison with dF, describing some of the work in progress in this laboratory.

ANORECTIC ACTIVITIES OF SEROTONINERGIC AGENTS IN VARIOUS ANIMAL SPECIES

The anorectic activity of dF, Fl and ST was compared in food-deprived mice, rats and guinea-pigs (Table 14.1). On a molar basis—and taking into account the ED_{50}—dF was more effective than Fl, which in turn was more effective than ST in rats. In mice Fl was more effective than dF or ST, while in guinea-pigs Fl showed potency similar to dF. It was impossible to determine an ED_{50} for ST in guinea-pigs because at doses between 10–20 mg/kg (i.e. the range of anorectic doses in rats and mice) the animals showed motor behavioural abnormalities which obviously interfered with the measurement of food intake (Table 14.1).

Differences in active doses of a given compound across animal species are related to kinetics or to intrinsic sensitivity. It was therefore interesting to investigate the brain concentrations of the parent drug and its known

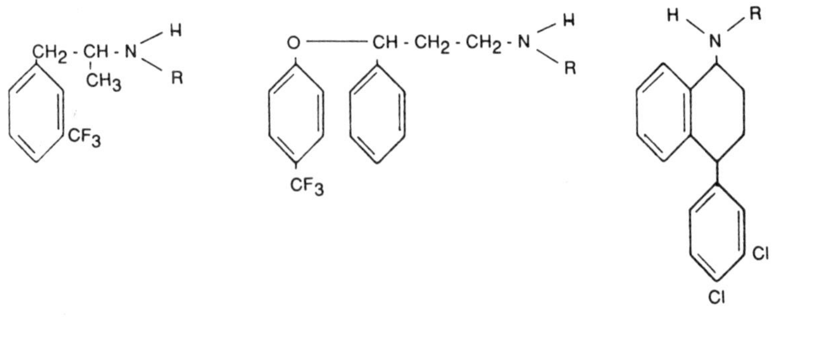

R = CH_2CH_3	d-FENFLURAMINE	R = CH_3	FLUOXETINE	R = CH_3	SERTRALINE
R = H	d-NORFENFLURAMINE	R = H	NORFLUOXETINE	R = H	NORSERTRALINE

Figure 14.1 Chemical structures of d-fenfluramine, fluoxetine and sertraline and their nor-metabolites

Table 14.1 Brain concentrations of d-fenfluramine (dF), fluoxetine (Fl) and sertraline (ST) and their *N*-dealkylated metabolites after equi-active anorectic doses (ED_{50}) to mice, rats and guinea-pigs. ED_{50} is the dose reducing food intake by 50% in animals trained to eat their meal in the morning, fasted overnight and presented with food for 1 h on the day of the experiment

Species	Compound	ED_{50} (95% confidence limits) (μmol/kg, i.p.)	Brain concentration (nmol/g)†	
			Parent drug	Nor-metabolite
Mouse	dF	60.9 (53.4–70.2)	138.5 ± 6.6	14.9 ± 0.1
	Fl	39.10 (30.7–47.2)	28.5 ± 8.9	2.2 ± 0.4
	ST	74.70 (71.8–77.6)	38.8 ± 5.9	12.1 ± 1.6
Rat	dF	4.90 (3.4–7.5)	8.8 ± 0.5	3.6 ± 0.1
	Fl	34.80 (19.7–49.6)	32.0 ± 9.8	7.1 ± 0.3
	ST	48.70 (23.3–74.4)	72.8 ± 11.4	8.3 ± 2.7
Guinea-pig	dF	18.30 (8.2–28.4)	1.9 ± 0.6	2.2 ± 0.4
	Fl	22.00 (4.3–39.4)	6.1 ± 0.6	1.5 ± 0.2
	ST	n.d.	3.0 ± 0.8	1.1 ± 0.1

†Each value is the mean ± SD of 4–5 animals. The animals were killed 30 min after drug injection.
n.d., not determinable (brain concentrations were measured after i.p. injection of 20 mg/kg (58.4 mmol/kg) ST.

metabolite after administration of the anorectic ED_{50} (Table 14.1). The brain concentrations of unchanged drug varied widely and were poorly correlated with the dose. Attempts to correlate brain concentrations with the anorectic potency are also limited by the fact that d-norfenfluramine (dNF) is more potent than dF in causing anorexia while no information is available on the anorectic potency of the nor-metabolites of Fl and ST. In relation to brain concentrations of the parent compounds, guinea-pigs appeared to be the most sensitive species to dF and Fl. ST was more effective in mice than in rats, while the opposite was true for dF; Fl was almost equally effective in rats and mice.

Importance of the nor-metabolites

It is interesting that the three drugs considered in this report are all metabolized through *N*-dealkylation giving rise to active nor-metabolites (see Figure 14.1 for chemical structures). As shown in Table 14.1, the nor-metabolites of dF, Fl and ST are found in the brain of the three species considered even after an acute ED_{50} when brain levels were determined 30 min after the dose of parent drug. However, the proportions of the nor-metabolites to the parent drug differed: for dF the ratio between the parent drug and dNF was 9, 2.4 and 0.8 in mice, rats and guinea-pigs; for Fl the ratio was 13.0, 4.5 and 4.0, and for ST the ratio was 3.2, 8.7 and 2.7.

dNF had greater anorectic activity than dF in all three species (Table 14.2). Brain levels of dNF compared after an ED_{50} of dF or dNF show clearly that dNF plays a negligible anorectic role in mice, it may contribute in rats, while it may have great importance for the anorectic effect of the parent drug in guinea-pigs. dNF is certainly more important during chronic treatment because it accumulates in the rat brain more than the parent compound because of its longer $t_{1/2}$ ($t_{1/2}$ about 2 h for dF and about 13 h for dNF in rats) (Caccia *et al.*, 1982; Fracasso *et al.*, 1988).

For Fl and ST too, the $t_{1/2}$ of the respective nor-metabolite is longer than for the parent compound in rats (Caccia *et al.*, 1990; Tremaine *et al.*, 1989). Hence, the repeated treatment will again result in higher rat brain levels of the nor-metabolite than of the parent drug.

BIOCHEMICAL EFFECTS OF dF, Fl AND ST AND THEIR NOR-METABOLITES

Table 14.3 summarizes the effects of the three serotoninergic anorectic agents and their nor-metabolites on the uptake (IC_{50}) and release (SC_{20}) of 5HT, DA and NA. ST is the most potent inhibitor or 5HT uptake followed by Fl, nor-ST and dF, nor-Fl and dNF; this effect is relatively specific,

Table 14.2 Anorectic activity of d-fenfluramine (dF) and d-norfenfluramine (dNF) expressed as ED_{50} (mg/kg, i.p.) and brain concentrations after equi-active anorectic doses (ED_{50}) to mice, rats and guinea-pigs

Species	Compound	ED_{50} (95% confidence limits) (μmol/kg, i.p.)	Time (min)	Brain concentration (μmol/g)	
				dF	dNF
Mouse	dF	60.9 (53.4–70.2)	5	50.5 ± 7.8	<0.1
			30	138.5 ± 6.6	14.9 ± 0.1
			60	106.8 ± 21.3	16.9 ± 2.5
	dNF	28.5 (23.0–33.9)	30	—	124.8 ± 14.9
Rat	dF	4.9 (3.4–7.5)	30	8.8 ± 0.5	3.6 ± 0.1
			90	4.6 ± 0.4	5.7 ± 0.4
	dNF	3.4 (2.1– 5.0)	30	—	7.4 ± 0.9
Guinea-pig	dF	18.3 (8.2–28.4)	30	1.9 ± 0.6	2.2 ± 0.4
			90	1.7 ± 0.6	12.6 ± 2.8
	dNF	12.6 (7.5–17.2)	30	—	2.0 ± 0.2

Each value is the mean ± SEM of 4–5 animals.

Table 14.3 Effect of anorectic agents on uptake and release of 5HT, DA, NA

	Drug (μM)	Nor-metabolite (μM)
d-Fenfluramine		
Uptake (IC_{50}) of		
5HT	0.4	0.6
DA	16.3	4.2
NA	2.7	0.5
Release (SC_{20}) of		
5HT	2.5	9.5
DA	>10.0	1.8
NA	>20.0	>20.0
Fluoxetine		
Uptake (IC_{50}) of		
5HT	0.07	0.5
DA	4.5	2.9
NA	2.4	2.3
Release (SC_{20}) of		
5HT	6.3	19.7
DA	3.8	3.4
NA	>20.0	>20.0
Sertraline		
Uptake (IC_{50}) of		
5HT	0.01	0.15
DA	0.5	2.4
NA	3.1	8.7
Release (SC_{20}) of		
5HT	2.3	4.2
DA	1.9	4.1
NA	>20.0	>20.0

These data can be compared with Table 14.1 showing the brain levels of the various drugs and nor-metabolites after the anorectic ED_{50}. Drugs or their metabolites were added *in vitro* to isolated synaptosomes, and uptake and release were studied as described by Mennini *et al.* (1981).

except for dNF which inhibits NA uptake at the dose acting on 5HT.

dF and ST have more effect on 5HT release than Fl. Among the metabolites nor-ST is more effective than dNF which in turn is more effective than nor-Fl. The effect on 5HT release is specific on NA release while DA release is mostly affected by dNF, ST, Fl, nor-Fl and nor-ST.

Comparing these *in vitro* data with the concentrations found in the brain after an anorectic ED_{50} of the three serotoninergic agents shows that in rats all three agents can inhibit 5HT uptake and enhance 5HT release.

Release of 5HT by dF and Fl

The fact that dF, Fl and ST at anorectic doses are all able to release 5HT besides inhibiting 5HT uptake helps sustain the hypothesis that these drugs enhance central serotoninergic transmission. However, it should be underlined that the ratios between the effects on 5HT release and uptake are quite different for the two drugs, Fl being about six times more potent in inhibiting 5HT uptake than dF (Table 14.3). In addition the mechanism of 5HT release induced by dF appears to differ from that of Fl. As shown in Figure 14.2, the concentration–effect relationship is quite different for the two drugs. The kinetics of release reported in Figure 14.3 are also different, dF being faster-acting than Fl. The release does not seem related to the leakage of synaptosomes because both drugs release 5HT at the second addition too (Figure 14.3); the fact that the second stimulation is less effective than the first is common to other releasers of 5HT such as KCl (data not shown).

Finally, Figure 14.4 shows another interesting difference: indalpine, which is a specific inhibitor of 5HT uptake completely blocks the 5HT releasing effect of dF but has no such effect on the 5HT release induced by various concentrations of Fl. This difference may be related to the fact that [^{3}H]dF is actively taken up by synaptosomes, probably through the same carrier as 5HT (Garattini *et al.*, 1989).

The importance of the 5HT releasing activity of dF and Fl can be shown

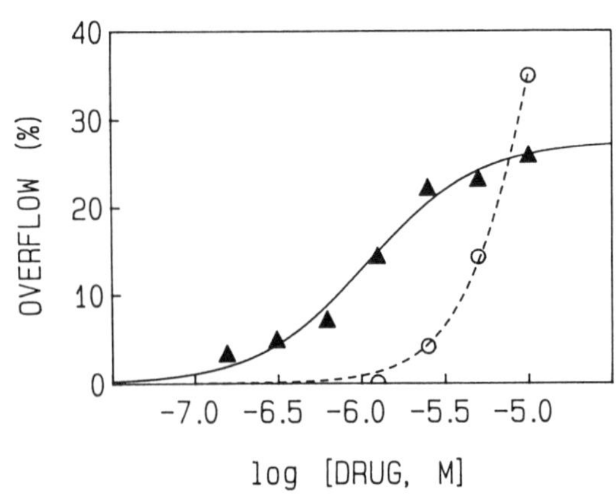

Figure 14.2 Dose–effect relationship of [^{3}H]5HT overflow induced by d-fenfluramine (▲) or fluoxetine (○). Data are obtained from two experiments, each in triplicate

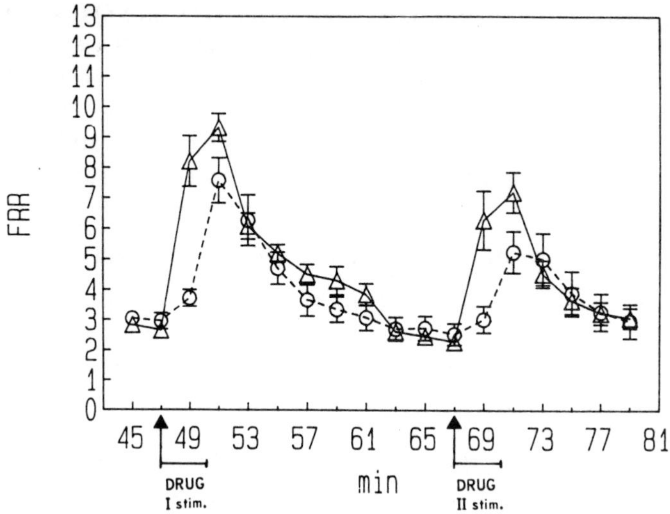

Figure 14.3 Fractional releasing rates of [^{3}H]5HT in the presence of dF (10 μM), △, or fluoxetine (5 μM), ○. Drugs were added twice for 3 min, as indicated in the figure. Each point is the mean ± SD of triplicate samples

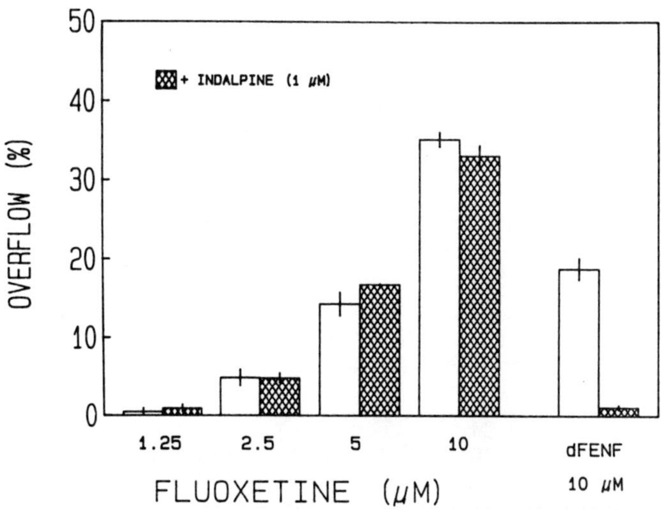

Figure 14.4 Fluoxetine was added to superfusion medium at 47 min, and removed at 50 min. [^{3}H]5HT overflow was calculated as the difference between the fractional releasing rate in the presence and absence of the drug. Indalpine was present in the superfusion medium from the beginning. Data are mean ± SD of triplicate samples

in vivo. A single dose of dF lowers both brain 5HT and 5HIAA 4 h later, an effect which is not shared by Fl which only lowers 5HIAA (Fuller *et al.*, 1988). However, during repeated treatment Fl lowers both 5HT and 5HIAA in the hippocampus and cortex, 24 h after the last dose. These results, summarized in Table 14.4, are seen when brain concentrations of Fl and nor-Fl are 2–4 times higher than those effective *in vitro* as releasers of 5HT.

Relative importance of dF and dNF for lowering brain 5HT

Kinetic studies have indicated that, particularly during repeated treatment, dNF accumulates in the brain where its levels may largely exceed those of dF (Caccia *et al.*, 1982). In addition dNF releases 5HT with a different mechanism from dF; in fact it releases 5HT even when synaptosomes are obtained from reserpinized rats, while under the same experimental conditions dF is inactive (Borroni *et al.*, 1983; Mennini *et al.*, 1981). It was therefore interesting to find out how much dNF contributes to the brain 5HT-lowering effect of dF: dF or dNF were infused intracerebroventricularly (i.c.v.) at various concentrations for 2 h; and 4 h; after the end of the infusion animals were killed to measure 5HT, dF and/or its metabolite (Table 14.5). Only dF infused at 500 μg/h reduced 5HT although dNF reached levels similar to those producing the same decrease of 5HT when dNF was infused directly.

In another experiment the metabolism of dF to dNF was blocked by SKF

Table 14.4 Brain indole and fluoxetine (Fl) and norfluoxetine (norFl) concentrations after administration of Fl hydrochloride to male rats

	Indole concentrations (% of control)		Drug concentrations (nmol/g)	
Day of treatment	5HT	5HIAA	Fl	norFl
Cortex				
Day 1	102 ± 33	65 ± 13**	1.0 ± 0.2	14.8 ± 2.8
Day 7	104 ± 13	54 ± 11**	8.3 ± 2.2‡	45.5 ± 2.5‡
Day 14	61 ± 13*†	39 ± 5**‡	14.5 ± 2.8‡	56.2 ± 4.2‡
Hippocampus				
Day 1	92 ± 24	54 ± 4**	1.0 ± 0.5	18.6 ± 5.1
Day 7	56 ± 10**†	39 ± 3**†	10.1 ± 1.7‡	51.7 ± 13.5‡
Day 14	49 ± 9**‡	37 ± 8**†	23.1 ± 4.8‡	78.6 ± 16.4‡

Fl was dissolved in water and injected i.p. at a dose of 15 mg/kg, b.i.d. The rats were killed 24 h after the last dose. Each value is the mean ± SD of 5 animals.
** $p<0.01$; * $p<0.05$ vs. vehicle.
‡ $p<0.01$; † $p<0.05$ vs. 1st day group.

Table 14.5 Effect of SKF 525A on brain serotonin depletion induced by fenfluramine (F) and drug levels in brain

Drug	Dose (μg/h)	Brain levels: 5HT (μg/g ± SEM)	dF (μg/g ± SD)	dNF (μg/g ± SD)
Saline		0.41 ± 0.01	—	—
SKF 525A		0.41 ± 0.03	—	—
dF	250	0.42 ± 0.02	0.17 ± 0.01	1.30 ± 0.03
dF	500	0.24 ± 0.01*	0.57 ± 0.12	4.26 ± 1.59
dNF	500	0.20 ± 0.02*	—	3.44 ± 0.73
dF+SKF 525A	250	0.40 ± 0.03	1.85 ± 0.21	0.22 ± 0.05
dF+SKF 525A	500	0.22 ± 0.00*	3.49 ± 0.62	0.37 ± 0.08

Rats received 50 mg/kg SKF 525A 1 h before and 3 h after the intracerebroventricular infusion of various doses of dF in 2 h and were killed 4 h after the end of the infusion (Invernizzi *et al.*, 1989).
*$p<0.01$ vs. saline (Dunnett's *t*-test).

525A, a compound known to inhibit liver *N*-dealkyl oxidation. The results indicated that dF was also effective in reducing brain 5HT when a brain concentration of 3.49 ± 0.62 μg/g dF was reached. It can thus be concluded that dNF is certainly very important in relation to the depletion of brain 5HT induced by i.c.v. dF but dF is also active at appropriate concentrations (Invernizzi *et al.*, 1989). Similar studies are in progress to assess the importance of nor-Fl and nor-ST in the action of Fl and ST on brain indole concentrations.

Binding and uptake of [^{3}H]dF

[^{3}H]dF binds to brain membranes with high affinity (Gobbi *et al.*, 1989; Mennini *et al.*, 1988) and may be actively taken up by brain synaptosomes (Garattini *et al.*, 1989b). The characteristics of binding and uptake of [^{3}H]dF have been described and these two processes are now known to be partly localized in serotoninergic nerve terminals (Garattini *et al.*, 1990; Garattini *et al.*, 1989a). ST and Fl show nM affinity for the binding of [^{3}H]dF but the nor-metabolites are less effective than the parent drugs (Table 14.6). Other inhibitors of 5HT uptake such as indalpine and paroxetine show good affinity for [^{3}H]dF binding too. It is of interest that ST and norST are very good inhibitors of the uptake of [^{3}H]dF while Fl and norFl are less active though at concentrations compatible with those reached in the brain after anorectic ED_{50} in rats (see Table 14.1 for comparison). The inhibition exerted by Fl on [^{3}H]dF uptake may explain why *in vivo* Fl antagonizes the

Table 14.6 Effects of various 5HT uptake inhibitors on [^{3}H]dF binding and uptake by brain synaptosomes

Drug	[^{3}H]dF binding IC_{50} (μM)	[^{3}H]dF uptake IC_{50} (μM)
Sertraline	0.007	0.153
Norsertraline	0.027	0.320
Fluoxetine	0.034	4.705
d-Fenfluramine	0.039	0.232
Indalpine	0.124	2.785
d-Norfenfluramine	0.532	3.961
Norfluoxetine	0.621	3.962
Paroxetine	1.158	1.177
5-Hydroxytryptamine	>100 000	0.485

[^{3}H]dF binding was studied by incubating 10 nM [^{3}H]dF for 90 min at 37°C with 4–6 drug concentrations with rat crude membrane preparation, as described by Mennini *et al.* (1988). [^{3}H]dF uptake was measured in rat brain synaptosomes, incubating 50 nM [^{3}H]dF for 5 min at 30°C in the absence or presence of different concentrations of inhibitors. Passive diffusion was measured at 4°C (Garattini *et al.*, 1989b).

depletion induced by dF (Fuller *et al.*, 1978). It is noteworthy that 5HT is effective in competing with [^{3}H]dF uptake but has no effect on the binding of [^{3}H]dF (Table 14.6). The significance of [^{3}H]dF binding and uptake in relation to the anorectic activity remains to be established.

RELATION BETWEEN THE SEROTONINERGIC AND ANORECTIC EFFECTS

Having established that the three anorectic agents do modify 5HT transmission in the brain, acting as indirect 5HT agonists, it was important to establish whether this effect was responsible for their anorectic activity. Previous studies indicated that the anorectic effect of dF was abolished by an electrolytic lesion of the median raphe (Samanin *et al.*, 1972), an area rich in 5HT neurons, but not by a lesion of the ventral noradrenergic bundle which instead blocked the anorectic effect of d-amphetamine (Garattini and Samanin, 1976). In addition the effect of dF was completely antagonized by treatment with a non-selective 5HT antagonist such as metergoline (Garattini and Samanin, 1976). Table 14.7 indicates that metergoline does not antagonize ST and—in agreement with other findings (Wong *et al.*, 1988)—it was impossible even with different schedules of treatment to block the anorectic effect of Fl. Further studies indicate that $5HT_2$ inhibitors such as ritanserin (Samanin *et al.*, 1989) or $5HT_3$ blockers such as odansetron do not affect the anorectic activity of dF (A. Bizzi, unpublished results).

Table 14.7 Effects of various 5HT antagonists on anorectic activity of dF, Fl and ST

	Effect on the anorectic activity of		
5HT antagonist	dF	Fl	ST
Metergoline (non-selective)	A	0	0
Ritanserin ($5HT_2$, $5HT_{1C}$)	0	0	0
Odansetron ($5HT_3$)	0	0	–
Xylamidine (5HT peripheral)	0	0	0

A, significant reduction of the effect of the agonist; 0, no effect; –, not tested.

Therefore the action of dF and ST appears to involve $5HT_1$ receptors. $5HT_{1A}$ receptor agonists such as 8-OH-DPAT (Dourish *et al.*, 1985; Bendotti and Samanin, 1986) and buspirone (Garattini *et al.*, 1988) enhance food intake; the action of dF is probably mediated by an effect on $5HT_{1B}$ and/or $5HT_{1C}$ receptor subtypes (Table 14.7) but for the time being the lack of selective antagonists for these subtypes makes it impossible to draw any firm conclusion. The fact that xylamidine, a peripheral 5HT antagonist, does not antagonize the action of dF shows that anorexia is mediated by central 5HT transmission (Borsini *et al.*, 1982).

CONCLUDING REMARKS

The accumulation of data showing that indirect and direct 5HT agonists reduce food intake in rats adds evidence for the importance of 5HT in controlling the complex mechanisms involved in appetite and satiety. The fact that some of these drugs—such as dF, Fl and ST—reduce food intake and body weight in man too reinforces the interest in animal studies to clarify the mechanism of action of these drugs.

More quantitative studies are needed to establish the roles of the compound administered and its metabolite(s) in modifying brain chemical mediators. The importance of serotoninergic mechanisms (uptake inhibition and enhanced release) needs to be established *in vivo* and in brain areas important in controlling food intake. More selective 5HT antagonists are needed to identify which 5HT receptor subtype mediates the anorectic effect of 5HT and of 5HT agonists. The question of the specificity of 5HT agonists at anorectic doses must be tackled more critically. Attention must also focus on the mechanisms that follow up on the enhanced serotoninergic transmission: the recent finding (Cooper *et al.*, 1990) that a cholecystokin antagonist, MK 329, reverses the anorectic activity of dF suggests there may be important interactions between 5HT and this peptide (Stallone *et al.*, 1989).

Finally, since anorectic drugs are prescribed for repeated administration, emphasis must be laid on the fact that they modify brain chemical mediators and cause changes in their receptors after chronic treatment.

REFERENCES

Apfelbaum, M. (ed.) (1988). Brain and nutrition. Dexfenfluramine weight control and regulation of eating patterns. *Clin. Neuropharmacol.*, **11**, Suppl. 1.

Bendotti, C., Borsini, F., Zanini, M.G., Samanin, R. and Garattini, S. (1980). Effect of fenfluramine and norfenfluramine stereoisomers on stimulant effects of d-amphetamine and apomorphine in the rat. *Pharmacol. Res. Commun.*, **12**, 567–74.

Bendotti, C. and Samanin, R. (1986). 8-Hydroxy-2-(di-n-propylamino) tetralin (8-OH-DPAT) elicits eating in free-feeding rats by acting on central serotonin neurons. *Eur. J. Pharmacology*, **121**, 147–50.

Berger, H.J., Brown, C.C. and Krantz, J.C. Jr. (1973). Fenfluramine blockade of CNS stimulant effects of amphetamines. *J. Pharm. Sci.*, **62**, 788–91.

Borroni, E., Ceci, A., Garattini, S. and Mennini, T. (1983). Differences between d-fenfluramine and d-norfenfluramine in serotonin presynaptic mechanisms. *J. Neurochem.*, **40**, 891–3.

Borsini, F., Bendotti, C., Aleotti, A., Samanin, R. and Garattini, S. (1982). d-Fenfluramine and d-norfenfluramine reduce food intake by acting on different serotonin mechanisms in the rat brain. *Pharmacol. Res. Commun.*, **14**, 671–8.

Caccia, S., Ballabio, M., Guiso, G., Rocchetti, M. and Garattini, S. (1982). Species differences in the kinetics and metabolism of fenfluramine isomers. *Arch. Int. Pharamocodyn.*, **258**, 15–28.

Caccia, S., Cappi, M., Fracasso, C. and Garattini, S. (1990). Influence of dose and route of administration on the kinetics of fluoxetine and its metabolite norfluoxetine in the rat. *Psychopharmacology*, **100**, 509–14.

Cooper, S.J., Dourish, C.T. and Barber, D.J. (1990). Reversal of the anorectic effect of (+)-fenfluramine in the rat by the selective cholecystokinin receptor antagonist MK-329. *Br. J. Pharmacol.*, **99**, 65–70.

De Simoni, M.G., Juraszczyk, Z., Fodritto, F., De Luigi, A. and Garattini, S. (1988). Different effects of fenfluramine isomers and metabolites on extracellular 5-HIAA in nucleus accumbens and hippocampus of freely moving rats. *Eur. J. Pharmacol.*, **153**, 295–9.

Dourish, C.T., Hutson, P.H. and Curzon, G. (1985). Low doses of the putative serotonin agonist 8-hydroxy-2-(di-n-propylamino) tetralin (8-OH-DPAT) elicit feeding in the rat. *Psychopharmacology*, **86**, 197–204.

Fantino, M., Boucher, H., Faion, F. and Mathiot, P. (1988). Dexfenfluramine and body weight regulation: Experimental study with hoarding behavior. *Clin. Neuropharmacol.*, **11**, (Suppl. 1), S97–104.

Fracasso, C., Guiso, G., Garattini, S. and Caccia, S. (1988). Disposition of D-fenfluramine in lean and obese rats. *Appetite*, **10**, 45–55.

Fuller, R.W., Snoddy, H.D. and Hemrick, S.K. (1978). Effects of fenfluramine and norfenfluramine on brain serotonin metabolism in rats. *Proc. Soc. Exp. Biol.*, **157**, 202–5.

Fuller, R.W., Snoddy, H.D. and Robertson, D.W. (1988). Mechanisms of effects of d-fenfluramine on brain serotonin metabolism in rats: Uptake inhibition versus release. *Pharmacol. Biochem. Behav.*, **30**, 715–21.

Garattini, S. and Samanin, R. (1976). Anorectic drugs and brain neurotransmitters. In: *Appetite and Food Intake* (ed. T. Silverstone), Dahlem Konferenzen, Berlin, pp. 83–108.

Garattini, S., Bendotti, C., Invernizzi, R. and Samanin, R. (1986). Neurochemical mechanism of action of drugs which modify feeding via the serotonergic system. *Appetite*, **7**, (Suppl.), 15–38.

Garattini, S., Bizzi, A., Caccia, S., Mennini, T. and Samanin, R. (1988). Progress in assessing the role of serotonin in the control of food intake. *Clin. Neuropharmacol.*, **11** (Suppl. 1), S8–32.

Garattini, S., Mennini, T. and Samanin, R. (1989a). Central serotonergic mechanisms and feeding regulation. In: *Neurochemical Pharmacology. A Tribute to B.B. Brodie* (ed. E. Costa), Raven Press, New York, pp. 89–106.

Garattini, S., Mennini, T. and Samanin, R. (1989b). Reduction of food intake by manipulation of central serotonin: Current experimental results. *Br. J. Psychiatry*, **155**, suppl. 8, 41–51.

Garattini, S., Caccia, S., Mennini, T. and Samanin, R. (1990). Serotonin transmission and food intake. In: *Serotonin. From Cell Biology to Pharmacology and Therapeutics* (eds R. Paoletti, P.M. Vanhoutte, N. Brunello and F.M. Maggi), Kluwer, Dordrecht, pp. 193–202.

Gobbi, M., Taddei, C. and Mennini, T. (1989). *In-vivo* (+)-[^{3}H]fenfluramine binding to rat brain: Biochemical and autoradiographic studies. *J. Pharm. Pharmacol.*, **41**, 253–6.

Invernizzi, R., Fracasso, C., Caccia, S., Garattini, S. and Samanin, R. (1989). Serotonin depletion caused by direct administration of d-fenfluramine and d-norfenfluramine in the rat brain. Presented at: The Neuropharmacology of Serotonin, Int. Conf. New York, July 10-13, 1989, *NY Acad. Sci.*, Abstr. 78.

Le Douarec, J.C. and Neveu, C. (1970). Pharmacology and biochemistry of fenfluramine. In: *Amphetamines and Related Compounds* (eds E. Costa and S. Garattini), Raven Press, New York, pp. 75–105.

Lucki, I., Kreider, M.S. and Simansky, K.J. (1988). Reduction of feeding behavior by the serotonin uptake inhibitor sertraline. *Psychopharmacology*, **96**, 289–95.

Mennini, T., Borroni, E., Samanin, R. and Garattini, S. (1981). Evidence of the existence of two different intraneuronal pools from which pharmacological agents can release serotonin. *Neurochem. Int.*, **3**, 289–94.

Mennini, T., Gobbi, M., Taddei, C. and Garattini, S. (1988). Characterization of high affinity and stereospecific ^{3}H-d-fenfluramine binding to rat brain. *Neurochem. Int.*, **13**, 345–51.

Samanin, R., Ghezzi, D., Valzelli, L. and Garattini, S. (1972). The effects of selective lesioning of brain serotonin or catecholamine containing neurons on the anorectic activity of fenfluramine and amphetamine. *Eur. J. Pharmacol.*, **19**, 318–22.

Samanin, R., Mennini, T., Bendotti, C., Barone, D., Caccia, S. and Garattini, S. (1989). Evidence that central 5-HT_2 receptors do not play an important role in the anorectic activity of d-fenfluramine in the rat. *Neuropharmacology*, **28**, 465–9.

Stallone, D., Nicolaidis, S. and Gibbs, J. (1989). Cholecystokinin-induced anorexia depends on serotoninergic function. *Am. J. Physiol.*, **256**, R1138–41.

Tremaine, L.M., Welch, W.M. and Ronfeld, R.A. (1989). Metabolism and disposition of the 5-hydroxytryptamine uptake blocker sertraline in the rat and dog. *Drug Metab. Disp.*, **17**, 542–50.

Weiss, G.F., Papadakos, P., Knudson, K. and Leibowitz, S.F. (1986). Medial

hypothalamic serotonin: Effects on deprivation and norepinephrine-induced eating. *Pharmacol. Biochem. Behav.*, **25**, 1223–30.
Wong, D.T., Reid, L.R. and Threlkeld, P.G. (1988). Suppression of food intake in rats by fluoxetine: Comparison of enantiomers and effects of serotonin antagonists. *Pharmacol. Biochem. Behav.*, **31**, 475–9.

Chapter Fifteen

β-Adrenoceptor-mediated Control of Thermogenesis, Body Composition and Glucose Homeostasis

J.R.S. Arch, M.A. Cawthorne, K.A. Coney†, B.A. Gusterson‡, V. Piercy, M.V. Sennitt, S.A. Smith, J. Wallace‡ and S. Wilson

SmithKline Beecham Pharmaceuticals, Great Burgh, Yew Tree Bottom Road, Epsom, Surrey KT18 5XQ, UK, †SmithKline Beecham Animal Health, Walton Oaks, Dorking Road, Tadworth, Surrey KT20 7NT, UK, and ‡The Institute of Cancer Research, Haddow Laboratories, 15 Cotswold Road, Belmont, Sutton, Surrey SM2 5NG, UK

INTRODUCTION

The recognition by Ahlquist (see Ahlquist, 1980) that adrenaline and noradrenaline stimulate two types of adrenoceptor (α and β) and the subsequent realization that each type can be further divided into at least two subtypes (α_1, α_2, β_1, β_2) (Lands *et al.*, 1967; Langer, 1977; Starke, 1977) has provided a theoretical framework for the development of a number of valuable drugs. Selective agonists or antagonists of the various receptor subtypes have proved of immense benefit, primarily in cardiovascular disorders and (in the case of β_2-adrenoceptor agonists) as bronchodilators in respiratory diseases.

These utilities are consequences of the influence of adrenoceptor stimulation on smooth and cardiac muscle contractility; but it has long been known that adrenoceptor—especially β-adrenoceptor—stimulation also has a profound influence upon the major pathways of lipid and carbohydrate

Obesity and Cachexia. Edited by N.J. Rothwell and M.J. Stock

metabolism (see Himms-Hagen, 1967; Haylett, 1979) and yet the potential of β-adrenoceptor agonists as agents for the treatment of metabolic disorders is only now beginning to be appreciated. Two reasons for this new awareness can be identified. One is the finding that *repeated* administration of certain β-adrenoceptor agonists has a marked influence upon body composition and insulin sensitivity (Arch and Ainsworth, 1983; Cawthorne *et al.*, 1984). The second is the discovery of compounds with a novel balance of β-adrenoceptor activities; for example, those that are especially potent as lipolytic agents (Arch *et al.*, 1984a), and those that are especially effective in promoting protein accretion (Asato *et al.*, 1984). The identification of these novel compounds has shown that the Lands *et al.* (1967) classification of β-adrenoceptors into just two subtypes is incomplete, suggesting new molecular targets for adrenoceptor ligands.

At SmithKline Beecham Pharmaceuticals our work with β-adrenoceptor agonists has involved us in three therapeutic areas: obesity, type 2 diabetes (also known as non-insulin-dependent or maturity onset diabetes) and protein wasting conditions. For obesity, our objective was to raise metabolic rate and thereby stimulate the combustion of body fat stores. In type 2 diabetes insulin levels are normal or even raised but the tissues are unresponsive to the hormone, and so our objective was to improve insulin sensitivity. The third area relates to the ability of some β-adrenoceptor agonists to increase body protein content. This area has attracted most attention from those interested in feed additives for increasing the production of lean meat in farm animals (Hanrahan, 1987), but these compounds may also prove useful for treating human conditions in which there is protein loss, for example trauma, infection, limb immobilization, muscular dystrophies (originating within the muscle or due to neuronal failure) or cancer cachexia (Rennie 1985; Booth 1977; Harris and Slater, 1980; Kern and Norton, 1988). By reviewing our work in each of these three areas we hope to provide some insights into the role of β-adrenoceptors in the control of thermogenesis, body composition and glucose homeostasis. A more general review has been provided by Yang and McElligott (1989).

OBESITY

Thermogenic β-adrenoceptor agonists

Most of the literature on the potential of β-adrenoceptor agonists as anti-obesity agents is to be found in publications emanating from four pharmaceutical companies: Beecham, Eli Lilly, Hoffman-La Roche and ICI. The chemical structures of the compounds that these companies have described are shown in Figure 15.1.

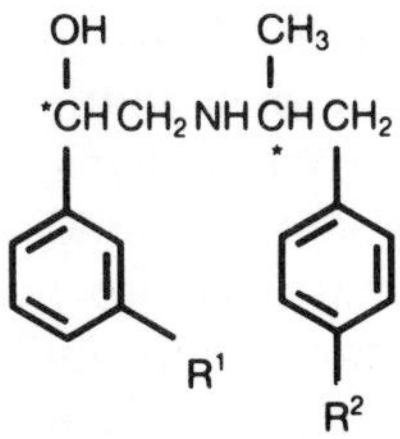

	R^1	R^2
BRL 26830	H	CO_2CH_3
BRL28410	H	CO_2H
BRL35135	Cl	$OCH_2CO_2CH_3$
BRL37344	Cl	OCH_2CO_2H

Racemates of RR and SS enantiomers

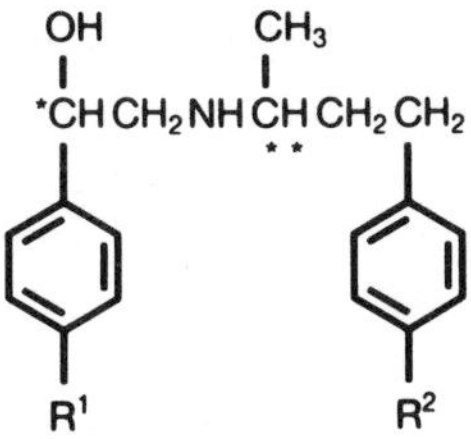

	Configuration *	**	R^1	R^2
LY79771	R	S	H	OH
LY104119	R	S	H	$CONH_2$
LY99134	R	R	OH	OH

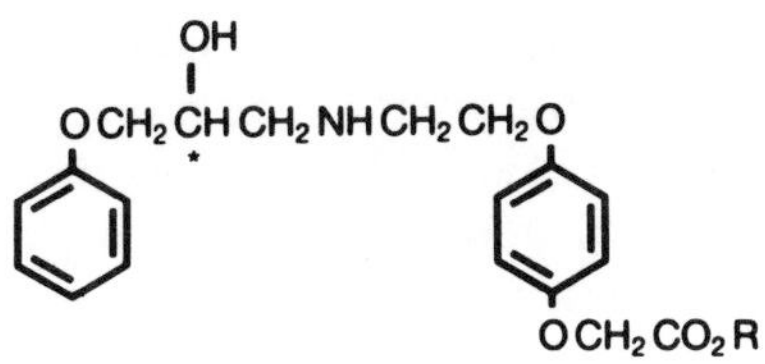

	R
ICI 198,157	CH_3
ICI 201,651	H

Racemates of R and S enantiomers

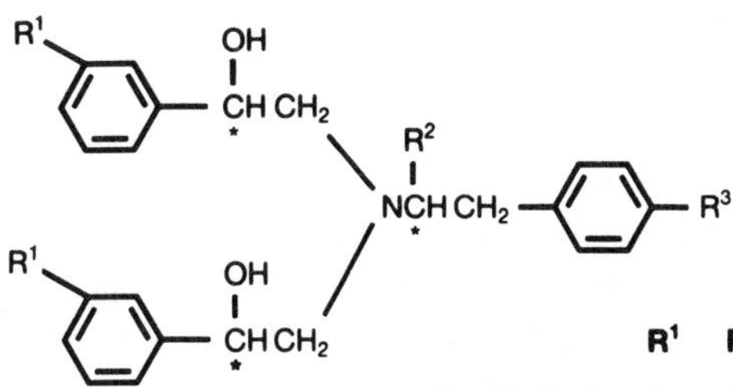

	R^1	R^2	R^3
Ro 16-8714	H	CH_3	$CO\,NH_2$
Ro 40-2148	Cl	H	$OCH_2CH_2OCH_2CH_3$

R Stereochemistry at all points of asymmetry

Figure 15.1 Chemical structure of β-adrenoceptor agonists described as having anti-obesity activity, and of the 'physical conditioning agent' LY 99134. Asterisks denote asymmetric centres

The phenethanolamine BRL 26830 was identified by its ability to cause a marked stimulation of metabolic rate in rodents whilst having little effect on heart rate compared with known β-adrenoceptor agonists (Arch *et al.*, 1984b, 1987). BRL 35135 was first identified as a potential anti-diabetic agent (Sennitt *et al.*, 1986; Cawthorne *et al.*, 1986), but it is also twenty-fold more potent than BRL 26830 as an anti-obesity agent with similar or better selectivity from chronotropic activity (Arch *et al.*, 1984a). BRL 26830 and BRL 35135 are methyl esters, and whilst they do behave as β-adrenoceptor agonists *in vitro*, when they are given orally to animals or man none of the parent compound is found in the systemic circulation. It is the parent acids BRL 28410 and BRL 37344 that appear and, since these compounds are selective β-adrenoceptor agonists *in vitro* (Arch *et al.*, 1984a), it is presumed that they mediate the effects of oral BRL 26830 and BRL 35135, respectively.

The Eli Lilly compounds (Shaw *et al.*, 1981; Yen, 1984; Yen *et al.*, 1984) were 'spin-offs' from a programme on inotropic agents for congestive heart failure. Like the Beecham compounds they contain two asymmetric centres, but whereas BRL 26830 and BRL 35135 are racemates containing the most potent RR stereoisomer and its almost inactive SS mirror image, LY 79771 and LY 104119 are single enantiomers having the RS stereochemistry. Of the four stereoisomers the RS is the second most potent. The more potent RR stereoisomers were patented only as inotropic and 'physical conditioning' agents (Mills *et al.*, 1979), but whether the RS stereoisomers are any more selective than the RR stereoisomers as thermogenic anti-obesity agents is dubious. The 'physical conditioning' property of the RR stereoisomer LY 99134 is perhaps relevant to the progression of ractopamine, a mixture of LY 99134 and its three stereoisomers, as a feed additive for meat-producing animals (see below).

The Hoffman-La Roche compounds, Ro 16-8714 and Ro 40-2148 (Meier *et al.*, 1984, 1989) differ from the others in being tertiary amines. It is possible that the full activity of these compounds is dependent upon their metabolism to secondary amines. Thus, Ro 16-8714 is only a weak stimulant of brown adipocyte oxygen consumption *in vitro* (EC_{50} 8 μM; maximum effect 55% of that of noradrenaline)(Meier *et al.*, 1984), and its failure to affect rat atrial rate *in vitro* seems inconsistent with its potent chronotropic activity *in vivo* (0.086 mg/kg, p.o. increased heart rate by more than 100 beats/min; unpublished results, C. Wilson and J. Arch), suggesting that metabolism to a more active compound occurs. Alternatively, Ro 16-8714 might have a marked indirect chronotropic effect *in vivo* (see discussion of the cardiovascular effects of Ro 40-2148 by Meier *et al.*, 1989). Studies in man revealed that Ro 16-8714 had inadequate thermogenic relative to chronotropic potency (Henny *et al.*, 1987, 1988), but Ro 40-2148, which has no effect on heart rate in pithed rats (Meier *et al.*, 1989), appears to hold more promise.

The ICI programme that led to ICI 198,157 was specifically designed to identify compounds that stimulate thermogenesis in brown adipose tissue but are weak stimulants of β_1- and β_2-adrenoceptor-mediated responses (Holloway *et al.*, 1989; Holloway 1989). ICI 198,157 differs from the other compounds shown in Figure 15.1 in that it is a phenoxypropanolamine—a structure usually associated with β-adrenoceptor antagonists rather than agonists. However, like BRL 26830 and BRL 35135, ICI 198,157 is an ester and it is metabolized to its parent acid ICI 201,651 *in vivo*. It seems probable that the carboxylic acid function endows these molecules with good selectivity for thermogenesis.

Apart from differences in potency, the effects of all these compounds on body weight, body composition and metabolic rate in rats and mice appear to be remarkably similar and the following description will refer mainly to BRL 26830.

Body weight

Daily oral administration of BRL 26830 causes a dose-related reduction in body weight or weight gain in obese rodents (Arch and Ainsworth, 1983; Arch *et al.*, 1984b), even though food intake is not decreased. Figure 15.2 illustrates the effects of BRL 26830A (the hemifumarate salt of BRL 26830) in genetically obese (ob/ob) mice and (fa/fa) rats. The control animals were gaining weight in these experiments and BRL 26830 reduced the rate of weight gain. In experiments where mice have been made obese by other manipulations and have reached a stable body weight, BRL 26830 causes weight loss (Arch and Ainsworth, 1983; Arch *et al.*, 1984b). Weight loss has also been produced in dogs (Arch *et al.*, 1989), and some studies in man have shown an enhancement of the effect of a low energy diet (Zed *et al.*, 1985; Connacher *et al.*, 1988). Figure 15.2 illustrates another fascinating feature of β-adrenoceptor agonists, which is that they have little effect on body weight in lean animals. Other workers have argued that this is because these compounds increase food intake in lean rodents, but this may not be the full explanation, at least in the case of BRL 26830 (see thermogenesis section).

Body composition

The reductions in body weight or weight gain produced by thermogenic β-adrenoceptor agonists are entirely due to reductions in body lipid content—lean tissue is either unaffected or may even increase in amount. Figure

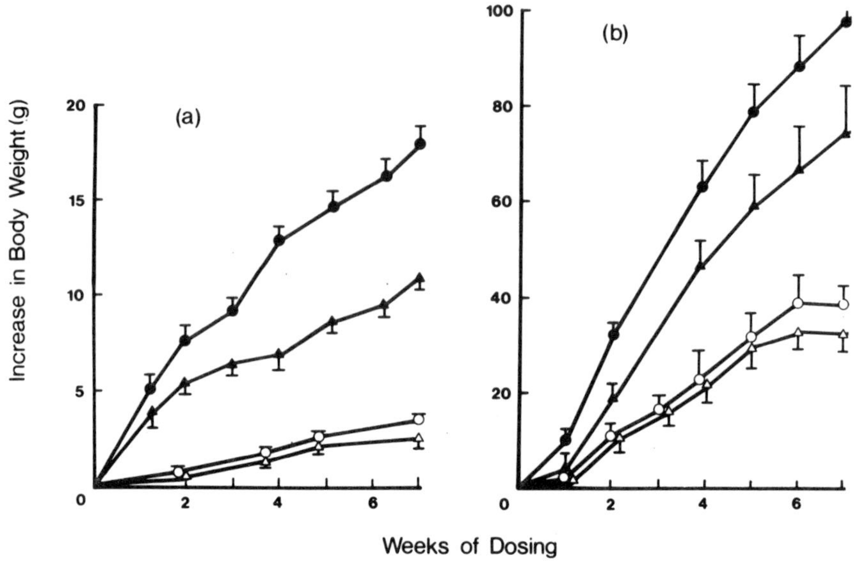

Figure 15.2 Effect of BRL 26830 on body weight gain in (a) ob/ob (n=24) and +/? (n=18) Zucker rats. BRL 26830 (△,▲) or water (○,●) was dosed daily to the obese (●,▲) and lean (○,△) animals. The dose level of BRL 26830 was 0.17 mg/mouse and 1.6 mg/kg body weight for the rats. Starting mean body weights were (a) obese 30.5 g, lean 16.8 g; (b) obese 310 g, lean 183 g. The mice were aged 8–9 weeks and the rats 10–12 weeks at the start of dosing. From week 2 the reduction in weight gain due to BRL 26830 was significant ($p<0.05$) in the obese mice and rats. Reproduced by permission from Arch and Ainsworth (1983). © *Am. J. Clin. Nutr.* American Society for Clinical Nutrition

15.3a shows the body composition of genetically obese (ob/ob) mice following 4 weeks' administration of BRL 26830 once daily by oral gavage. The reductions in body weight and body lipid content were almost identical and there was no effect of BRL 26830 on lean body mass. Diet restriction, in contrast, depleted both lipid and lean tissue, whilst administration of BRL 26830 to mice fed on a restricted diet caused further loss of lipid but not lean tissue. In this experiment the treated animals lost about 5 g lipid compared with the controls. Lean litter mates of ob/ob mice usually contain less than 4 g of lipid in total. Thus if β-adrenoceptor stimulation can only deplete lipid, it is not surprising that there is little effect on body weight in lean mice.

Even though it is difficult to demonstrate a reduction in body weight in lean mice, small but significant depletions of total body lipid content have been detected (Arch *et al.*, 1984b). Furthermore, inclusion of BRL 26830 in the diet of normal mice for 3 days consistently produces a significant reduction in the weight of the perigenital fat pads (Table 15.1).

Administration of thermogenic β-adrenoceptor agonists in the diet of genetically obese mice reveals a slightly different picture from administration

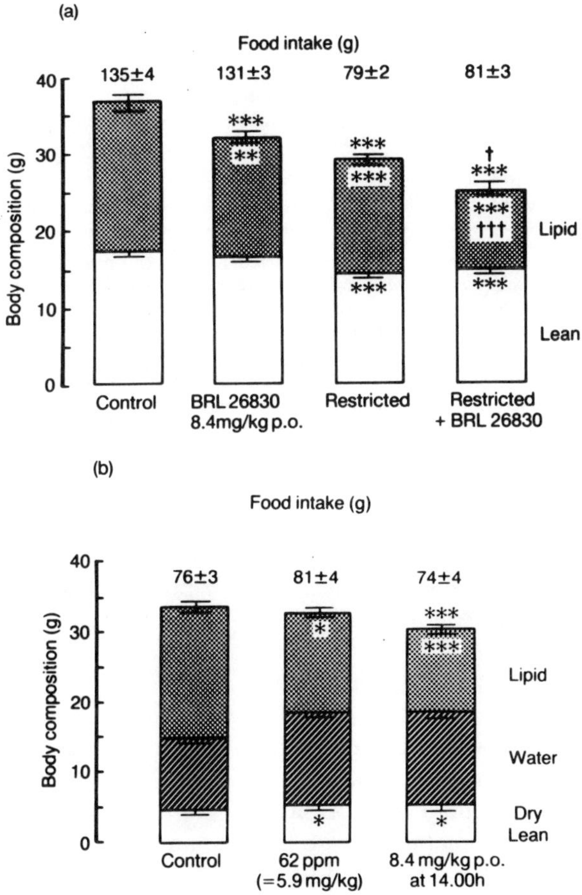

Figure 15.3 Effects of BRL 26830 on body composition of ob/ob mice. (a) The mice were fed *ad libitum* or restricted to 60% of *ad lib* food intake and dosed once daily at 11.00–12.00 h for 4 weeks. (b) The mice were trained to feed from 10.00–16.00 h for a week and BRL 26830 was then included in their diet or given once daily by oral gavage at 14.00 h for 4 weeks. Statistical significances are shown for weight gain, final lipid content and final lipid-free weight (a) or lipid-free dry weight (b). *$p<0.05$; **<0.01, ***<0.001 relative to controls; †††$p<0.001$ relative to restricted group. $n=8$; bars show SEM. Reproduced by permission of the publisher from Arch *et al.* (1989). Copyright 1989 by Elsevier Science Publishing Co., Inc

by gavage during the day. First, there is a tendency for food intake to increase (Yen *et al.*, 1984) and this may lessen the depletion of lipid. Second, even in experimental situations where food intake does not increase, lean body mass tends to increase, at least with the BRL series of compounds. It seems that absorption of the compound with food is critical to this effect, because lean tissue mass also increased when genetically obese (ob/ob) mice were trained to feed between 10.00 and 16.00 h, and BRL 26830 was

Table 15.1 Effect of 3 days administration of BRL 26830 in the diet of normal mice

	Weight loss (g/mouse)	Food intake (g/3 mice or % control*)	Parametrial fat pad weight (mg or % control*)
Controls	0.64 ± 0.30	26.5 ± 5.2	323 ± 65
BRL 26830	0.79 ± 0.29	100 ± 8.6*	62.1 ± 7.7*

Female BALB/c mice were housed in threes and fed for 3 days on a control powdered diet or the same diet containing BRL 26830 (47 mg/kg diet). Food was removed at 09.00 h and the mice were killed 2–4 h later. Results are means ± SD of mean values from 20 experiments, each involving 9 control and 9 treated mice. *Results for the effects of BRL 26830 on food intake and fat pad weight have been expressed as a percentage of the control value in each experiment. BRL 26830 significantly ($p<0.05$) decreased fat pad weight in 19 out of 20 experiments but increased body weight loss in only 3 experiments.

administered by oral gavage at 14.00 h—about two-thirds of the way through the feeding period (Figure 15.3b). Similar results have been reported in genetically obese diabetic (db/db) mice (Almond *et al.*, 1988), whilst in man BRL 26830 appears to reduce the loss of body protein that accompanies an energy-restricted diet (Abraham *et al.*, 1987; Connacher *et al.*, 1988).

Thermogenesis

Single doses of the β-adrenoceptor agonists shown in Figure 15.1 stimulate metabolic rate in both obese and lean rodents. BRL 26830 (unpublished results of P. Thurlby and J.R.S. Arch) and LY 79771 (Shaw *et al.*, 1981) are known to stimulate metabolic rate in dogs, and BRL 26830, BRL 35135 and Ro 16-8714 have been reported to stimulate metabolic rate in lean and obese man (Arch *et al.*, 1989; Connacher *et al.*, 1988; Zed *et al.*, 1985; Smith *et al.*, 1989; Mitchell *et al.*, 1989; Henny *et al.*, 1987, 1988).

The thermogenic effect of BRL 26830 is similar in fa/fa rats and their lean littermates. It is also similar in ob/ob mice and their lean littermates provided the animals are maintained at thermoneutrality, but below thermoneutrality BRL 26830 is more effective in the ob/ob mice (Arch and Ainsworth, 1983; Wilson *et al.*, 1984). Similar results have been obtained with other sympathomimetic agents (Arch, 1983). It appears that below thermoneutrality sympathetic activity is elevated in the lean but not the obese mice, reducing the capacity for stimulation of metabolic rate by an exogenous agent (Arch, 1983; Arch *et al.*, 1989).

There is a marked contrast between the effect of repeat dosing BRL 26830 in obese and lean rodents. In ob/ob mice and fa/fa rats the peak thermic effect of BRL 26830 is enhanced by repeat dosing, whereas in the lean littermates of these animals there is at most a slight increase in the peak response. The main effect of repeat dosing in lean animals is a reduction in the duration of the response followed by a fall in metabolic rate below the control level (Arch and Ainsworth, 1983; Arch *et al.*, 1984b). In consequence, after 8 days, once daily dosing of BRL 26830 fails to elicit an increase in total daily energy expenditure in lean rats and mice, whereas the effect in the obese animals is greater than on day 1 (Table 15.2). This mechanism may account for the failure of β-adrenoceptor agonists to influence the body weight of lean animals, even when there is no increase in food intake.

Selectivity and brown adipose tissue

The selectivity of BRL 26830 as a stimulant of thermogenesis relative to heart rate was originally demonstrated in conscious rodents, but more conclusive evidence of its novel selectivity was obtained in pithed rats. This model has the advantage that it is possible to measure thermogenesis, β_1- and β_2-adrenoceptor-mediated effects simultaneously. To obtain a good thermogenic response it is necessary to respire the rats with a 1:1 oxygen:nitrogen mixture, since otherwise oxygen supply seems to limit oxygen consumption. Pithing ablates reflex control of heart rate, notably that due to β_2-adrenoceptor-mediated vasodilation, so that heart rate provides a measure of primarily β_1-activity, though some direct β_2-activity may remain (Kaumann, 1986). Finally, by suspending a uterine horn outside the abdomen it is possible to measure β_2-activity (Piercy, 1988; Arch *et al.*, 1987). Using this procedure, we have shown that BRL 26830 is more potent than the β_1-agonist denopamine or the β_2-agonist salbutamol as a stimulant of thermogenesis, but it is less potent than the β_1-agonist as a stimulant of heart rate and less potent than the β_2-agonist as a relaxant of uterine tension (Table 15.3). Other approaches to demonstrating selectivity have been described by Holloway *et al.* (1989) and Meier *et al.* (1989), but they have the disadvantage that thermogenesis, β_1- and β_2-adrenoceptor-mediated effects are not measured in the same experimental situation (or even the same species), or that heart rate in intact animals is taken as a measure of both β_1- and β_2-adrenoceptor stimulation.

The novel selectivity of the thermogenic β-agonists suggests that they selectively stimulate a non-β_1-, non-β_2-adrenoceptor. This view is strongly supported by the finding that the thermogenic effect of BRL 26830 is totally unaffected by doses of propranolol and other β-blockers that can completely prevent any increase in heart rate. Thus 10 mg/kg, p.o. propranolol blocks the effect of 10 mg/kg, p.o. BRL 26830 on heart rate but does not reduce its thermogenic effect (Fig 15.4). Only when the dose of propranolol is raised to 100 mg/kg is there a consistent inhibition of thermogenesis (Arch and Ainsworth, 1983). Similar findings have been reported for other

Table 15.2 Effect of single and repeated daily doses of BRL 26830 on energy expenditure in genetically obese (ob/ob) mice and (fa/fa) rats, and their lean littermates

		Percentage increase in energy expenditure due to BRL 26830A			
		C57Bl/6 mice		Zucker rats	
	Dose of BRL 26830A:	ob/ob 0.2 mg/mouse	Ob/? 0.2 mg/mouse	fa/fa 3.7 mg/kg	Fa/? 1.9 mg/kg
	Time after dosing (h)				
Single dose	0–3	67 ± 8	18 ± 4	30 ± 3	33 ± 8
	0–21	25 ± 3	10 ± 2	22 ± 3	12 ± 4
Repeat dosing	0–3	170 ± 11	32 ± 3	56 ± 7	27 ± 2
	0–21	52 ± 5	2 ± 2	29 ± 7	3 ± 4

Energy expenditure was measured after the 1st or 8th daily dose of BRL 26830A (1st or 18th dose for ob/ob mice). Results are means of four values ± SEM and percentage increases in metabolic rate are relative to animals dosed with water. All the increases were statistically significant except those measured over 21 h in the lean animals following repeat dosing. The mice were 16 weeks and the rats 10 to 12 weeks old.

Table 15.3 Effects of BRL 26830 on metabolic rate, heart rate and uterine relaxation in pithed rats

	Dose eliciting response (μmol/kg, i.p.) (geometric mean, 95% confidence interval, n = 4–6)		
	25% increase O_2 consumption	70 beats/min increase heart rate	50% relaxation uterine tension
BRL 26830	1.2 (0.4–4.1)	6.1 (3.1–11.9)	2.0 (1.1–3.6)
Denopamine	2.1 (1.7–2.6)	0.55 (0.48–0.63)	No effect
Salbutamol	9.8 (5.8–16.3)	4.0 (3.4–4.9)	0.032 (0.025–0.043)

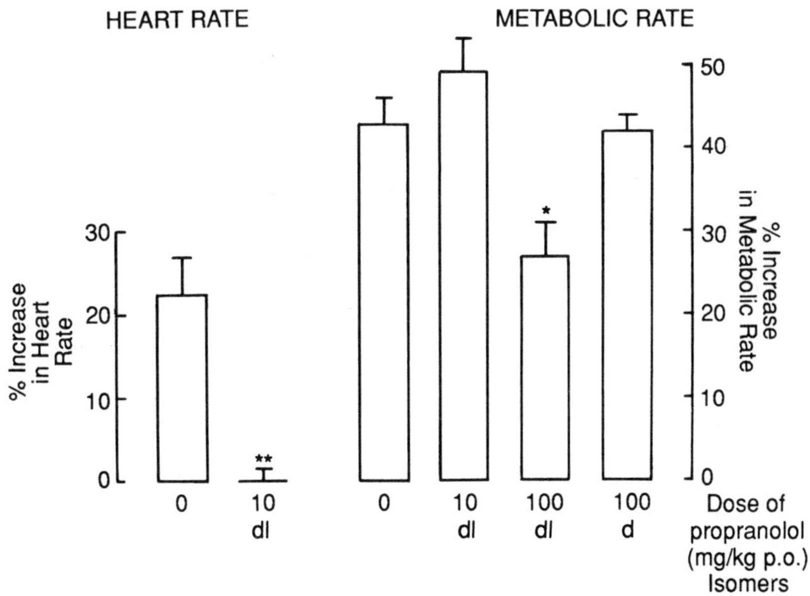

Figure 15.4 Effect of propranolol on thermogenic and chronotropic responses to BRL 26830 (8.6 mg/kg, p.o.) in male Sprague-Dawley rats. Heart rate (n=4) was measured 30 min and metabolic rate (n=3) was averaged over 3 h after dosing BRL 26830 and propranolol. These are times or periods when responses to BRL 26830 are near maximal. *p<0.05; **p<0.01 relative to BRL 26830 alone

thermogenic β-agonists by Yen *et al.* (1984) and Meier *et al.* (1989).

These results raise the issue of the location of the β-adrenoceptor that is selectively stimulated by BRL 26830. The work of Foster and Frydman (1978), demonstrating an important role for brown adipose tissue in noradrenaline-stimulated thermogenesis prompted us to focus our attention on this tissue. We found that BRL 28410 and more especially BRL 37344—

the metabolites of BRL 26830 and BRL 35135—selectively stimulate brown adipocyte lipolysis *in vitro*, whereas standard β-adrenoceptor agonists are more potent as stimulants of atrial rate or tension (β_1-adrenoceptor-mediated), or as relaxants of tracheal or uterine tension (primarily β_2-adrenoceptor-mediated). Furthermore, consistent with the *in vivo* results, standard β-adrenoceptor antagonists, such as propranolol, were much less potent as antagonists of brown adipocyte lipolysis than as antagonists of β_1- or β_2-adrenoceptor-mediated effects (Arch *et al.*, 1984a). Similar findings have been reported for ICI 201,651, the metabolite of ICI 198,157, and it has further been shown that ICI 198,157 selectively activates brown adipose tissue thermogenesis *in vivo* (Holloway, 1989).

This work has led to the proposal that the rat brown adipocyte β-adrenoceptor should be designated as being of the β_3-subtype (Arch, 1989). The pharmacological identification of this subtype has recently been supported by the cloning of a β-adrenoceptor gene from a human c-DNA library, which, when expressed in Chinese hamster ovary cells, shows many (though not all) of the features of the rat brown adipocyte receptor (Emorine *et al.*, 1989; Zaagsma and Nahorski, 1990). In particular, it exhibits a high sensitivity to BRL 37344 and a resistance to standard β-blockers. The sequence identity of the human β_3-adrenoceptor with β_1- or β_2-adrenoceptors is not greater than that between β_1- or β_2-adrenoceptors, emphasizing its distinctive character.

Although the pharmacology of the rat brown adipocyte β_3-adrenoceptor is consistent with that of the receptor that mediates thermogenesis *in vivo*, the contribution of brown adipose tissue to BRL 26830-induced thermogenesis remains unclear. Measurement of blood flow to the tissue suggested that it contributed no more than 8% of the acute thermogenic effect of BRL 26830 in mature rats (Thurlby and Ellis, 1986) and brown adipose tissue may be of even less importance in man. It is therefore of interest that BRL 28410 appears to act via β_3-adrenoceptors to stimulate hind limb thermogenesis in the rat, suggesting that skeletal muscle is an important site of thermogenesis (Thurlby and Ellis, 1986; Arch *et al.*, 1989).

DIABETES

Influences of β-adrenoceptor agonists on glucose homeostasis

Selective β_3-adrenoceptor agonists show potential in the treatment of not only obesity, but also type 2 diabetes—a convenient combination of indications since type 2 diabetics are frequently obese (Zimmet, 1982). To understand why β_3-selectivity endows a compound with such properties, it is necessary to appreciate the various ways in which β-adrenoceptor agonists influence glucose homeostasis.

Two acute effects and one chronic effect are important. One of the acute effects is to release insulin from the pancreas. The extent to which this increases glucose uptake and inhibits hepatic glucose release depends upon the sensitivity of the tissues to insulin, which in the type 2 diabetic is depressed. The second, opposing, acute effect is to stimulate glycogenolysis in the liver. Whilst this may seem a deleterious effect, in fact it serves to ensure that the normal counter-regulatory mechanisms that prevent severe hypoglycaemia can operate. Thus β-adrenoceptor agonists do not produce severe hypoglycaemia even in insulin-resistant animals or humans, whilst sulphonylureas, which are predominantly insulin secretagogues, can precipitate hypoglycaemic coma.

The chronic effect of β-adrenoceptor agonists is to improve the sensitivity of the tissues to insulin. This results in both glucose uptake being enhanced and hepatic glucose release being reduced. The interplay of these effects in different experimental situations will be illustrated using results obtained with the potent β_3-adrenoceptor agonist BRL 35135 (Figure 15.1), the properties of which have been reviewed in more detail by Smith *et al.* (1990). Similar results have been obtained (at 20-fold higher dose levels) using BRL 26830 (Cawthorne *et al.*, 1984) and to a limited extent with Ro 16-8714 (Meier *et al.*, 1984).

Single dose effects

A single oral dose of BRL 35135 improves glucose tolerance in 24 h-fasted normal lean mice and rats. This effect is dependent on the insulin secretagogue activity of BRL 35135. Thus, plasma insulin concentrations are elevated within 30 min of administration of the compound to fasted rats, and if insulin stores are destroyed by streptozotocin pretreatment BRL 35135 is ineffective (Smith *et al.*, 1990).

Figure 15.5 illustrates the improvement in glucose tolerance when BRL 35135 is administered to normal lean Zucker rats. By contrast, glucose tolerance is, if anything, exacerbated in genetically obese fa/fa Zucker rats. Measurement of glucose turnover using [6-^{3}H]glucose radiotracer has shown that this is because hepatic glucose output is stimulated in fa/fa rats and glucose utilization is not increased, despite an elevation in blood insulin concentration. The extra insulin has no effect because the fa/fa rat is very insulin resistant and peripheral glucose metabolism is already maximally stimulated by the basal hyperinsulinaemia (Smith *et al.*, 1990). Similar reasoning probably explains why a single dose of BRL 35135 fails to improve glucose tolerance in genetically obese ob/ob mice.

The receptor that mediates the insulin secretagogue effect of β-adrenoceptor agonists appears to be of the same general type—β_3—as that involved in thermogenesis and lipolysis. Thus, co-administration of BRL 35135 and

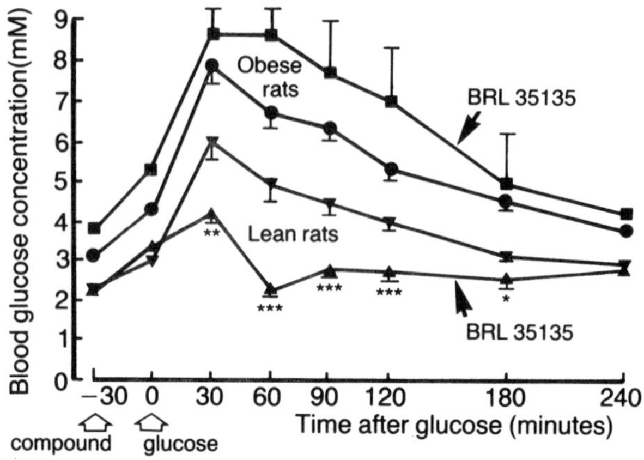

Figure 15.5 Effect of a single dose of BRL 35135 on glucose tolerance in lean and obese Zucker rats. The rats were fasted for 24 h before an oral dose of water or BRL 35135 (38 μg/kg to lean rats and 23.4 μg/kg to obese rats). Glucose (1.2 g/kg, s.c.) was administered 30 min later and blood samples were taken at intervals. Results are means of 8 values; vertical lines show SEM $^{*}p<0.05$; $^{**}<0.01$; $^{***}p<0.001$ relative to controls. Reproduced with permission from Smith *et al.* (1990)

the β_1/β_2-adrenoceptor antagonist propranolol (10 mg/kg, p.o.) did not shift the dose–response curve for the effect of BRL 35135 on glucose tolerance in normal rats but it abolished the increase in heart rate (Smith *et al.*, 1990).

The β-adrenoceptor that mediates liver glycogenolysis is, on the other hand, generally believed to be of the β_2-subtype. In fact, in the adult rat, adrenergic activation of hepatic glycogenolysis is predominantly α_1- rather than β-adrenoceptor mediated, and it is probable that the main effect of BRL 35135 in this species is indirect, via glucagon release (see Smith *et al.*, 1990). Our results are, nevertheless, more consistent with the involvement of a β_2- rather than a β_3-adrenoceptor because a moderate dose (10 mg/kg, p.o.) of propranolol reduces the elevation of blood glucose produced by BRL 35135 in both fa/fa rats and ob/ob mice (S.A. Smith and M.V. Sennitt, unpublished results).

The conclusion that the receptor that mediates insulin secretion is of the β_3-subtype, whereas the receptor that mediates glycogenolysis is of the β_2-subtype, is consistent with the effects of various selective β-adrenoceptor agonists on glucose tolerance in normal rats. Thus BRL 35135 and other β_3-selective agonists improve glucose tolerance, but the β_2-agonists fenoterol, terbutaline and salbutamol all impair glucose tolerance, and the β_1-agonist denopamine has no effect (Smith *et al.*, 1990).

Repeat dose effects

There is a remarkable difference between the effects of a single dose of BRL 35135 and once daily dosing for periods in excess of 7 days on glycaemic control in insulin-resistant rodents, such as fa/fa rats, ob/ob mice or genetically diabetic db/db mice. A single dose exacerbates glucose tolerance in fa/fa rats but repeat dosing normalizes glucose tolerance to that found in lean (Fa/?) Zucker rats (Figure 15.6). Similarly, glucose tolerance is normalized in ob/ob mice and blood glucose levels are greatly reduced throughout the day in db/db mice. These effects of BRL 35135 are achieved using dose levels that are too low or administered over too short a time to affect body weight or body composition (Smith *et al.*, 1990).

By using the hyperinsulaemic–euglycaemic clamp procedure in fa/fa rats, it has been demonstrated that the improvement in glucose tolerance is due to an increased whole body sensitivity to insulin. When insulin was infused into control obese rats to maintain a plasma concentration of 300 μU/ml only 3.9 μmol/min of exogenous glucose was required to maintain euglycaemia, but following 56 days of treatment with BRL 35135 (0.023 mg/kg/day, p.o.) the glucose infusion rate required was 15.1 μmol/min. Simultaneous measurement of tissue glucose uptake rates using [^{14}C]2-deoxyglucose showed that BRL 35135 increased glucose uptake significantly in the heart, diaphragm and brown adipose tissue. Further evidence that BRL 35135 improves insulin sensitivity comes from the finding that it reduces both fasting hyperglycaemia and plasma insulin concentrations in ob/ob mice (Smith *et al.*, 1990).

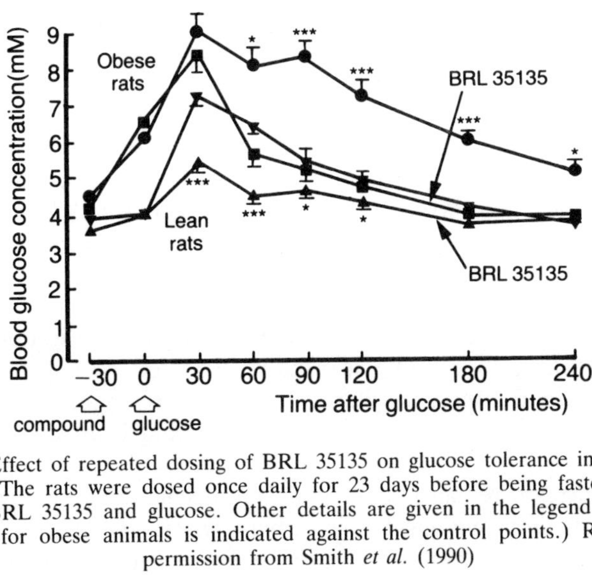

Figure 15.6 Effect of repeated dosing of BRL 35135 on glucose tolerance in lean and obese Zucker rats. The rats were dosed once daily for 23 days before being fasted for 24 h and dosed with BRL 35135 and glucose. Other details are given in the legend to Figure 15.5. (Significance for obese animals is indicated against the control points.) Reproduced with permission from Smith *et al.* (1990)

Clinical studies

Two 10 day studies have shown that repeated administration of BRL 35135 also improves insulin sensitivity in obese, insulin-resistant man (Mitchell *et al.*, 1989; Smith *et al.*, 1990). In the first, treatment of obese patients with BRL 35135 (2 mg q.d.s. for 5 days; 6 mg q.d.s. for 5 days) both improved glucose tolerance and reduced the post-glucose load hyperinsulinaemia. The greatest effect occurred in those patients who were markedly glucose intolerant or diabetic. It was calculated that the improvement in glucose tolerance was entirely due to improved glucose storage (rather than increased oxidation), which is where the major defect in the obese subject lies.

In the second study, insulin sensitivity was measured directly by the hyperinsulinaemic–euglycaemic clamp technique in obese type 2 diabetics. BRL 35135 (2 mg q.d.s. for 2 days; 4 mg q.d.s. for 2 days; 6 mg q.d.s. for 6 days) increased the rate of glucose infusion required to maintain euglycaemia by 91%. In a similar study in healthy lean men, BRL 26830 elicited a 31% increase in the rate of glucose infusion required to maintain euglycaemia and also reduced fasting plasma insulin without affecting glucose concentrations (Smith *et al.*, 1987).

Although fasting insulin and glucose concentrations were not affected by BRL 35135, these studies bode well for full clinical efficacy studies, which would probably involve dosage for at least 3 months.

Mechanism of chronic effect

Further work is required to define the nature of the β-adrenoceptors responsible for the chronic effect on insulin sensitivity, but it appears that, in rats and mice at least, the receptor belongs to the β_3-subtype. Thus only β_3-selective adrenoceptor agonists have been found to improve insulin sensitivity or glucose tolerance in insulin-resistant rodents, and β_2-selective agonists, such as salbutamol or fenoterol, are inactive. Furthermore, our attempts to block the chronic effects of BRL 35135 and BRL 26830 with propranolol have so far failed. In man, β_3-adrenoceptor agonists may not be the only effective agents. Scheidegger *et al.* (1984) reported that 2 weeks treatment of healthy lean men with terbutaline enhanced the rate of glucose infusion required to maintain euglycaemia during insulin infusion by 29%. However, although terbutaline is usually classified as a β_2-adrenoceptor agonist, there is some evidence that it stimulates a receptor that is insensitive to propranolol (Mena *et al.*, 1978), suggesting that it affects both β_2- and β_3-adrenoceptors.

A major question remains concerning the mechanism whereby β_3-adrenoceptor agonists improve insulin sensitivity. BRL 26830 increases insulin receptor number and insulin receptor tyrosine kinase activity in brown adipose tissue and skeletal muscle of mice made obese using gold thioglucose (Rochet *et al.*, 1988), but it is unclear whether these are its primary sites of action. It is also unclear how hepatic glucose output is modified, or how β-adrenoceptor stimulation and insulin receptor number are linked at the receptor level.

Finally, it must be emphasized that though the term 'β_3-adrenoceptor' has been used for the receptors mediating thermogenesis, lipolysis, insulin secretion and insulin sensitivity because all are especially sensitive to BRL 35135 (or BRL 37344) and insensitive to propranolol, these receptors may not be identical. Indeed, our unpublished work shows that it is possible to design compounds that have similar efficacy to BRL 26830 and BRL 35135 (or their metabolites) as stimulants of thermogenesis and lipolysis, but do not affect insulin secretion.

PROTEIN WASTING CONDITIONS

Repartitioning agents and their mechanism of action

It has been described how, in some experimental situations in rodents and in clinical studies, BRL 26830 not only reduces body lipid content but also increases protein. Loss of protein, other than from adipose tissue, is an undesirable consequence of eating an energy-restricted diet, and so this property is beneficial in an anti-obesity drug, provided protein accretion does not outweigh lipid loss and prevent a net loss of body weight (see Arch *et al.*, 1989).

In some β-adrenoceptor agonists the balance between the protein accretion and lipid loss effects is far more in favour of protein accretion than it is in β_3-adrenoceptor agonists. These compounds have been dubbed 'repartitioning agents', because in growing animals they repartition energy deposition from lipid to protein, the protein accretion being almost exclusively in skeletal muscle (Asato *et al.*, 1984). Since lean tissue contains about one-seventh the energy content of adipose tissue, repartitioning agents can enhance the efficiency of growth (feed efficiency) in farm animals, even though energetic efficiency may be reduced. Consequently, repartitioning agents produce a leaner and often a bigger carcass, and the commercial interest in them as potential feed additives for meat-producing animals has fostered a large amount of literature in the last 5 years (see Yang and McElligott, 1989, for review).

The mechanisms linking β-adrenoceptor stimulation to protein accretion remain, however, a mystery. To begin with, some workers have argued that protein degradation is decreased, whilst others argue that synthesis is increased. It now appears that both may be altered but the balance varies according to the experimental situation. Another fundamental question is whether the compounds act directly on skeletal muscle or via a humoral intermediate. Protein accretion in response to these agents has not been consistently demonstrated *in vitro*, but equally, changes in the concentrations of hormones, such as insulin, growth hormone, thyroxine, glucocorticoids or gonadal steroid hormones do not appear to mediate the responses. It has therefore been suggested that skeletal muscle preparations fail to respond to β-adrenoceptor agonists *in vitro* because they lack normal contractile activity (Yang and McElligott, 1989).

Two aspects of the effects of β-adrenoceptor agonists on protein metabolism are discussed in more detail below: the nature of the β-adrenoceptor that mediates this activity, and the potential of these agents in one particular protein wasting condition—cancer cachexia.

β-Adrenoceptor subtype influencing skeletal muscle protein metabolism

The first β-adrenoceptor agonist that was reported to promote protein accretion was the β_2-selective compound clenbuterol (Asato *et al.*, 1984; Emery *et al.*, 1984). Another β_2-selective compound, cimaterol, has also been extensively studied (see Figure 15.7 for structures). The differing profiles of clenbuterol, cimaterol and the β_3-selective compounds BRL 35135 and BRL 26830 are clearly apparent when the compounds are included in

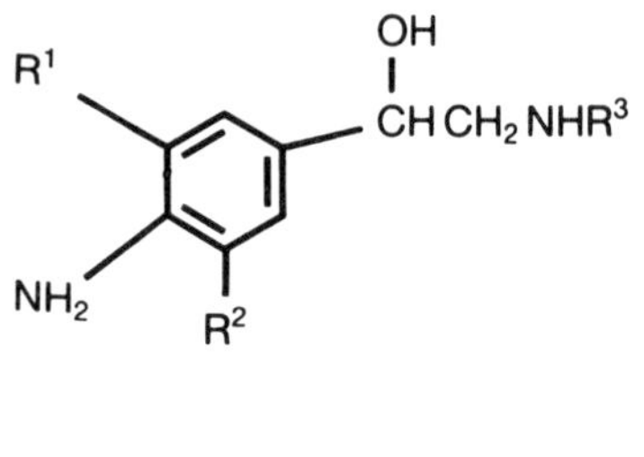

	R^1	R^2	R^3
clenbuterol	Cl	Cl	$C(CH_3)_3$
cimaterol	CN	H	$CH(CH_3)_2$

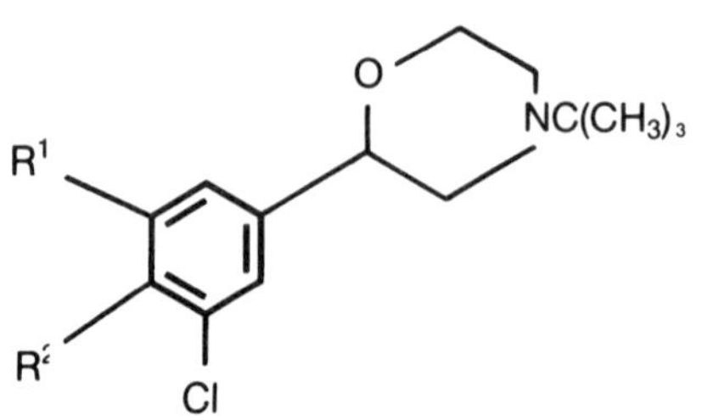

	R^1	R^2
BRL 46104	H	H
BRL 47672	Cl	NH_2

Figure 15.7 Chemical structures of compounds that promote skeletal muscle protein accretion

the diet of young growing rats (Table 15.4). Clenbuterol and cimaterol cause a marked increase in the weights of both the gastrocnemius and plantaris (mainly white) and the soleus (red) muscles within 2 days, the maximal percentage increase being achieved by day 6. Protein accumulation is in proportion to the increased muscle mass (results not shown). BRL 26830 was ineffective in this model; BRL 35135 elicited an increase in soleus muscle weight, but it had no effect on the weight of the gastrocnemius and plantaris muscles, and its potent β_3-agonist activity was apparent from the large decrease in the weight of the epididymal fat pads.

The efficacy of clenbuterol and cimaterol and the lack of effect of the β_3-adrenoceptor agonists would suggest the involvement of β_2-adrenoceptors in promoting protein accretion. However, a number of factors warn against this conclusion. First, we find that clenbuterol and cimaterol are only about 30-fold more potent as relaxants of rat uterine tension (β_2) than as stimulants of atrial rate (β_1). Cimaterol ($EC_{50} = 2.6 \times 10^{-7}$ M), though not clenbuterol, is also quite potent as a stimulant of rat white adipocyte lipolysis (β_3). Second, not all β_2-adrenoceptor agonists are effective. For example, salbutamol and terbutaline were ineffective in the studies of Reeds *et al.* (1988), though it is possible that this was because the dose level was too low, or they are not as long-acting as clenbuterol. Third, some compounds that are not obviously β_2-selective promote protein accretion. Ractopamine (LY 99134 and its three stereoisomers—Figure 15.1) is a possible example. In our own studies we have identified the morpholines BRL 46104 and BRL 47672 (Figure 15.7), neither of which elicit a significant β_2-adrenoceptor-mediated relaxation of rat uterine tension *in vitro* and yet they display similar potency to clenbuterol and cimaterol as stimulants of skeletal muscle growth (Table 15.4). These compounds have in fact shown no β-adrenoceptor agonist properties in any of the *in vitro* systems in which they have been tested, and are of interest because they are less potent stimulants of heart rate than clenbuterol or cimaterol. BRL 46104, in particular, bears little structural resemblance to known β_2-adrenoceptor agonists.

A fourth consideration is the ability of β-adrenoceptor antagonists to prevent the stimulation of protein accretion. Reeds, Maltin and co-workers have reported that propranolol fails to prevent the effect of clenbuterol on protein accretion and have suggested that the effect is not directly mediated by any type of β-adrenoceptor (Reeds *et al.*, 1988; Maltin *et al.*, 1989). Maclennan and Edwards (1989), however, found that propranolol did block clenbuterol when both compounds were given parenterally, whilst Choo *et al.* (1989), using the dietary route of administration, agreed that propranolol was ineffective but found that the selective β_2-adrenoceptor antagonist ICI 118,551 blocked clenbuterol.

In our own studies we have found that oral propranolol is not very effective against clenbuterol: propranolol (50 mg/kg diet) may have partially

Table 15.4 Effects of β-adrenoreceptor agonists and morpholine derivatives on growth and organ weights in young rats

	Percentage difference from controls					
	Clenbuterol	Cimaterol	BRL 26830	BRL 35135	BRL 46104	BRL 47672
Dose (mg/kg diet):	4.4	5.0	21	4.1	4.4	4.5
Number of experiments:	4	6	1	1	24	1
Body weight gain	$+13^{2}$	$+12^{5}$	0 ± 3	$+6 \pm 4$	$+15^{18}$	$+8 \pm 4$
Gastrocnemius plus plantaris muscle weight	$+26^{4}$	$+27^{6}$	$+2 \pm 2$	$+8 \pm 3$	$+27^{24}$	$+23 \pm 2^{***}$
Soleus muscle weight	$+23^{4}$	$+26^{6}$	$+10 \pm 4$	$+20 \pm 4^{*}$	$+28^{24}$	$+14 \pm 2^{***}$
Fat pad weight	-4^{0}	-5^{0}	-9 ± 8	-27 ± 6	$+3^{0}$	-5 ± 8
Food intake	$+2^{0}$	$+1^{0}$	$+7 \pm 2$	-4 ± 3	$+1^{1}$	-2 ± 4

Compounds were included in the diets of male Sprague-Dawley rats (initially 45–50 g) for 6 days. There were 9 rats per treatment, housed in threes, in each experiment. Where more than one experiment was performed, superscripts indicate the number of experiments in which p was <0.05. Results from single experiments are shown with SEM. $^{*}p<0.05$; $^{**}p<0.01$; $^{***}p<0.001$.

inhibited the increase in weight of the gastrocnemius and plantaris muscles produced by clenbuterol (4.4 mg/kg diet), but it had no effect on the increase in weight of the red soleus muscle. However, the effects of cimaterol and BRL 46106 were inhibited (Table 15.5), and in the case of BRL 46104 analysis of the gastrocnemius and plantaris muscles demonstrated that protein accretion was blocked (results not shown).

Further work was conducted with l- and d-propranolol. These have identical physical properties, but the l-enantiomer has about 100-fold higher affinity for β_1- and β_2-adrenoceptors and 20-fold higher affinity for the β_3-adrenoceptor (Harms *et al.*, 1977). l-Propranolol was more effective than d-propranolol in antagonizing BRL 46104, but the enantiomers were equally effective (or ineffective) against clenbuterol (Table 15.6).

It appears, therefore, that clenbuterol does not promote muscle accretion by binding to typical β_2-adrenoceptors. The relationship between the clenbuterol receptor and that stimulated by BRL 46104 and other repartitioning agents is unclear. 'Atypical β_2' may be the best designation for the receptor until it is better understood.

Evaluation of BRL 47672 in a model of cancer cachexia

One of the morpholine repartitioning agents, BRL 47672, has been evaluated in a model of cachexia in which a human hypernephroma is implanted into immune-suppressed mice. The tumour caused considerable weight loss in the patient from whom it was taken and it produces marked weight loss in mice when it comprises no more than 6% of body weight. Weight loss is not associated with reduced food intake or absorption and is rapidly reversed following removal of the tumour (Strain *et al.*, 1980).

Tumour-bearing mice lost weight after tumour implantation, whereas non-tumour-bearing mice continued to gain weight. Administration of BRL 47672 from week 3 caused a rapid increase in body weight and appeared to prevent weight loss between weeks 9 and 11 (Figure 15.8). The fall in carcass (minus tumour) and muscle weights elicited by the tumour was prevented by BRL 47672 (Table 15.7). Experiments are currently being conducted to confirm and extend these findings.

SUMMARY

The recognition that β-adrenoceptor agonists influence thermogenesis, body composition and glucose homeostasis by stimulating atypical β-adrenoceptors is opening up new possibilities for therapeutic intervention in metabolic diseases. Certain novel β-adrenoceptor agonists promote lipid loss or prevent lipid gain in obese rodents by stimulating metabolic rate. They are largely

Table 15.5 Effect of dl-propranolol (50 mg/kg diet) on weight gain and skeletal muscle weight responses to clenbuterol (4.4 mg/kg diet), cimaterol (5 mg/kg diet) and BRL 46104 (4.4 mg/kg diet) in young rats

	Percentage above controls					
	Body weight gain		Gastrocnemius plus plantaris muscle weight		Soleus muscle weight	
dl-Propranolol:	–	+	–	+	–	+
Control	(34.0 ± 1.2)	0 ± 4	(380 ± 2)	0 ±4	(35.6 ± 2.5)	1 ± 3
Clenbuterol	8 ± 4	2 ± 2	23 ± 4***	14 ± 4**	27 ± 7***	28 ± 8**
Cimaterol	13 ± 3	−1 ± 5‡	32 ± 2***	2 ± 3†‡	29 ± 4***	10 ± 4*‡
BRL 46104	14 ± 2**	4 ± 4†	31 ± 2***	9 ± 2*†‡	30 ± 3***	17 ± 2**†

Methods etc. are described in the footnote to Table 15.4. Asterisks indicate significance compared with groups not treated with repartitioning agents (i.e. relative to 0); daggers indicate significance relative to groups not given propranolol. The figures in parentheses show the absolute weight gain (g) and muscle weights (mg) for controls treated with neither β-agonists/morpholine nor propranolol (mean body weight 51.6 g at start).

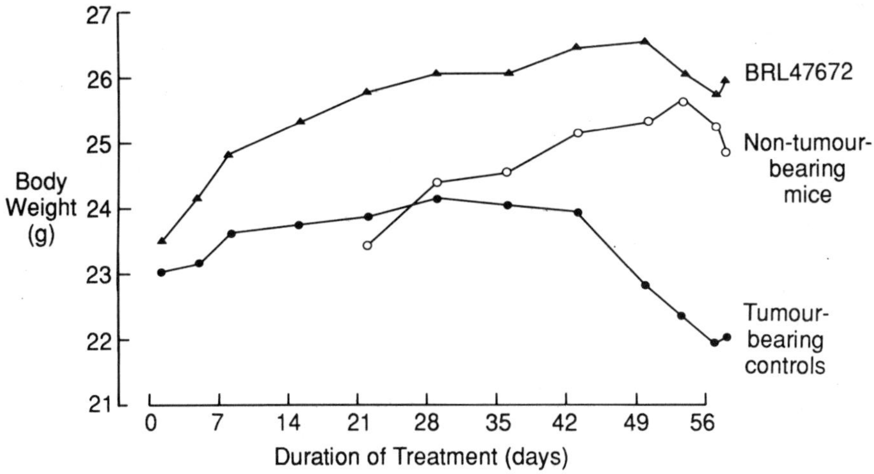

Figure 15.8 Effect of BRL 47672 on body weight in tumour-bearing mice. Small pieces (about 1 mm^3) of a human (XK1) hypernephroma were implanted subcutaneously into the flanks of female immune-suppressed Cba/ca mice. BRL 47672 (13 mg/kg diet) was included in the powdered diet of some of these mice from the third week after tumour implantation. A group of mice in which the tumour failed to grow and which had not been treated with BRL 47672 was selected to serve as negative controls

Table 15.6 Effects of d- and l-propranolol (5 mg/kg diet) on weight gain and muscle weight responses to clenbuterol and BRL 46104 (each 4.4 mg/kg diet) in young rats

	Percentage above controls			
	Enantiomer of propranolol	Body weight gain	Gastrocnemius plus plantaris muscle weight	Soleus muscle weight
Clenbuterol	—	8 ± 5	21 ± 2**	14 ± 5*
	d	5 ± 3	15 ± 4**	12 ± 5
	l	8 ± 9	15 ± 2**	10 ± 6
BRL 46104	—	12 ± 4	27 ± 3***	26 ± 5**
	d	5 ± 4	12 ± 5*††	18 ± 3*
	l	−11 ± 3††	−4 ± 2†††	−7 ± 5

Methods etc. are described in the footnotes to Tables 15.4 and 15.5.

Table 15.7 Effects of BRL 47672 on tissue weights and food intake in tumour-bearing mice

	Non-tumour-bearing controls $n = 8$	Tumour-bearing controls $n = 6$	BRL 47672 $n = 6$
Tumour weight (g)	—	0.92 ± 0.18	0.88 ± 0.17
Carcass weight (g)	24.9 ± 0.7**	21.1 ± 0.9	25.1 ± 0.8**
Gastrocnemius plus plantaris weight (mg)	130 ± 4***	98 ± 6	130 ± 8**
Soleus weight (mg)	5.9 ± 0.3	5.2 ± 0.3	6.8 ± 0.5*
Parametrial fat pads weight (mg)	359 ± 15**	184 ± 59	270 ± 40
Food intake (g/mouse)	—	253 ± 5	257 ± 5

Methods are described in the legend to Figure 15.8. Carcass weight is after removal of the tumour. $n = 3$ for food intake since the mice were housed in twos or threes. *$p<0.05$; **$p<0.01$; ***$p<0.001$ compared with the tumour-bearing controls.

ineffective in lean rodents, however, because during repeat-dosing their effect on daily energy expenditure wanes and there is a tendency for food intake to increase. One of these compounds, BRL 26830, has proved effective in some anti-obesity clinical trials in man.

Although there are exceptions, generally the same type of β-adrenoceptor agonist that stimulates thermogenesis also stimulates insulin secretion. In insulin-sensitive animals this results in improved glucose tolerance after a single dose of compound, but in insulin-resistant animals the predominant influence on glucose homeostasis after a single dose is a stimulation of glycogenolysis, so that glucose tolerance tends to decrease. During repeat dosing, however, insulin sensitivity improves and there is a remarkable improvement in blood glucose control. Improvements in insulin sensitivity and glucose tolerance have been reported following treatment of insulin-resistant men with BRL 35135.

Other β-adrenoceptor agonists are particularly effective in promoting protein accretion in skeletal muscle and these may have potential both as feed additives for meat-producing animals and for treating protein wasting conditions in man. BRL 47672, which appears to act via a β-adrenoceptor mechanism, delayed the onset of cachexia in tumour-bearing mice.

The receptors that mediate the effects on body lipid, metabolic rate, insulin secretion and probably insulin sensitivity are so different from β_1- or β_2-adrenoceptors that they should be described as β_3. They resemble a receptor whose gene has been cloned from a human c-DNA library. The receptor responsible for the protein accretion effect is ill-defined, though it may be a type of β_2-adrenoceptor. It would appear, therefore, that the clinical potential of β-adrenoceptor agonists has yet to be fully exploited.

REFERENCES

Abraham, R., Zed, C., Mitchell, T., Parr, J. and Wynn, V. (1987). The effect of a novel β-agonist BRL 26830A on weight and protein loss in obese patients. *Int. J. Obesity*, **11**, 306A.

Ahlquist, R.P. (1980). Historical perspective: classification of adrenoreceptors. *J. Auton. Pharmac.*, **1**, 101–6.

Almond, R.E., Cawthorne, M.A. and Enser, M. (1988). Muscles of diabetic (db/db) mice: fibre size, fibre type and the effects of a thermogenic, β-adrenoceptor agonist. *Int. J. Obesity*, **12**, 81–91.

Arch, J.R.S. (1983). Defective thermoregulatory thermogenesis in genetically obese (ob/ob) mice is not due to insensitivity to noradrenaline. In: *Environment, Drugs and Thermoregulation* (eds P. Lomax and E. Schöbaum), S. Karger, Basel, pp. 35–9.

Arch, J.R.S. (1989). The brown adipocyte β-adrenoceptor. *Proc. Nutr. Soc.*, **48**, 215–23.

Arch, J.R.S. and Ainsworth, A.T. (1983). Thermogenic and antiobesity activity of a novel β-adrenoceptor agonist (BRL 26830A) in mice and rats. *Am. J. Clin. Nutr.*, **38**, 549–58.

Arch, J.R.S., Ainsworth, A.T., Cawthorne, M.A., Piercy, V., Sennitt, M.V., Thody, V.E., Wilson, C. and Wilson, S. (1984a). Atypical β-adrenoceptor on brown adipocytes as target for anti-obesity drugs. *Nature*, **309**, 163–5.

Arch, J.R.S., Ainsworth, A.T., Ellis, R.D.M., Piercy, V., Thody, V.E., Thurlby, P.L., Wilson, C., Wilson, S. and Young, P. (1984b). Treatment of obesity with thermogenic β-adrenoceptor agonists: studies on BRL 26830A in rodents. *Int. J. Obesity*, **8**, (Suppl. 1), 1–11.

Arch, J.R.S., Piercy, V., Thurlby, P.L., Wilson, C. and Wilson, S. (1987). Thermogenic and lipolytic drugs for the treatment of obesity: old ideas and new possibilities. In: *Recent Advances in Obesity Research: V* (eds E.M. Berry, S.H. Blondheim, H.E. Eliahou and E. Shafrir), John Libbey, London, pp. 285–96.

Arch, J.R.S., Bywater, R.J., Coney, K.A., Ellis, R.D.M., Thurlby, P.L., Smith, S.A. and Zed, C. (1989). Influences on body composition and mechanism of action of the β-adrenoceptor agonist BRL 26830A. In: *Hormones, Thermogenesis and Obesity* (eds H.A. Lardy and F.W. Stratman), Elsevier, New York, pp. 465–76.

Asato, G., Baker, P.K., Bass, R.T., Bentley, T.J., Chari, S., Dalrymple, R.H., France, D.J., Ginger, P.E., Lences, B.L., Pascavage, J.J., Pensack, J.M. and Ricks, C.A. (1984). Repartitioning agents: 5-[1-hydroxy-2-(isopropylamino)ethyl]-anthranilonitrile and related phenethanolamines; agents for promoting growth, increasing muscle accretion and reducing fat deposition in meat-producing animals. *Agric. Biol. Chem.*, **48**, 2883–8.

Booth, F.W. (1977). Time course of muscular atrophy during immobilization of hindlimbs in rats. *J. Appl. Physiol.*, **43**, 656–61.

Cawthorne, M.A., Carroll, M.J., Levy, A.L., Lister, C.A., Sennitt, M.V., Smith, S.A. and Young, P. (1984). Effects of novel β-adrenoceptor agonists on carbohydrate metabolism: relevance for the treatment of non-insulin-dependent diabetes. *Int. J. Obesity*, **8** (Suppl. 1), 93–102.

Cawthorne, M.A., Young, P. and Smith, S.A. (1986). Effect of BRL 35135, a novel β-adrenoceptor agonist, on glucose tolerance and insulin sensitivity in obese Zucker rats. *Diabetes*, **35** (Suppl. 1), abstr. 263.

Choo, J.J., Horan, M.A., Little, R.A. and Rothwell, N.J. (1989). The anabolic action of a β_2-adrenoceptor agonist clenbuterol, on muscle, is mediated through β_2-adrenoceptor stimulation. *Br. J. Pharmacol.*, **98**, 856P.

Connacher, A.A., Jung, R.T. and Mitchell, P.E.G. (1988). Weight loss in obese subjects on a restricted diet given BRL 26830A, a new atypical β-adrenoceptor agonist. *Br. Med. J.*, **296**, 1217–20.

Emery, P.W., Rothwell, N.J., Stock, M.J. and Winter, P.D. (1984). Chronic effects of β_2-adrenergic agonists on body composition and protein synthesis in the rat. *Biosci. Rep.*, **4**, 83–91.

Emorine, L.J., Marullo, S., Briend-Sutren, M.-M., Patey, G., Tate, K., Delavier-Klutchko, C. and Strosberg, A.D. (1989). Molecular characterization of the human β_3-adrenergic receptor. *Science*, **245**, 1118–1121.

Foster, D.O. and Frydman, M.L. (1978). Nonshivering thermogenesis in the rat. II. Measurements of blood flow with microspheres point to brown adipose tissue as the dominant site of the calorigenesis induced by noradrenaline. *Can. J. Physiol. Pharmacol.*, **56**, 110–22.

Hanrahan, J.P. (1987). *Beta-agonists and their Effects on Animal Growth and Carcass Quality*, Elsevier Applied Science, London.

Harms, H.H., Zaagsma, J. and de Vente, J. (1977). Differentiation of β-adrenoceptors in right atrium, diaphragm and adipose tissue of the rat, using stereoisomers of propranolol, alprenolol, nifenalol and practolol. *Life Sci.*, **21**, 123–8.

Harris, J.B. and Slater, C.R. (1980). Animal models: what is their relevance to the pathogenesis of human muscular dystrophy? *Br. Med. Bull.*, **36**, 193–7.

Haylett, D.G. (1979). The identification of the metabolic receptors for catecholamines. *Trends Auton. Pharmacol.*, **1**, 309–45.

Henny, C., Schutz, Y., Bückert, A., Meylan, M., Jéquier, E. and Felber, J.-P. (1987). Thermogenic effect of the new β-adrenoreceptor agonist Ro 16-8714 in healthy male volunteers. *Int. J. Obesity*, **11**, 473–83.

Henny, C., Bückert, A., Schutz, Y., Jéquier, E. and Felber, J.-P. (1988). Comparison of the thermogenic activity induced by the new sympathomimetic Ro 16-8714 between normal and obese subjects. *Int. J. Obesity*, **12**, 227–36.

Himms-Hagen, J. (1967). Sympathetic regulation of metabolism. *Pharmacol. Rev.*, **19**, 367–461.

Holloway, B.R. (1989). Selective β-agonists of brown fat and thermogenesis. In: *Hormones, Thermogenesis and Obesity* (eds H.A. Lardy and F.W. Stratman), Elsevier, New York, pp. 477–84.

Holloway, B.R., Howe, R., Rao, B.S. and Stribling, D. (1989). ICI 198157: a novel selective agonist of brown fat and thermogenesis. In: *Obesity in Europe 88* (eds P. Björntorp and S. Rössner), John Libbey, London, pp. 323–328.

Kaumann, A.J. (1986). The β_1-adrenoceptor antagonist CGP20712A unmasks β_2-adrenoceptors activated by (-)-adrenaline in rat sinoatrial node. *Naunyn Schmiedebergs Arch. Pharmacol.*, **332**, 406–9.

Kern, K.A. and Norton, J.A. (1988). Cancer cachexia. *JPEN*, **12**, 286–98.

Lands, A.M., Arnold, A., McAuliff, J.P., Luduena, F.P. and Brown, T.G. (1967). Differentiation of receptor systems activated by sympathomimetic amines. *Nature*, **214**, 597–8.

Langer, S.Z. (1977). Presynaptic receptors and their role in the regulation of transmitter release. *Br. J. Pharmacol.*, **60**, 481–97.

Maclennan, P.A. and Edwards, R.H.T. (1989). Effects of clenbuterol and propranolol on muscle mass. Evidence that clenbuterol stimulates muscle β-adrenoceptors to induce hypertrophy. *Biochem. J.*, **264**, 573–9.

Maltin, C.A., Hay, S.M., Delday, M.I., Reeds, P.J. and Palmer, R.M. (1989). Evidence that the hypertrophic action of clenbuterol on denervated rat muscle is not propranolol-sensitive. *Br. J. Pharmacol.*, **96**, 817–22.

Meier, M.K., Alig, L., Burgi-Saville, M.E. and Müller, M. (1984). Phenethanolamine derivatives with calorigenic and antidiabetic qualities. *Int. J. Obesity*, **8** (Suppl. 1), 215–25.

Meier, M.K., Blum-Kaelin, D., Gerold, M., Isler, D. and Mueller, M. (1989). Ro 40-2148, a novel thermogenic β-agonist with anti-obesity activity. In: *Obesity in Europe 88* (eds P. Björntorp and S. Rössner), John Libbey, London, pp. 329–38.

Mena, H., Talley, J., Lewis, L. and Avner, B.P. (1978). The terbutaline site of action may not be the adrenergic β_2-receptor. *Proc. West. Pharmacol. Soc.*, **21**, 197–202.

Mills, J., Schmiegel, K.K. and Tuttle, R.R. (1979). Optically active phenethanolamines, their formulations, use and preparation. *European Patent* 0 007 205.

Mitchell, T.H., Ellis, R.D.M., Smith, S.A. and Cawthorne, M.A. (1989). Effects of BRL 35135, a β-adrenoceptor agonist with novel selectivity, on glucose tolerance and insulin sensitivity in obese subjects. *Int. J. Obesity*, **13**, 757–66.

Piercy, V. (1988). Methods for assessing the activity of drugs at β_1- and β_2-adrenoceptors in the same animal. *J. Pharmacol. Methods*, **20**, 125–33.

Reeds, P.J., Hay, S.M., Dorward, P.M. and Palmer, R.M. (1988). The effect of β-agonists and antagonists on muscle growth and body composition of young rats (*Rattus* sp.). *Comp. Biochem. Physiol.*, **89C**, 337–41.

Rennie, M.J. (1985). Muscle protein turnover and the wasting due to injury and disease. *Br. Med. Bull.*, **41**, 257–64.

Rochet, N., Tanti, J.-F., Greméaux, T., Van Obberghen, E. and Le Marchand-Brustel, Y. (1988). Effect of a thermogenic agent, BRL 26830A, on insulin receptors in obese mice. *Am. J. Physiol.*, **255**, E101–9.

Scheidegger, K., Robbins, D.C. and Danforth, E. (1984). Effects of chronic beta receptor stimulation on glucose metabolism. *Diabetes*, **33**, 1144–9.

Sennitt, M.V., Smith, S.A. and Cawthorne, M.A. (1986). Anti-hyperglycaemic activity in rats and mice of BRL 35135, a novel β-adrenoceptor agonist. *Diabetes* **35**, (Suppl. 1), abstr. 263.

Shaw, W.N., Schiegel, K.K., Yen, T.T., Toomey, R.E., Meyers, D.B. and Mills, J. (1981). LY 79771: a novel compound for weight control. *Life Sci.*, **29**, 2091–101.

Smith, S.A., Cawthorne, M.A., Fay, L.C., McCullough, D.A. and Mitchell, T.H. (1987). Effect of a novel β-adrenoceptor agonist on insulin sensitivity in lean healthy male volunteers. *Diabetes*, **36**, (Suppl. 1), 15A.

Smith, S.A., Zed, C., McCullough, D., Harris, G. and Cawthorne, M.A. (1989). Thermogenic activity in man of BRL 35135: a potent and selective atypical β-adrenoceptor agonist. *Int. J. Obesity*, **13**, (Suppl. 1), 33.

Smith, S.A., Sennitt, M.V. and Cawthorne, M.A. (1990). BRL 35135: an orally active anti-hyperglycaemic agent with weight reducing effects. In: *New Anti-diabetic Drugs* (eds C.J. Bailey and P.R. Flatt), Smith-Gordon, London, 177–189.

Starke, K. (1977). Regulation of noradrenaline release by presynaptic receptor systems. *Rev. Physiol. Biochem. Pharmac.*, **77**, 1–124.

Strain, A.J., Easty, G.C. and Neville, A.M. (1980). An experimental model of cachexia induced by a xenografted human tumour. *J. Nat. Cancer Inst.*, **64**, 217–21.

Thurlby, P.L. and Ellis, R.D.M. (1986). Differences between the effects of noradrenaline and the β-adrenoceptor agonist BRL 28410 in brown adipose tissue and hind limb of the anaesthetized rat. *Can. J. Physiol. Pharmacol.*, **64**, 1111–14.

Wilson, S., Arch, J.R.S. and Thurlby, P.L. (1984a). Genetically obese C57B1/6 ob/ob mice respond normally to sympathomimetic compounds. *Life Sci.*, **35**, 1301–9.

Wilson, C., Wilson, S., Piercy, V., Sennitt, M.V. and Arch, J.R.S. (1984b). The rat lipolytic β-adrenoceptor: studies using novel β-adrenoceptor agonists. *Eur. J. Pharmacol.*, **100**, 309–19.

Yang, Y.T. and McElligott, M.A. (1989). Multiple actions of β-adrenergic agonists on skeletal muscle and adipose tissue. *Biochem. J.*, **261**, 1–10.

Yen, T.T. (1984). The antiobesity and metabolic activities of LY 79771 in obese and normal mice. *Int. J. Obesity*, **8**, 69–78.

Yen, T.T., Mckee, M.M. and Stamm, N.B. (1984). Thermogenesis and weight control. *Int. J. Obesity*, **8**, (Suppl. 1), 65–78.

Zaagsma, J. and Nahorski, S.R. (1990). Is the adipocyte β-adrenoceptor a prototype for the recently cloned atypical 'β_3-adrenoceptor'? *Trends Pharmacol. Sci.*, **11**, 3–7.

Zed, C.A., Harris, G.S., Harrison, P.J. and Robb, G.H. (1985). Anti-obesity activity of a novel β-adrenoceptor agonist (BRL 26830A) in diet-restricted obese subjects. *Int. J. Obesity*, **9**, 231.

Zimmet, P. (1982). Type 2 (non-insulin-dependent) diabetes—an epidemiological overview. *Diabetologia*, **22**, 399–411.

Chapter Sixteen

Cytokines and Nutritional Aspects of the Acute Phase Response

A. Shenkin
The University Department of Chemical Pathology, Duncan Building, Royal Liverpool Hospital, Prescot Street, PO Box 147, Liverpool L69 3BX, UK

INTRODUCTION

Serious illness, whether acute or chronic, gives rise to a variety of changes in metabolism, which in turn lead to alterations in nutritional requirement, and in body composition. The object of this brief review is to consider the inter-relationship of the acute phase reaction and the nutritional consequences of illness.

WHAT IS THE ACUTE PHASE REACTION?

The acute phase reaction (APR) is generally defined as the response of the body to a variety of injuries, including microbial infection, physical or thermal trauma, ischaemic necrosis and malignancy. Thus virtually any illness associated with inflammation or damage to tissue will give rise to an APR.

The APR is a complex coordinated series of physiological and metabolic alterations. The main features can all be regarded as beneficial, at least in the short term, these benefits accounting for the highly conserved nature of the response in virtually all higher organisms.

Obesity and Cachexia. Edited by N.J. Rothwell and M.J. Stock

Fever

Endogenous pyrogens, which include interleukin-1, interleukin-6, and tumour necrosis factor amongst other cytokines, are the main mediators of fever in both infectious and non-infectious illnesses (Dinarello, 1989). Although very high fevers may be harmful in causing central nervous abnormalities or metabolic problems, moderate fevers are of some benefit to metabolic processes, since an elevation in core temperature will lead to increased speed of biochemical reactions (Q_{10} effect) and therefore enhanced mobilization of defence mechanisms.

Leukocytosis

There is an increase in both the number and activity of white blood cells (Dinarello, 1989). A number of cytokines, including G-CSF (granulocyte colony stimulating factor) and GM-CSF (granulocyte macrophage colony stimulating factor) stimulate production of polymorphonuclear leukocytes. The metabolic activity and properties of these cells are also increased by cytokines.

Lymphocyte activity is also markedly increased. T-cell activity is particularly stimulated by IL-1 and IL-2, whereas B-cell activity is stimulated by IL-6 and IL-1, although other cytokines have some additive or synergistic properties.

The benefits of these increases in white cell activity are self-evident in protection against microorganisms and in removal of necrotic tissue.

Activation of clotting and fibrinolytic mechanisms

Damage to body tissues stimulates the activation of both extrinsic and intrinsic coagulation pathways, this being enhanced by the increased synthesis of coagulation factors including fibrinogen (an acute phase protein). The benefits of this are apparent in minimizing blood loss.

On the other hand, the fibrinolytic mechanism is also stimulated, with an increase in plasminogen activator. This stimulates the removal of blood clot, and the revascularization of tissue.

Activation of complement system

The concentration of complement C3 and C4 increases as part of the acute phase response, and this promotes removal of necrotic tissue or microorganisms via the complement mechanism.

Changes in trace elements

The serum concentrations of both iron and zinc fall very early in infection or trauma, apparently under the action of interleukin-1 and other cytokines. Serum copper concentration on the other hand increases, primarily due to the increase in concentration of caeruloplasmin in serum. The fall in serum iron appears to be due to reduced release of iron from macrophages (Weinberg, 1984), and also to increased synthesis of ferritin in the liver, which then extracts iron from the plasma (Konijn and Hershko, 1977). The beneficial effect of this appears to be to remove iron from the circulation and hence make it less available to stimulate the growth of microorganisms.

The fall in plasma zinc is due to binding the zinc in liver to increased amounts of the protein metallothionein, whose synthesis is induced by IL-1 and other cytokines (Cousins and Leinart, 1988). The precise benefit of this is not clear, but one explanation is to provide the zinc necessary for the increased protein synthesis in the liver following injury. Alternatively, the value may be similar to that of the fall in serum iron in reducing the availability of plasma zinc and hence reducing bacterial growth.

Alteration in plasma proteins

The acute phase protein response (APPR) is a particularly well-studied feature of the APR. Some 20–30 proteins increase rapidly in concentration in acute illness, as a result of increased synthesis (Kushner, 1982) usually in the liver. The acute phase proteins are heterogeneous in terms of molecular weight, glycosylation and function. The functions do, however, indicate why these proteins are of benefit during illness.

1. Antiproteases—e.g. alpha-1-antitrypsin, alpha-1-antichymotrypsin. These proteins will help to neutralize any intracellular proteases released during damage to tissues.
2. Haemostatic—e.g. fibrinogen (see above).
3. Free radical scavenging—e.g. caeruloplasmin. Caeruloplasmin has ferro-oxidase activity, and helps to minimize the excessive oxidative damage due to free radicals.
4. Complexing of potentially harmful substances—e.g. haptoglobin, which will bind to free haemoglobin; complement fractions, which will bind antigen—antibody complexes.
5. Complement activation—e.g. C-reactive protein. CRP will bind a variety of types of cell debris, especially cell membranes, and this complex activates the complement system.

Monitoring of the APPR has been used clinically in the diagnosis of acute inflammatory processes, such as acute infection, and in monitoring the response to therapy. The best studied of the acute phase proteins is CRP, since its concentration changes quite dramatically in man during an inflammatory process. In normal individuals the serum concentration is usually less than 1 mg/l, and this can rise to 200–300 mg/l within 48 h of serious injury. The precise control of the CRP response is not clear, although increased production of messenger RNA for CRP under the control of IL-6 and other cytokines is probably the main mechanism.

Control of the acute phase reaction

It is now clear that the acute phase response results from a complex interaction between cytokines and classical hormones. For many years attention focused on the acute metabolic effects of the counter regulatory hormones, adrenaline, cortisol and glucagon (Bessey *et al.*, 1984). It was observed that the combined effects of these hormones, which increased substantially after a serious injury, led to an alteration in the normal balance between them and insulin, so that net catabolism took place. However, many of the effects of serious illness could not be explained by these hormones alone, and the properties of certain cytokines, including fever, leukocytosis, and the APPR, has led to a clearer understanding of this. Although some of the effects of cytokines can directly explain features of the acute phase response, it seems clear that these two mediating systems are closely inter-related. Thus an increased production of IL-1 or TNF gives rise to increased release of adrenocortical hormones (Roh *et al.*, 1987), whereas an increase in corticosteroids reduces the output of IL-1 and TNF from isolated monocytes (Lew *et al.*, 1988). The precise modulation of this inter-relationship remains to be established, but it is important to consider the response as a whole, and not as two independent control mechanisms.

It is now recognized that the main protein mediating the increased synthetic rate of APPR is interleukin-6 (Gauldie *et al.*, 1987). The other cytokines, especially IL-1 and TNF, may also have a role in this, although the sensitivity and the profile of proteins induced appears to be more restricted (Baumann *et al.*, 1987).

Cruickshank *et al.* (1990), found that IL-6 could be detected in serum post-surgery within 1 h of skin incision, and usually reached a peak by 4–8 h, a time course which would be consistent with a mediating effect on the induction of CRP synthesis. However, the relatively poor correlation between serum IL-6 and serum CRP suggested that other factors may be responsible in mediating the CRP response apart from IL-6.

Nutritional implications of the acute phase response

The greater the injury, the greater is the associated APR. The magnitude of the catecholamine and cortisol response to burn injury relates to the area of body surface damaged, and this is greater than in elective surgery (Wilmore *et al.*, 1974). Similarly, serum TNF may be high in severe septicaemia (Waage *et al.*, 1987) or burn injury (Marano *et al.*, 1990), but may be undetectable in other serious conditions.

We have recently demonstrated that patients undergoing elective surgery have a graded response of CRP and IL-6 to operations of varied severity. The largest serum IL-6 response was found with colorectal surgery and aortic bifurcation graft surgery, which was larger than hip replacement and cholecystectomy. A small but detectable IL-6 response was also observed in minor surgery (Cruickshank *et al.*, 1990).

We have also recently observed that patients with complications to acute pancreatitis had much higher serum IL-6 and CRP within 24–36 h of admission, than patients who went on to an uncomplicated recovery. Again this is evidence that the more serious the disease or inflammatory process, the greater is the cytokine response and the acute phase reaction.

Since the hormonal and cytokine changes are related to the severity of the disease, it is therefore inevitable that the metabolic and nutritional sequelae will respond in a similar 'dose-related' manner.

Effects of the acute phase reaction on nutritional status

The range of systemic responses brought about by the APR will have major nutritional implications. However, these are difficult to quantify, since the disease itself will usually also have a direct bearing on nutritional status. Most forms of injury or chronic inflammatory disease will bring about variable anorexia, perhaps together with malabsorption or diarrhoea.

Effects of APR on whole body protein status

The acute phase response gives rise to an increase in whole body protein turnover. There is increased catabolism of protein in skeletal muscles, as evidenced by the increased release of 3-methylhistidine from skeletal muscle, and also from isotope infusion studies of muscle protein turnover (Bessey and Wilmore, 1988). On the other hand there is increased uptake of amino acids from the splanchnic circulation into the liver, with increased hepatic protein synthesis (Clowes *et al.*, 1985).

In studies of patients with colon cancer we have demonstrated that those with more severe disease and hepatic metastases have a higher whole body protein synthetic rate and also a greater acute phase protein response than those with less severe disease (Fearon *et al.*, 1988). Such patients also have a higher circulating plasma IL-6 than non-cancer patients, and a small proportion of them have detectable circulating TNF (Fearon *et al.*, 1990).

Effect of APR on energy expenditure

Fever. It is generally accepted that an increase in body temperature of 1°C gives rise to an increase in resting energy expenditure of about 13%. Thus in severe acute fevers, an increase of energy expenditure by 25–35% might well be anticipated, and lead to the progressive mobilization of energy stores from adipose tissue.

Protein synthesis. The energy cost of protein synthesis remains debatable, but a figure of about 3.6 Kcal/g protein seems reasonable (Reeds *et al.*, 1985). If the whole body protein synthetic rate is therefore increased during the acute phase response, this should give rise to a substantial increase in energy expenditure. However, most studies have failed to demonstrate a good correlation between the increase in whole body protein turnover and increase in resting energy expenditure. This probably reflects the fact that body protein metabolism accounts for less than 20% of total body energy expenditure, and hence it is difficult to see the specific changes due to protein metabolism against a background in which other factors may also affect total body energy expenditure. Although quantitative evidence is therefore lacking, it is nonetheless still reasonable to conclude that the increase in protein turnover during the acute phase response will lead to increased energy requirements.

Cell synthesis and control mechanisms. It is even more difficult to quantitate the energy implications of tissue repair resulting from tissue damage during illness. All of the biosynthetic pathways involved in nucleic acid, protein, lipid and carbohydrate metabolism have an energy cost, as have the activation of systems which control metabolism, and hence if substantial new cell synthesis is required, this will also have an energy cost on body energy stores. As Reeds *et al.* (1985) pointed out, each additional mole ATP directly expended in synthesizing protein is associated with an increase in total energy expenditure of 3.2 moles ATP, that is, an additional 2.2 moles ATP related to, but not necessarily directly expended in protein synthesis.

Effects of the acute phase reaction on trace elements

As indicated above, the acute phase response gives rise to reproducible changes in the serum concentrations of iron, zinc and copper. The redistribution of iron and zinc in particular from serum into the liver, is an active process, and supplements of iron tend to have no effect on serum iron concentration over several days, although zinc supplements intravenously lead to a slightly more rapid return to normal zinc concentrations (Shenkin, 1988). This response appears to be highly conserved, and is not modified by cyclo-oxygenase inhibitors under conditions which block the pyrogenic effects of IL-1 and endotoxin (Sobrado *et al.*, 1983).

Effect of nutritional status on the acute phase reaction

Despite the undoubted nutritional cost of mounting an acute phase response, severe malnutrition appears to have only a small effect on the magnitude of the response produced by any particular stimulus. Some studies have failed to show any effect. For example, we have demonstrated that in patients with an acute exacerbation of Crohn's disease there was no difference in the magnitude of the CRP response whether patients commenced either malnourished (with low body weight, skinfold thickness and muscle circumference), or relatively well nourished (Park *et al.*, 1988).

On the other hand, we have also demonstrated that in patients undergoing elective upper abdominal surgery, patients with low ideal body weight (less than 80%) together with reduced triceps skinfold thickness, have a significantly attenuated CRP response (Cruickshank *et al.*, 1989). No differences, however, were found between malnourished and well-nourished groups with respect to other acute phase proteins. Moreover, recent studies in rats have indicated that protein malnutrition in acute inflammation does lead to a reduced peak alpha-2-macroglobulin (the main acute phase protein in the rat) and also a delayed response in comparison with control animals (Jennings *et al.*, 1990). These changes are also associated with an altered plasma IL-6 response. Protein deprivation in rabbits also reduces the extent of the fall in serum iron in response to bacterial infection (Hoffman-Goetz and Kluger, 1979). Other studies have also suggested that malnutrition gives rise to a reduced monocyte ability to secrete cytokines (Keenan *et al.*, 1982; Kauffman *et al.*, 1986). On the other hand, we have found that malnourished cancer patients undergoing surgery secrete a normal amount of IL-6 into the serum.

It therefore appears that malnutrition has different effects on both the cytokine response and the APR, perhaps depending upon the clinical situation and therefore the nature of the stimulus to produce an APR.

Relationship between acute phase response and outcome

There are few studies which have directly compared the magnitude of the APR with outcome. From a theoretical point of view, it would seem clear that the more severely ill the patient, the greater is the APR likely to be—hence a large APR would be associated with poor outcome. On the other hand, since the APR is broadly beneficial, the higher the APR, the better might be the mobilization of the patient's defence mechanisms, and hence the more likely he would be to survive. One study which bears on this is a report by Dominioni *et al.* (1987), who showed that in severely septic patients in Intensive Care, those who subsequently died had a lower APR in terms of alpha-l-acid glycoprotein, and alpha-l-antitrypsin than those who survived. Since the non-survivors tended to have a higher sepsis score than the survivors, it is not clear why they had a lower acute phase protein response. Nonetheless this study does provide some evidence that a full APR may at least be part of the overall survival process.

Modification of the acute phase response—therapeutic or harmful?

If the acute phase response is beneficial, might it be possible to stimulate the acute phase response by judicious use of certain stimuli? For example, some of the early studies which led to the identification of tumour necrosis factor indicated that injection of bacterial toxins sometimes can give rise to reduction in tumour mass—this proved to be due to the induction of tumour necrosis factor (Balkwill, 1989). This might be especially beneficial if particular patient groups are identified with a reduced APR. On the other hand this would of course lead to increased mobilization of nutritional reserves, and therefore speed up the process of nutritional depletion.

There may also be clinical situations where the acute phase response appears to be either excessive, or continuing for a prolonged period. This would particularly be the situation in chronic inflammatory disease processes such as Crohn's disease, rheumatoid arthritis, or neoplastic disease. Strategies may become available to modify the magnitude of the acute phase response. The sequence of events can be considered to be as follows.

Disease—leading to stimulation of macrophages—which release cytokines—which subsequently cause the acute phase reaction.

It therefore seems clear that the most effective method of reducing the acute phase response should be by control of disease activity. Nonetheless, dietary or pharmacological manipulation might either reduce the extent of macrophage activation, or the effectiveness of the cytokines on their receptors. Should this become possible, it will become necessary to ask the question whether such manipulation is likely to do more harm than good.

It would probably be appropriate to bear in mind the thoughts of John Hunter on the body's response to disease in his classic work *A Treatise on the Blood, Inflammation and Gunshot Wounds* almost 200 years ago. He stated:

> It will be proper to observe here that there is a circumstance attending accidental injury which does not belong to disease, vis. that the injury done, has in all cases a tendency to produce both the disposition and the means of cure.
>
> The operations of restoration arise naturally out of the accident itself; for when there is only a mechanical alteration in the structure, the stimulus of imperfection taking place, immediately calls forth the action of restoration; but this is contrary to what happens in disease; for disease is a disposition producing a wrong action and it must continue this wrong action till the disposition is stopped or wears itself out; when this salutary effect, however, has once taken place, the state of the body becomes similar to that in the simple accident, vis. *The consciousness of imperfection is excited, which produces the action of restoration'.*

Hunter's insight into the response to injury is remarkable. Any attempt to modify the acute phase response should therefore be seen against the background that the APR is a highly conserved mechanism which has evolved to bring about restoration and cure. Much work remains to be done to identify the beneficial, rather than the harmful aspects of the APR and determine whether any alteration of this will improve outcome.

REFERENCES

Balkwill, F.R. (1989). Tumour necrosis factor. *Br. Med. Bull.*, **45**, 389–400.

Baumann, H., Onorato, V., Gauldie, J. *et al.* (1987). Distinct sets of acute phase plasma proteins are stimulated by separate human hepatocyte stimulating factors and monokines in rat hepatoma cells. *J. Biol. Chem.*, **262**, 9756–68.

Bessey, P.Q., Watters, J.M., Aoki, T.T. and Wilmore, D.W. (1984). Combined hormonal infusion stimulates the metabolic response to injury. *Ann. Surg.*, **200**, 264–81.

Bessey, P.Q. and Wilmore, D.W. (1988). The burned patient. In: *Nutrition and Metabolism in Patient Care* (eds J.M. Kinney, K.N. Jeejeebhoy, G.L. Hill and O.E. Owen), W.B. Saunders, Philadelphia, pp. 672–700.

Clowes, G.H., Hirsch, E., George, B.C., Bigatello, L.M., Mazuksi, J.E. and Villee, C.A. (1985). Survival from sepsis: The significance of altered protein metabolism regulated by proteolysis inducing factor, the circulating cleavage product of interleukin 1. *Ann. Surg.*, **202**, 446–58.

Cousins, R.J. and Leinart, A.S. (1988). Tissue-specific regulation of zinc metabolism and metallothionein genes by interleukin 1. *FASEB J.*, **2**, 2884–90.

Cruickshank, A.M., Fraser, W.D., Burns, H.J.G., Van Damme, J. and Shenkin, A. (1990). Serum interleukin-6 response in patients undergoing elective surgery of varying severity. *Clin. Sci.*, **79**, 161–5.

Cruickshank, A.M., Hansell, D., Burns, H.J.G. and Shenkin, A. (1989). Effect of nutritional status on acute phase response to elective surgery. *Br. J. Surg.*, **76**, 165–8.

Dinarello, C.A. (1989). Interleukin 1 and its related cytokines. *The Cytokine*, Vol 1 (ed. C. Sorg), Karger, Basel, pp. 105–54.

Dominioni, L., Dionigi, R., Zanello, M. *et al.*, (1987). Sepsis score and acute phase protein response as predictors of outcome in septic surgical patients. *Arch. Surg.*, **122**, 141–6.

Fearon, K.C.H., Hansell, D.T., Preston, T. *et al.* (1988). Influence of whole body protein turnover rate on resting energy expenditure in patients with cancer. *Cancer Res.*, **48**, 2590–5.

Fearon, K.C.H., McMillan, D.C., Preston, T., Winstanley, F.P., Cruickshank, A.M. and Shenkin, A. (1990). Reduced rates of hepatic protein synthesis, the acute phase protein response and elevated circulating IL-6 and tumour necrosis factor in advanced malignancy. *Proc. Nutr. Soc.*, **49**, 167A.

Gauldie, J., Richards, C., Harnish, D. *et al.* (1987). Interferon B_2/B-cell stimulatory factor type 2 shares identity with monocyte-derived hepatocyte-stimulating factor and regulates the major acute phase protein response in liver cells. *Proc. Natl Acad. Sci. USA*, **84**, 7251–6.

Hoffman-Goetz, L. and Kluger, M.J. (1979). Protein deprivation: its effects on fever and plasma iron during bacterial infection in rabbits. *J. Physiol.*, **295**, 419–30.

Hunter, J. (1794). *A Treatise on the Blood, Inflammation and Gunshot Wounds*, George Nicol, London, p. 190.

Jennings, G., Cruickshank, A.M., Shenkin, A. and Elia, M. (1990). Effect of protein deficiency on the IL-6 and acute phase protein response (APPR) in rats injected with turpentine. *Clin. Nutr.*, **9**, (Suppl.), p. 52.

Kauffman, C.A., Jones, P.G. and Kluger, M.J. (1986). Fever and malnutrition: endogenous pyrogen/interleukin-1 in malnourished patients. *Am. J. Clin. Nutr.*, **44**, 449–52.

Keenan, R.A., Moldawar, L.L., Young, R.D., Kawamura, I., Blackburn, G.L. and Bistrian, B.R. (1982). An altered response by peripheral leukocytes to synthesise or release leukocyte endogenous mediator in critically ill, protein malnourished patients. *J. Lab. Clin. Med.*, **100**, 844–57.

Konijn, A.M. and Hershko, C. (1977). Ferritin synthesis in inflammation. 1. Pathogenesis of impaired iron release. *Br. J. Haematol.*, **37**, 7–16.

Kushner, I. (1982). The phenomenon of the acute phase response. *Ann. NY Acad. Sci.*, **389**, 39–48.

Lew, W., Oppenheim, J.J. and Matsushima, K. (1988). Analysis of the suppression of IL-1 alpha and IL-1 beta production in human peripheral blood mononuclear adherent cells by glucocorticoid hormone. *J. Immunol.*, **140**, 1895–902.

Marano, M.A., Fong, Y., Moldawar, L.L. *et al.* (1990). Serum cachectin, tumour necrosis factor in critically ill patients with burns, correlates with infection and mortality. *Surg. Gynecol. Obstet.*, **170**, 32–8.

Park, R.H.R., Khasawneh, S., Duncan, A., Russell, R.I., Hutchison, A.S. and Shenkin, A. (1988). Acute phase response in malnourished patients with Crohn's disease. *Clin. Nutr.*, **7**, 207–11.

Reeds, P.J., Fuller, M.F. and Nicholson, B.A. (1985). Metabolic basis of energy expenditure with particular reference to protein. In: *Substrate and Energy Metabolism in Man* (eds J.S. Garrow and D. Halliday), John Libbey, London, pp. 46–57.

Roh, M.S., Drazenovich, K.A., Jeffrey, B.S. *et al.* (1987). Direct stimulation of the adrenal cortex by interleukin 1. *Surgery*, **102**, 140–6.

Shenkin, A. (1988). Trace elements and vitamins in perioperative nutrition. *Br. J. Clin. Pract.*, **42**, (Suppl. 63), 138–43.

Sobrado, J., Moldawar, L.L., Bistrian, B.R., Dinarello, C.A. and Blackburn, G.L. (1983). Effect of ibuprofen on fever and metabolic changes induced by continuous infusion of leukocytic pyrogen (interleukin 1) or endotoxin. *Infect. Immun.*, **42**, 997–1005.

Waage, A., Halstenen, A. and Espevik, T. (1987). Association between tumour necrosis factor in serum and fatal outcome in patients with meningococcal disease. *Lancet*, **i**, 355–7.

Weinberg, E.D. (1984). Iron withholding: a defense against infection and neoplasia. *Physiol. Rev.*, **64**, 65–102.

Wilmore, D.W., Long, J.M., Mason, A.D., Skreen, R.W. and Pruitt, B.A. (1974). Catecholamine mediation of the hypermetabolic response to thermal injury. *Ann. Surg.*, **180**, 653–69.

Chapter Seventeen

Summary and the Way Forward in Obesity and Cachexia

D. Stribling
ICI Pharmaceuticals, Alderley Park, Macclesfield, Cheshire SK10 4TG, UK

INTRODUCTION

This symposium is being held at an opportune moment. Following the identification of brown adipose tissue as the key site for diet-induced thermogenesis, research in obesity is becoming more fragmented as the control of BAT and its integration with the control of food intake are unravelled. Running in parallel the newer research into cachexia and hypermetabolism is in a position to benefit from obesity research—even if only to exclude processes or mediators which can be identified as possible candidates by reference to the obesity field.

Without reiterating the points made in the individual contributions to the symposium, this review will aim to draw parallels between the two areas and highlight differences. The ultimate target is the identification of novel therapies and the article concludes with a consideration of how these could be assessed against the treatment objectives.

ENERGY BALANCE

From the vast body of experimental data in man and animals, the homeostatic mechanisms controlling energy balance are interlocking with both a hierarchy

Obesity and Cachexia. Edited by N.J. Rothwell and M.J. Stock

and matrix of controls to defend the important function of maintaining energy reserves.

In a wide range of animal models, of both obesity and cachexia, when energy balance is perturbed, there are reciprocal changes in energy intake and energy expenditure which contribute to the change. One of the problems which still faces workers in both obesity and cachexia is establishing cause and effect. At a superficial level, obesity and cachexia seem to be inverse situations. However, this is inevitably an oversimplification as revealed by consideration of the individual components of the energy balance equation.

CONTROL OF APPETITE

From the earliest observations of the gross control of food intake exerted by the feeding centres in the brain, analysis has continued using increasingly sophisticated techniques. This has continued to reveal an extensive matrix management system employing multiple neurotransmitters and alternate pathways to control not only appetite, satiety and nutrient selection but total feeding behaviour through the feeding cycle. Once again, there are problems establishing cause and effect when changes in any particular mediator are detected.

With this complexity, there are now signs that even within inbred strains of animals the balance of controls employed in a given situation varies. In the search for novel therapies therefore, one can expect difficulties extrapolating across the species. Equally one can also begin to understand how many interventions which achieve short-term changes in nutrient intake are rapidly overcome by adaptive changes within the overall control system.

CONTROL OF ENERGY EXPENDITURE

A similar interdigitated series of control systems appear to control energy expenditure with the common efferent pathway of the sympathetic nervous supply to brown adipose tissue as a primary site of adaptive thermogenesis in both diet-induced thermogenesis and hypermetabolism.

Reflecting the reciprocal changes in energy intake and expenditure, feedback loops, common pathways and neurotransmitters have been identified. In fact, the unravelling of the central control of non-shivering thermogenesis has been facilitated by the earlier work on central control of appetite.

FAT STORAGE

Whilst the control of brown adipose tissue has become closely drawn into the web of control of energy balance, white adipose tissue appears to have been designed as the 'servant' of the system. As such, reciprocal changes in lipid deposition and mobilization occur in obesity and cachexia. However, there are powerful overriding controls—for example during lactation.

Over the past decade the location of fat deposition has been shown to have a major impact on the consequences of obesity. Although this is an area of great interest, there is only limited information on how this might be mediated. Research into this issue could lead to a reassessment of white adipose tissue as an active participant in the control system as opposed to a passive depot. As our understanding increases, there may well be opportunities for reducing the risks of android obesity to those associated with a gynoid distribution. This would be a novel therapeutic target which may be more attainable than attempting an overall reduction in body weight, yet achieve a comparable impact on morbidity and mortality.

LEAN BODY MASS

Progress in understanding physiological control systems is critically dependent on the availability of sensitive methodology for quantifying changes and this has been a key factor in the study of the control of lean body mass.

It is evident that insulin and adrenaline play critical roles in the control of protein deposition and breakdown—with consequent effects on energy expenditure in obesity and the prognosis for patients suffering from hypermetabolism/cachexia. However, the regulation of protein deposition and breakdown can be separated and may well respond differently in different disease states. This offers a future area for investigation and an opportunity for novel therapies. There are signs that atypical β-adrenoreceptors are involved at the tissue level offering a short-term target for intervention.

AFFERENT SIGNALS (Obesity and Cachexia)

The area of afferent signals perhaps not surprisingly reveals the greatest dichotomy between obesity and cachexia. In diet-induced thermogenesis, insulin and nutrients play a critical role along with sensory inputs of taste, smell, gastric distension, etc., ably integrated by the complex matrix of controls. The involvement of insulin is complex when one considers the changes in insulin sensitivity associated with obesity. Once again, it is hard to establish primary and secondary events but insulin resistance may play a role in the differential effect of android and gynoid obesity.

In the control of hypermetabolism, release of interleukins, IL-1, IL-6 and TNF feed into different aspects of the cascade of peptidergic/serotoninergic and eicosonoid control pathways. There would appear to be overlap between the systems triggered in response to different stimuli perhaps accounting for differential responses—and offering an opportunity for selective manipulation in novel therapies.

In the case of tumours, evidence is emerging for specific factors released from tumours with powerful effects on energy balance. As these are identified and synthesized, their relationship with the other afferent signals can be established—and may also offer selective interventions for the future. At a simplistic level the same factor which debilitates the cancer patient may prove to be an elixir for the obese, although the chances that it would offer anything other than a lead for medicinal chemistry must be small.

EFFERENT SIGNALS

The earlier observation that adrenalectomy prevented the evolution of obesity related to a wide range of mechanisms has moved on to a consideration of the role of glucocorticoid receptors and their regulation. It has also led back to the identification of a key role for CRF in transducing the integrated signal to stimulate thermogenesis. However, there is no evidence that a lack of CRF is the cause of obesity. Further down the hierarchy of efferent signals, reduced sympathetic outflow appears to correlate with the propensity to obesity and may be critical in determining the response to dieting. This offers a more attractive target for therapeutic intervention which is currently under evaluation.

The changes in metabolic activity after injury are complex. Whilst an increase in sympathetic activity controls the early hypermetabolism, the identity of the efferent signal in the long term is less clear although changes in adrenal and pituitary function would seem to be possible candidates.

APPLICATIONS TO THERAPY

Appetite control

Fenfluramine, dextrofenfluramine, and the newer generation of 5HT uptake inhibitors such as fluoxetine and sertraline increase synaptic 5HT concentrations and have been shown to inhibit food intake in a wide range of models including obese patients. With the identification of subtypes of

5HT receptor, the possibility of a selective appetite stimulant to disinhibit food intake in cachexia could be envisaged.

Basic research has moved on to identify a new generation of neurotransmitters and mediators. Some of these may offer more tenacious or selective modulators of food intake but it remains to be seen which of these will offer a new generation of therapeutic agents giving an acceptable profile of CNS effects not liable to abuse.

An alternative approach to reducing food intake is to reduce the efficiency of digestion or absorption of either carbohydrate or fat. This can be achieved either by inhibition of gut enzymes or substitution of dietary constituents. In both cases, the problem is achieving a balance between the reduction in absorption and the impact of the residual nutrients on lower gut function. However, lipase inhibitors, alpha-glucosidase inhibitors and indigestible nutrient substitutes are all available for evaluation.

Energy expenditure

The presence of a specific category of β_3-receptors on brown adipose tissue has led to the identification of selective thermogenic agents which are being evaluated for the treatment of obesity. Whilst in the short term the effects may not be as dramatic as those achieved by very low calorie diets or anorexiants; in the longer term, these compounds may offer a means to reverse individual propensity to obesity which appears to be related to differences in sympathetic tone.

The same receptor mechanism may offer the opportunity of selective β_3-antagonists to inhibit an adrenergically mediated hypermetabolism in cachexia/injury. However, the long-term efficacy could be in doubt since other mediators would appear to take over after the initial increase in sympathetic outflow subsides.

Lean body mass

The preservation of lean body mass during dieting and during cachexia is an important target for therapy. Whilst in the therapy of obesity the emphasis is on weight loss, experience from inappropriately formulated diets has shown the importance of avoiding an associated loss of lean body mass. One attraction of the β-adrenergic thermogenic agents is that, in both animal models and man, lean body mass is preserved. This would appear to link with the proposed role for adrenaline in the physiological control of muscle metabolism.

This could well lead on to the identification of selective agents for the control of lean body mass independent of direct effects on energy expenditure as indicated by the work on clenbuterol, BRL 47672 and related compounds.

Current research on the modulation of insulin sensitivity as an approach to the therapy of non-insulin dependent diabetes may well lead to the identification of alternative mechanisms to control the loss of muscle mass in trauma or cachexia.

CLINICAL ENDPOINTS IN THERAPY

Obesity

In the Western world, obesity is currently unfashionable and supports a burgeoning industry in diets, health foods and dieting manuals. Fashion may yet change but the serious medical consequences of significant obesity (>125% IBW) will remain. Obesity is a major long-term risk factor both directly and by association with hyperlipidaemia, hypertension and diabetes.

Weight loss (by dieting) has been shown to be an effective treatment reversing the risks to the age related norm and reducing lipids, blood pressure and glucose intolerance.

Most treatments of the obese will achieve a short-term weight loss. However, success rates in maintaining the lower weight at 12 months are lamentably poor. Whilst a short-term weight loss may meet a particular social objective or even facilitate an operative procedure, the major risks from obesity are apparently long term and therefore long-term weight loss must be a criterion for assessing novel therapies. Given the poor success at 12 months, it should be possible to demonstrate an improvement in long-term weight loss by an assessment over this timescale.

As discussed, much of the risk associated with obesity is due to the associated diabetes, hypertension and hyperlipidaemia. Novel therapies should certainly not produce a compensatory deterioration in these three sequelae and they should be reduced at least in line with the weight loss achieved. If long-term benefit on mortality is being projected, other features will prove important—such as the proportion of fat lost relative to lean body mass. In this respect the thermogenic β-stimulants may show benefit over the simple reduction of food intake by dieting or anorexiants.

It is doubtful whether any one therapy will prove to be the complete answer to the treatment of obesity. Reflecting the complexity of control mechanisms and the large genetic component determining propensity to obesity, different approaches are likely to be required for individual patients.

In obesity where the treatment objectives are to reduce long-term risks, the products will need to be safe and well tolerated. This will prove to be an important aspect once a spectrum of treatments is available—although for the present, any effective therapy seems to be enthusiastically accepted by patients.

Cachexia/hypermetabolism

In the treatment of cachexia or hypermetabolism associated with injury or surgery, the treatment objectives are much more short term and obvious. Therapies can be assessed for effects on survival, cure rate, duration of hospitalization or regain of weight.

At present, research is directed at establishing the mechanisms involved in the different components of the syndromes. Thereafter, model compounds to block or potentiate the different phases will need to be evaluated in animal models since there are already doubts over whether specific phases of the stress response play a therapeutic or dystherapeutic role. Similarly some components during one phase of a response may be beneficial, others detrimental. This would direct attention at selective blocking of efferent signals rather than the afferent stimulus triggering the complex of responses.

It is interesting to note that the novel β-stimulant BRL 47672a which has been shown to have effects on protein deposition in animal models has a beneficial effect on survival of mice with implanted tumours suggesting that sympathetic drive is deficient in cachexia. However, in the same model, there is evidence for enhanced sympathetic drive on brown adipose tissue.

At the present time, the catabolic states associated with tumours, surgery and different types of trauma are being grouped together as closely related syndromes. However, the availability of model compounds will help to expose differences and direct future clinical evaluation.

In acute disease, short-term benefit can be important and a higher level of side effects may be acceptable opposite to a clear gain on hard clinical endpoints.

CONCLUSION

Research into obesity, cachexia and hypermetabolism can clearly benefit from cross fertilization. Common techniques can be applied and common mechanisms and mediators identified. However, on closer analysis fundamental differences emerge both between the different syndromes as well as within each syndrome.

Moving on to the identification and evaluation of possible therapies, each condition is discrete. Whilst it is tempting to see any intervention which had a clinical effect as being a useful therapy, it is essential that these are assessed against meaningful clinical endpoints to ensure that they achieve a true benefit to the patient.

Index

NB: Page numbers in *italics* refer to figures and tables

Index compiled by Liza Weinkove

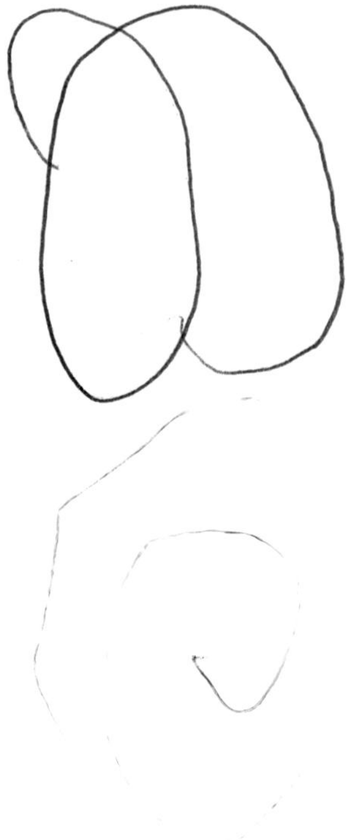